INORGANIC SYNTHESES

Volume 22

Board of Directors

HENRY F. HOLTZCLAW   *University of Nebraska*
JAY H. WORRELL   *University of South Florida*
BODIE E. DOUGLAS   *University of Pittsburgh*
DUWARD F. SHRIVER   *Northwestern University*
DARYLE H. BUSCH   *Ohio State University*

Future Volumes

23   STANLEY KIRSCHNER   *Wayne State University*
24   JEAN'NE M. SHREEVE   *University of Idaho*
25   HERBERT D. KAESZ   *University of California, Los Angeles*

International Associates

E. O. FISCHER   *Technische Universität München*
JACK LEWIS   *Cambridge University*
LAMBERTO MALATESTA   *University of Milan*
F. G. A. STONE   *University of Bristol*
GEOFFREY WILKINSON   *Imperial College of Science and Technology London*
AKIO YAMAMOTO   *Tokyo Institute of Technology Yokohama*
MARTIN A. BENNETT   *Australian National University*
R. POILBLANC   *National Center for Scientific Research Toulouse*
H. W. ROESKY   *University Göttingen*

Editor-in-Chief
SMITH L. HOLT, JR.
Department of Chemistry
Oklahoma State University

INORGANIC SYNTHESES

Volume 22

ELMHURST COLLEGE LIBRARY

A Wiley-Interscience Publication
JOHN WILEY & SONS
New York   Chichester   Brisbane   Toronto   Singapore

Published by John Wiley & Sons, Inc.

Copyright © 1983 by Inorganic Syntheses, Inc.

All rights reserved. Published simultaneously in Canada.

Reproduction or translation of any part of this work beyond that permitted by Section 107 or 108 of the 1976 United States Copyright Act without the permission of the copyright owner is unlawful. Requests for permission or further information should be addressed to the Permissions Department, John Wiley & Sons, Inc.

Library of Congress Catalog Number: 39-23015

ISBN 0-471-88887-7

Printed in the United States of America

10 9 8 7 6 5 4 3 2 1

PREFACE

Since the contents of this volume are adequately described in the Table of Contents they need no further elaboration here. This space does provide, however, the opportunity to recognize those individuals whose contributions are essential to producing a volume of *Inorganic Syntheses*. Without the submitters, there would be no syntheses. Without the checkers, there would be no independent verification that each synthesis proceeds as advertised. Without those individuals who contribute their time to help with the editing chores, our volume would lack necessary clarity and completeness. Without secretaries there would be no completed manuscripts. I would like to thank Duward Shriver for providing guidance and a sense of urgency. John Bailar, Therald Moeller, Warren Powell, and Thomas Sloan were particularly helpful during the editing process. Particular recognition must go to Otis Dermer for his editing of that material which I thought had been edited. This volume has gained much in style and clarity as a result of his critical eye. My secretary, Norma Sue Earp, provided invaluable assistance and exhibited considerable tolerance throughout the entire process. Debbie Shields exhibited great skill with the word processor when called upon to help salvage a battered schedule. I must also express my appreciation to those individuals who checked syntheses which ultimately did not perform as advertised: Ken Emmerson, Barry Garrett, and Pam Hood.

More than 20 years ago as an undergraduate at Northwestern University I first used *Inorganic Syntheses* to prepare "Ag_2F" in an inorganic laboratory course taught by Fred Basolo. It was an invaluable tool to me then, a time when "modern" inorganic chemistry was in its infancy. Though the world and inorganic chemistry are much more complex today, I hope this volume will serve to stir the interest of a newer generation of chemists.

SMITH L. HOLT

Stillwater, Oklahoma
July 1983

Previous volumes of *Inorganic Syntheses* are available. Volumes I–XVI can be ordered from R. E. Krieger Publishing Co., Inc., P.O. Box 9542, Melbourne, Florida 32901; Volume XVII is available from McGraw-Hill, Inc.; and Volumes XVIII–XXI can be obtained from John Wiley & Sons, Inc.

NOTICE TO CONTRIBUTORS AND CHECKERS

The *Inorganic Syntheses* series is published to provide all users of inorganic substances with detailed and foolproof procedures for the preparation of important and timely compounds. Thus the series is the concern of the entire scientific community. The Editorial Board hopes that all chemists will share in the responsibility of producing *Inorganic Syntheses* by offering their advice and assistance in both the formulation of and the laboratory evaluation of outstanding syntheses. Help of this kind will be invaluable in achieving excellence and pertinence to current scientific interests.

There is no rigid definition of what constitutes a suitable synthesis. The major criterion by which syntheses are judged is the potential value to the scientific community. An ideal synthesis is one that presents a new or revised experimental procedure applicable to a variety of related compounds, at least one of which is critically important in current research. However, syntheses of individual compounds that are of interest or importance are also acceptable. Syntheses of compounds that are readily available commercially at reasonable prices are not acceptable. Corrections and improvements of syntheses already appearing in *Inorganic Syntheses* are suitable for inclusion.

The Editorial Board lists the following criteria of content for submitted manuscripts. Style should conform with that of previous volumes of *Inorganic Syntheses*. The introductory section should include a concise and critical summary of the available procedures for synthesis of the product in question. It should also include an estimate of the time required for the synthesis, an indication of the importance and utility of the product, and an admonition if any potential hazards are associated with the procedure. The Procedure should present detailed and unambiguous laboratory directions and be written so that it anticipates possible mistakes and misunderstandings on the part of the person who attempts to duplicate the procedure. Any unusual equipment or procedure should be clearly described. Line drawings should be included when they can be helpful. All safety measures should be stated clearly. Sources of unusual starting materials must be given, and, if possible, minimal standards of purity of reagents and solvents should be stated. The scale should be reasonable for normal laboratory operation, and any problems involved in scaling the procedure either up or down should be discussed. The criteria for judging the purity of the final product should be delineated clearly. The section on Properties should supply and discuss those physical and chemical characteristics that are relevant to judging the purity of the product and to permitting its handling and use in an intelligent manner. Under References, all pertinent literature citations should be listed in order. A style sheet is available from the Secretary of the Editorial Board.

The Editorial Board determines whether submitted syntheses meet the general specifications outlined above. Every procedure will be checked in an independent laboratory, and publication is contingent upon satisfactory duplication of the syntheses.

Each manuscript should be submitted in duplicate to the Secretary of the Editorial Board, Professor Jay H. Worrell, Department of Chemistry, University of South Florida, Tampa, FL 33620. The manuscript should be typewritten in English. Nomenclature should be consistent and should follow the recommendations presented in *Nomenclature of Inorganic Chemistry*, 2nd Ed., Butterworths & Co, London, 1970 and in *Pure Appl. Chem.*, **28**, No. 1 (1971). Abbreviations should conform to those used in publications of the American Chemical Society, particularly *Inorganic Chemistry*.

Chemists willing to check syntheses should contact the editor of a future volume or make this information known to Professor Worrell.

TOXIC SUBSTANCES AND LABORATORY HAZARDS

Chemicals and chemistry are by their very nature hazardous. Chemical reactivity implies that reagents have the ability to combine. This process can be sufficiently vigorous as to cause flame, an explosion, or, often less immediately obvious, a toxic reaction.

The obvious hazards in the syntheses reported in this volume are delineated, where appropriate, in the experimental procedure. It is impossible, however, to foresee every eventuality, such as a new biological effect of a common laboratory reagent. As a consequence, *all* chemicals used and *all* reactions described in this volume should be viewed as potentially hazardous. Care should be taken to avoid inhalation or other physical contact with all reagents and solvents used in this volume. In addition, particular attention should be paid to avoiding sparks, open flames, or other potential sources which could set fire to combustible vapors or gases.

A list of 400 toxic substances may be found in the *Federal Register*, Vol. 40, No. 23072, May 28, 1975. An abbreviated list may be obtained from *Inorganic Syntheses*, Vol. 18, p. xv, 1978. A current assessment of the hazards associated with a particular chemical is available in the most recent edition of *Threshold Limit Values for Chemical Substances and Physical Agents in the Workroom Environment* published by the American Conference of Governmental Industrial Hygienists.

The drying of impure ethers can produce a violent explosion. Further information about this hazard may be found in *Inorganic Syntheses*, Vol. 12, p. 317.

CONTENTS

Chapter One SOLID STATE

1. Ternary Chlorides and Bromides of the Rare Earth Elements 1
   A. $A^I RE_2^{III} X_7$: Cesium Praesodymium Chloride ($CsPr_2Cl_7$) and Potassium Dysprosium Chloride (KDy_2Cl_7) 2
   B. $A_3^I RE_2^{III} X_9$, $A_2^I RE^{III} X_5$, $A_3^I RE^{III} X_6$: The Cesium Lutetium Chlorides, $Cs_3Lu_2Cl_9$, Cs_2LuCl_5, Cs_3LuCl_6 6
2. Quaternary Chlorides and Bromides of the Rare Earth Elements: Elpasolites $A_2^I B^I RE^{III} X_6$ ($r(A^I) > r(B^I)$) 10
   A. Cesium Lithium Thulium Chloride, ($Cs_2LiTmCl_6$) 10
   B. Other Chloro- and Bromo elpasolites 11
3. Tantalum as a High-Temperature Container Material for Reduced Halides 15
4. Cesium Scandium(II) Trichloride 23
   A. Synthesis with $CsCl$ 23
   B. Synthesis with $Cs_3Sc_2Cl_9$ 25
5. Zirconium Monochloride and Monobromide 26
   A. Synthesis with Zirconium Foil 27
   B. Synthesis Using Zirconium Turnings 28
6. Lanthanum Triiodide (And Other Rare Earth Metal Triiodides) . . . 31
   A. Synthesis Using HgI_2 32
   B. Synthesis from the Elements 33
7. Lanthanum Diiodide 36
8. Trichlorides of Rare Earth Elements, Yttrium, and Scandium 39
9. Single Crystal Growth of Oxides by Skull Melting: The Case of Magnetite (Fe_3O_4) 43
10. Lithium Nitride, Li_3N 48
    A. Polycrystalline Material 49
    B. Single Crystals of Li_3N 51
11. The Alkali Ternary Oxides A_xCoO_2 and A_xCrO_2 (A = Na, K) 56
    A. Sodium Cobalt Oxides: Na_xCoO_2 ($x \leq 1$) 56
    B. Potassium Cobalt Oxide Bronzes: $K_{0.50}CoO_2$, $K_{0.67}CoO_2$ . . . 57
    C. Potassium Cobalt Oxide: $KCoO_2$ 58
    D. Potassium Chromium Oxide: $KCrO_2$ 59
    E. Potassium Chromium Oxide Bronzes: K_xCrO_2 59

12. Zeolite Molecular Sieves 61
    A. Zeolite A 63
    B. Zeolite Y 64
    C. TMA Offretite 65
    D. ZSM-5 67
13. Lead Ruthenium Oxide, $Pb_2[Ru_{2-x}Pb_x^{4+}]O_{6.5}$ 69
14. Calcium Manganese Oxide, $Ca_2Mn_3O_8$ 73
15. Synthesis of Silver Tetratungstate 76
16. Cadmium Mixed Chalcogenides and Layers of Cadmium (Mixed) Chalcogenides on Metallic Substrates 80
    A. Cadmium Selenide Telluride ($CdSe_{0.65}Te_{0.35}$) 81
    B. Cadmium Selenide Thin Layers on Titanium 82
    C. Cadmium Selenide Telluride Layers on Molybdenum 84
17. Layered Intercalation Compounds 86
    A. Charge-Transfer-Type Intercalation Compounds: FeOCl(Pyridine derivative)$_{1/n}$ 86
    B. Grafted-Type Compound from FeOCl 87
    C. The Organic Intercalates of $HTiNbO_5$ 89
18. Iron Titanium Hydride ($FeTiH_{1.94}$) 90
19. Aluminum Lanthanum Nickel Hydride 96

Chapter Two TRANSITION METAL COMPLEXES AND COMPOUNDS

20. Purification of Copper(I) Iodide 101
21. Cobalt(III) Amine Complexes with Coordinated Trifluoromethanesulfonate 103
    A. Pentammine(trifluoromethanesulfonato)cobalt(III) Trifluoromethanesulfonate, [Co(NH$_3$)$_5$OSO$_2$CF$_3$](CF$_3$SO$_3$)$_2$ 104
    B. cis-Bis(1,2-ethylenediamine)di(trifluoromethanesulfonato)-cobalt(III) Trifluoromethanesulfonate, cis-[Co(en)$_2$(OSO$_2$CF$_3$)$_2$](CF$_3$SO$_3$) 105
    C. fac-Tris(trifluoromethanesulfonato)[N-(2-aminoethyl)-1,2-ethanediamine]cobalt(III), Co(dien)(OSO$_2$CF$_3$)$_3$ 106
22. 2,9-Dimethyl-3,10-diphenyl-1,4,8,11-tetraazacyclotetradeca-1,3,8,10-tetraene, (Me$_2$Ph$_2$[14]-1,3,8,10-tetraeneN$_4$) Complexes 107
    A. Bis(acetonitrile)(2,9-dimethyl-3,10-diphenyl-1,4,8,11-tetraazacyclotetradeca-1,3,8,10-tetraene)iron(II) Hexafluorophosphate 108
    B. (2,9,Dimethyl-3,10-diphenyl-1,4,8,11-tetraazacyclotetradeca-1,3,8,10-tetraene)copper(II) Hexafluorophosphate 110

	C. Chloro(2,9-dimethyl-3,10-diphenyl-1,4,8,11-tetraazacyclotetradeca-1,3,8,10-tetraene)zinc(II) Hexafluorophosphate	111
23.	Dichlorobis(3,3',3''-phosphinidynetripropionitrile)nickel(II) and Dibromobis(3,3',3''-phosphinidynetripropionitrile)nickel(II) Monomers and Polymers	113
	A. Dichlorobis(3,3',3''-phosphinidynetripropionitrile)-nickel(II)	114
	B. Dibromobis(3,3',3''-phosophinidynetripropionitrile)-nickel(II)	115
24.	Polymer-stabilized Divanadium	116
25.	Triamminechloroplatinum(II) Chloride	124
	A. *trans*-Diamminechloroiodoplatinum(II) Chloride	124
	B. *trans*-Diammineaquachloroplatinum(II) Nitrate	125
	C. Triamminechloroplatinum(II) Chloride	125
26.	Chlorotris(dimethyl sulfide)platinum(II) Compounds	126
27.	Tetrabutylammonium Trichloro(dimethyl sulfide)platinum(II)	128

Chapter Three MAIN GROUP, LANTHANIDE, ACTINIDE, AND ALKALI METAL COMPOUNDS

28.	Tributyl Phosphorotrithioite $(C_4H_3S)_3P$ from Elemental Phosphorus	131
29.	Preparation of Dimethylphenylphosphine	133
30.	Electrochemical Synthesis of Salts of Hexahalodigallate(II) and Tetrahalogallate(III) Anions	135
	A. Triphenylphosphonium Hexahalodigallate(II), $[(C_6H_5)_3PH]_2[Ga_2X_6]$ (X = Cl, Br, or I)	137
	B. Tetraphenylphosphonium Hexabromodigallate(II)	139
	C. Tetrabutylammonium Tetrachloro- and Tetrabromogallate(III)	139
	D. Tetrabutylammonium Tetraiodogallate(III)	140
	E. Tetraethylammonium Tetrabromogallate(III)	141
31.	Polymeric Sulfur Nitride (Polythiazyl), (SN_x)	143
32.	Pyridinium Hexachloroplumbate(IV)	149
33.	Bis[(4,7,13,16,21,24-hexaoxa-1,10-diazabicyclo[8.8.8.]hexacosane) potassium] Tetrabismuth (2-)	151
34.	[5,10,15,20-Tetraphenylporphyrinato(2-1)] Lanthanides and Some [5,10,15,20-Tetraphenylporphyrinato(2-1)] Actinides	156

Chapter Four ORGANOMETALLIC COMPOUNDS

35. [μ-Nitrido-bis(triphenylphosphorus)](1+) Tricarbonylnitrosylferrate(1-) and [μ-Nitrido-bis(triphenylphosphorus)](1+) Decacarbonyl-μ-nitrosyl-triruthenate(1-), [(Ph$_3$P)$_2$N] [Fe(CO)$_3$(NO)] and [(Ph$_3$P)$_2$N][Ru$_3$(CO)$_{10}$(NO)] ... 163
    A. μ-Nitrido-bis(triphenylphosphorus)(1+) Nitrite 164
    B. μ-Nitrido-bis(triphenylphosphorus)(1+) Tricarbonylnitrosylferrate (1-) 165
    C. μ-Nitrido-bis(triphenylphosphorus)(1+) Decacarbonyl-μ-nitrosyltriruthenate(1-) 165
36. Metallacyclopentane Derivatives of Palladium 167
    A. (1,4-Butanediyl)[1,2-ethanediylbis(diphenylphosphine)]-palladium(II) 167
    B. (1,4-Butanediyl)(N,N,N',N'-tetramethyl-1,2-ethanediamine)-palladium(II) 168
    C. (1,4-Butanediyl)bis(triphenylphosphine)palladium(II) 169
    D. (2,2'-bipyridine)(1,4-Butanediyl) palladium(II) 170
37. Metallacyclopentane Derivatives of Cobalt, Rhodium, and Iridium 171
    A. (1,4-Butanediyl)(η^5-cyclopentadienyl)-(triphenylphosphine)cobalt(III) 171
    B. (1,4-Butanediyl)(η^5-pentamethylcyclopentadienyl)-(triphenylphosphine)rhodium(III) 173
    C. (1,4-Butanediyl)(η^5-pentamethylcyclopentadienyl)-(triphenylphosphine)iridium(III) 174
38. Cycloolefin Complexes of Ruthenium 176
    A. (η^6-Benzene)(η^4-1,3-cyclohexadiene)ruthenium(0) 177
    B. (η^4-1,5-Cyclooctadiene)(η^6-1,3,5-cyclooctatriene)-ruthenium(0) 178
    C. Bis(η^5-2,4-cycloheptadienyl)ruthenium(II) 179
    D. Bis(η^5-cyclopentadienyl)ruthenium(II) (Ruthenocene) 180
39. Mononuclear Pentacarbonyl Hydrides of Chromium, Molybdenum, and Tungsten 181
    A. μ-Nitrido-bis(triphenylphosphorus)(1+) Pentacarbonylhydridotungsten(1-) 182
    B. μ-Nitrido-bis(triphenylphosphorus)(1+) Pentacarbonylchromium(1-) 183

Chapter Five COMPOUNDS OF BORON

40. [(2,2-Dimethylpropanoyl)oxy]diethylborane, ((Pivaloyloxy)diethylborane) 185

41.	Tetraethyldiboroxane	188
42.	Diethylmethoxyborane, (Methyl diethylborinate)	190
43.	Diethylhydroxyborane, (Diethylborinic acid)	193
44.	Bis[μ-(2,2-dimethylpropanoato-O,O)]-diethyl-μ-oxo-diboron	196
45.	Alkali Metal (Cyclooctane-1,5-diyl)dihydroborates(1-)	198
	A. Lithium (Cyclooctane-1,5-diyl)dihydroborate(1-)	199
	B. Sodium (Cyclooctane-1,5-diyl)dihydroborate(1-)	200
	C. Potassium (Cyclooctane-1,5-diyl)dihydroborate(1-)	200
46.	Decaborane(14)	202
47.	Dichlorophenylborane	207
48.	(Dimethylamino)diethylborane	209
49.	2,3-Diethyl-2,3-dicarba-$nido$-hexaborane(8)	211
50.	2,2',3,3'-Tetraethyl-1,1-dihydro-[1,1'-$commo$-bis(2,3-dicarba-1-ferra-$closo$-heptaborane)](12), [2,3-$(C_2H_5)_2C_2B_4H_4]_2FeH_2$, and 2,3,7,8-Tetraethyl-2,3,7,8-tetracarbadodecaborane(12), $(C_2H_5)_4C_4B_8H_8$	215
	A. 2,2',3,3'-Tetraethyl-1,1-dihydro-[1,1'-$commo$-bis(2,3-dicarba-1-ferra-$closo$-heptaborane)](12)	216
	B. 2,3,7,8-Tetraethyl-2,3,7,8-tetracarbadodecaborane(12)	217
51.	[^{10}B]-Labeled Boron Compounds	218
	A. [^{10}B]Boron Tribromide	219
	B. Dibromomethyl [^{10}B]Borane	223
	C. 3,5-Dimethyl-[$^{10}B_2$]-1,2,4,3,5-trithiadiborolane	225
52.	The Thiadecaboranes $arachno$-[6-SB$_9$H$_{12}$]$^-$, $nido$-6-SB$_9$H$_{11}$, and $closo$-1-SB$_9$H$_9$	226
	A. Cesium Dodecahydro-6-thia-$arachno$-decaborate(1-)	227
	B. 6-Thia-$nido$-decaborane(11)	228
	C. 1-Thia-$closo$-decaborane(9)	229
53.	Potassium Dodecahydro-7-8-dicarba-$nido$-undecaborate(1-), K[7,8-C$_2$B$_9$H$_{12}$], Intermediates, Stock Solution, and Anhydrous Salt	231
54.	3-(η^5-Cyclopentadienyl)-1,2-dicarba-3-cobalta-$closo$-dodecaborane(11), 3-(η^5-C$_5$H$_5$)-3-Co-1,2-C$_2$B$_9$H$_{11}$	235
55.	2,6-Dicarba-$nido$-nonaborane(11), 2,6-C$_2$B$_7$H$_{11}$	237
56.	9-(Dimethyl sulfide)-7,8-dicarba-$nido$-undecaborane(11), 9-[(CH$_3$)$_2$S]-7,8-C$_2$B$_9$H$_{11}$	239
57.	Direct Sulfhydrylation of Boranes and Heteroboranes, 1,2,-Dicarba-$closo$-dodecaborane(12)-9-thiol, 9-HS-1,2-C$_2$B$_{10}$H$_{11}$	241

Index of Contributors 245

Subject Index 249

Formula Index 263

INORGANIC SYNTHESES

Volume 22

Chapter One
SOLID STATE

1. TERNARY CHLORIDES AND BROMIDES OF THE RARE EARTH ELEMENTS

Submitted by GERD MEYER*
Checked by S.-J. HWU† and J. D. CORBETT†

Compound formation in the alkali metal halide/rare earth metal trihalide systems seems to be limited to the four following formula types:[1-4]

$$A^I RE_2^{III} X_7, \quad A_3^I RE_2^{III} X_9, \quad A_2^I RE^{III} X_5, \quad A_3^I RE^{III} X_6$$

Most of the existing ARE_2X_7 and A_3REX_6 phases (with A = K, Rb, Cs and X = Cl, Br) melt congruently, and those of A_2REX_5 and $A_3RE_2X_9$ type incongruently. They may be prepared by heating together appropriate amounts of the anhydrous halides AX and REX_3.

The synthetic route described here, starting with aqueous solutions of the appropriate amounts of alkali metal halides and rare earth metal halides, makes use of a high concentration of hydrogen halide gas as a driving force in a flow

---

*Institut für Anorganische und Analytische Chemie I, Justus-Liebig-Universitaet Giessen, Heinrich-Buff-Ring 58, 6300 Giessen, West Germany.

†Department of Chemistry and Ames Laboratory, Iowa State University, Ames, IA 50011. The Ames Laboratory is operated for the U.S. Department of Energy by Iowa State University under Contract No. W-7405-ENG-82.

system for the dehydration of a mixture that contains binary and ternary compounds (hydrates). It therefore avoids the complicated and time-intensive preparation of the anhydrous rare earth metal trihalides which have to be sublimed at least twice to get rid of their contamination with oxyhalides. For example, $Cs_3Sc_2Cl_9$ [5,6] has been used as a starting material for the preparation of $CsScCl_3$ with divalent scandium (reduction with Sc metal in sealed tantalum containers[7]).

A. $A^I RE_2^{III} X_7$: CESIUM PRASEODYMIUM CHLORIDE ($CsPr_2Cl_7$) AND POTASSIUM DYSPROSIUM CHLORIDE (KDy_2Cl_7)

$$CsCl + 2PrCl_3 \cdot 7H_2O \xrightarrow{HCl(aq)} (CsPr_2Cl_7)(hyd) \xrightarrow[-H_2O]{HCl(g)\ 500°} CsPr_2Cl_7$$

$$KCl + Dy_2O_3 \xrightarrow{HCl(aq)} (KDy_2Cl_7)(hyd) \xrightarrow[-H_2O]{HCl(g)\ 500°} KDy_2Cl_7$$

Procedure

■ **Caution.** *Inhalation of HCl or HBr gas is harmful to the respiratory system; all operations should be carried out in a well-ventilated hood.*

1. For the preparation of $CsPr_2Cl_7$, 168.4 mg CsCl (1 mmole) and 746.7 mg $PrCl_3 \cdot 7H_2O$ (2 mmole, commercially available from various sources; "Pr_6O_{11}" may also be used in appropriate amounts, but the Pr content should be determined by analysis) are dissolved by adding a small amount of concentrated hydrochloric acid, which should be at least reagent grade and in the case of optical investigations, completely free of transition metal ions, especially iron.

2. For the preparation of KDy_2Cl_7, 74.6 mg KCl (1 mmole) and 373.0 mg Dy_2O_3 (1 mmole) are used as starting materials. Here and in the following syntheses, the rare earth sesquioxides should be freshly ignited before weighing, because they are suspected of slowly forming carbonates when exposed to air for a long period of time. Concentrated hydrochloric acid (≈ 100 mL) is added to the reactants. A homogeneous solution is obtained after some minutes of boiling.

The clear solutions in 1 or 2 are slowly evaporated to dryness by using a heating plate or an infrared lamp. The residue is then transferred to a corundum crucible (size: 10×60 mm), and this is inserted into a fused quartz tube in the center of a tubular furnace. (To prepare larger quantities, e.g., 10 mmole, a $100 \times 15 \times 10$-mm Alundum boat is more convenient.) The sizes of the tubular furnace and the fused quartz tube should be sufficient that the sample can be

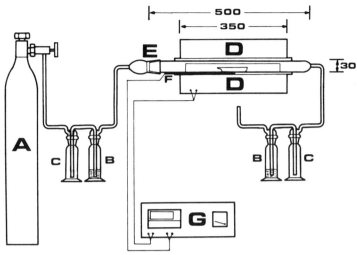

*Fig. 1. Apparatus for the synthesis of anhydrous ternary and quaternary chlorides and bromides. Sizes are given in millimeters; suitable for a mmolar scale. (**A**) Steel cylinder or lecture bottle with HCl or HBr gas; (**B**) washing bottles filled with conc. sulfuric acid (for HCl only) or P_2O_5 and paraffin oil, respectively; (**C**) empty washing bottles to prevent suck-back; (**D**) tubular furnace; (**E**) quartz tube; (**F**) thermocouple, e.g., Chromel-alumel or Pt-Pt/Rh; (**G**) temperature control and measuring device (a programming unit may be attached for slow heating and cooling).*

heated without a temperature gradient (see Fig. 1). A slow flow of HCl gas (available in steel cylinders or lecture bottles in high purity, e.g., "transistor grade"), one to two bubbles per second, dried with concentrated sulfuric acid or a trap cooled with dry ice/acetone, is then commenced and the sample heated to 500°. After 1–2 days, the HCl gas stream may be replaced by a dry inert gas stream (N_2 or Ar) and the sample cooled by turning off the power to the furnace. Yields: approximately 100%; 662.9 mg $CsPr_2Cl_7$ and 612.3 mg KDy_2Cl_7, respectively.

Properties

The products $CsPr_2Cl_7$ (light green) and KDy_2Cl_7 (colorless) are obtained as slightly sintered, moisture-sensitive powders. They should be handled in a dry box (dry N_2 or Ar atmosphere) and sealed in Pyrex tubes for storage.

For characterization, common X-ray techniques (Debye-Scherrer or, better, Guinier patterns) are useful. The compound $CsPr_2Cl_7$ crystallizes in the orthorhombic space group $P222_1$, with $a = 1657.1$, $b = 963.7$, and $c = 1486.6$ pm and 8 formula units per unit cell. Some characteristic d values are (relative intensities on a 1–100 scale are given in parentheses): 4.840, 4.832(16); 4.262, 4.260(46); 4.055, 4.050, 4.032, 4.026(100); 3.649, 3.644(16); 3.394, 3.393(9) Å. The crystal structure contains [$PrCl_8$] groups (partially disordered) linked together via common corners, edges, and faces.

The crystal structures of both KDy_2Cl_7 and $RbDy_2Cl_7$ have been determined from single-crystal X-ray data.[8] $RbDy_2Cl_7$ represents the orthorhombic aristotype: $a = 1288.1(3)$, $b = 693.5(2)$, $c = 1267.2(3)$ pm, space group Pnma, $Z = 4$. The compounds $RbRE_2Cl_7$, as well as the room-temperature modifications of $CsRE_2Cl_7$ that are indicated in Fig. 2 by filled circles, are isotypic to $RbDy_2Cl_7$. The other KRE_2Cl_7 compounds obtained so far crystallize as the monoclinic hettotype with (for KDy_2Cl_7) $a = 1273.9(8)$, $b = 688.1(5)$, $c = 1262.1(6)$ pm, $\beta = 89.36(3)°$, space group $P112_1/a$ (an unconventional setting of $P2_1/c$), $Z = 4$.

TYPE:	ARE_2X_7						$A_3RE_2X_9$						A_2REX_5						A_3REX_6					
RE\A	K		Rb		Cs		K		Rb		Cs		K		Rb		Cs		K		Rb		Cs	
	Cl	Br	Cl	Br	Cl	Br	Cl	Br	Cl	Br	Cl	Br	Cl	Br	Cl	Br	Cl	Br	Cl	Br	Cl	Br	Cl	Br
La	○				●	○	–				–	–	○		–	–	–						○	○
Ce	○						○	–			–		○				○		○				○	
Pr	–	○	○		●	○	○	–			–	–	○	○			–	–	○	○			○	○
Nd	–	○	◐	○		○	○	–	–	–	–	–	○	○	○	○	–	–	○	○	○	○	○	○
Sm	◐	○	●	○	○	○	–	–	–	–	–	●	○	○	○	○	–	○	○	○	○	○	○	○
Eu	◐		●					–			–	●	○			–			○					
Gd	●	○	●	○		○		–	–		–	●	○	○		○	–	–	○	○		○		○
Tb	●		●					–			–	●	○			–			○					
Dy	●	–	●	○	●			–	–		–	●	–	○	○	●			○	○		○	○	
Y	●		●		●			–			●	●	–			●			◐		◐		◐	
Ho	●		●		●						●	●				●			◐					
Er	●	–	●	–	●	–		–			●	●	–	○	●	○		–	◐	○		○	○	○
Tm	●		●		●						●	●			●				◐					
Yb	–	–	●	–	●	–		–			●	●	–	○		–	●		–	◐	○		○	○
Lu	–				●						●	●	–		●		●			●				
Sc	–	–	–	–	–	–	○	●	●	●	●	●	○	–	–	–	–	–	●	○	○	◐	○	○

Fig. 2. Compound formation in the alkali halide/rare-earth trihalide systems (chlorides and bromides); ● crystal structure known, see tables I and II (A_2REX_5 type compounds with $A = K, Rb$; $RE = La-Dy$; $X = Cl, Br$ are isotypic with K_2PrCl_5, see: G. Meyer and E. Hüttl, Z. Anorg. Allgem. (1983)); ○ observed in the phase diagram; ◐ phase investigated by x-ray diffraction, crystal structure not known so far; – not observed in the phase diagram; no entry means that the respective phase diagram was not investigated.

In both structures (Fig. 3), monocapped trigonal prims [DyCl$_7$] are connected via a triangular face and two edges with other prisms to form layers that are stacked in the [100] direction. These are held together by alkali metal cations (K$^+$, Rb$^+$), which are essentially 12-coordinated (bicapped pentagonal prism). Characteristic d values (in Å) are: for RbDy$_2$Cl$_7$, 6.440(70), 6.084(27), 4.423(100), 3.474(64); for KDy$_2$Cl$_7$, 6.374(100), 6.063(52), 4.417(36), 4.370(40), 3.471(34). Numbers in parentheses are relative intensities on a 1-100 scale.

The compound CsY$_2$Cl$_7$ may serve as an example for the cesium compounds with the heavier rare earths (those with La-Nd have the CsPr$_2$Cl$_7$ type structure). Lattice constants are: a = 1335.6(3), b = 696.6(2), c = 1266.1(4) pm,

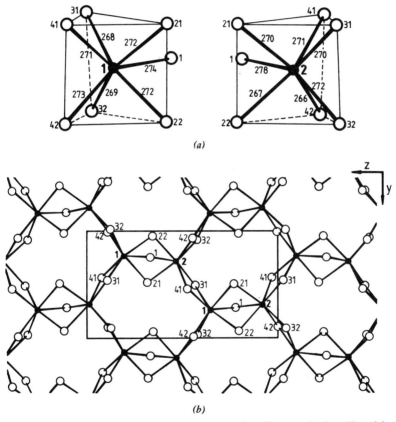

Fig. 3. Characteristic structural features of KDy$_2$Cl$_7$ and RbDy$_2$Cl$_7$. (a) The two crystallographically independent monocapped trigonal prisms [DyCl$_7$]; (b) part of one $^2_\infty$[Dy$_2$Cl$_7$] layer.

Pnma, Z = 4. Characteristic d values are: 6.678(16), 4.595(16), 4.505(100), 3.610(20), 3.485(35) Å.

For the preparation of the analogous bromides, all the chloride-containing starting materials and HCl gas as well are replaced by bromides and HBr gas, respectively.

B. $A_3^I RE_2^{III} X_9$, $A_2^I RE^{III} X_5$, $A_3^I RE^{III} X_6$: THE CESIUM LUTETIUM CHLORIDES, $Cs_3Lu_2Cl_9$, Cs_2LuCl_5, Cs_3LuCl_6

$$4CsCl + Lu_2O_3 \xrightarrow{HCl(aq)} 2Cs_2LuCl_5 \cdot H_2O \xrightarrow[-H_2O]{HCl(g)\ 500°} 2Cs_2LuCl_5$$

$$3CsCl + Lu_2O_3 \xrightarrow{HCl(aq)} 1.5Cs_2LuCl_5 \cdot H_2O\ (+0.5LuCl_3 \cdot hyd)$$

$$\xrightarrow[-H_2O]{HCl(g)\ 500°} Cs_3Lu_2Cl_9$$

$$6CsCl + Lu_2O_3 \xrightarrow{HCl(aq)} 2Cs_2LuCl_5 \cdot H_2O + 2CsCl$$

$$\xrightarrow[-H_2O]{HCl(g)\ 500°} 2Cs_3LuCl_6$$

Procedure

Appropriate molar amounts of CsCl and Lu_2O_3 (CsCl should be reagent grade or ultrapure, Lu_2O_3 at least 99.9%; both are available from various sources) are used as starting materials:

1. for Cs_2LuCl_5: 673.4 mg CsCl (4 mmole) and 397.9 mg Lu_2O_3 (1 mmole)
2. for $Cs_3Lu_2Cl_9$: 505.1 mg CsCl (3 mmole) and 397.9 mg Lu_2O_3 (1 mmole)
3. for Cs_3LuCl_6: 1010.1 mg CsCl (6 mmole) and 397.9 mg Lu_2O_3 (1 mmole)

Approximately 100 mL hydrochloric acid (reagent grade) is added to the reactants (1, 2, or 3). A homogeneous solution is obtained after some minutes of boiling. After that, a heating plate or an infrared lamp is used for slow evaporation to dryness. For large (e.g., 10 mmole) quantities it might be necessary to grind the product from the initial evaporation and put it in a 100° oven overnight to reduce the degree of hydration and considerably speed the final step. The procedure following this is identical to that described in A. The yields approach the theoretical values: 1.236 g (2 mmole) Cs_2LuCl_5, 1.068 g (1 mmole) $Cs_3Lu_2Cl_9$, 1.573 g (2 mmole) Cs_3LuCl_6.

Properties

The compounds Cs_2LuCl_5, $Cs_3Lu_2Cl_9$, and Cs_3LuCl_6 are obtained as slightly sintered, moisture-sensitive, colorless powders. They should be stored and handled in a dry box and for longer periods of time sealed under inert gas (N_2 or Ar) in Pyrex tubes to prevent hydration. Principal impurities in the samples obtained are most likely oxyhalides such as LuOCl. These are usually not detectable at a sensitivity level of a few percent with the Guinier method.

X-ray powder diffraction techniques are suitable for the characterization of Cs_2LuCl_5 and $Cs_3Lu_2Cl_9$: Cs_2LuCl_5 [9] is orthorhombic displaying the Cs_2DyCl_5 [10] structure type (Table I). The first six strongest lines (in Å) are: d = 4.178(100), 4.021(44), 3.995(32), 3.858(31), 3.700(54), 3.531(34); further strong lines are observed with d = 2.862(34), 2.720(57), 2.524(50); intensities are given in parentheses on a 1-100 scale.

The compound $Cs_3Lu_2Cl_9$ [6] belongs to the trigonal $Cs_3Tl_2Cl_9$ [11-13] structure type. The strongest X-ray lines have d values equal to 4.443(45), 3.734(88), 2.856(100), 2.758(36), 2.219(48), 1.867(40) Å.

Little is known about Cs_3LuCl_6 and other A_3REX_6 type compounds. Preliminary investigations show a reversible but slow phase transition above 300° for Cs_3LuCl_6. The compound $Cs_2LuCl_5 \cdot H_2O$,[9] detected as an intermediate in all three cases, crystallizes with the erythrosiderite type ($K_2FeCl_5 \cdot H_2O$-type[14], orthorhombic, space group Pnma) with a = 1456.4(4), b = 1046.6(3), c = 752.0(3) pm. The d values of the strongest X-ray lines are: 6.099(45), 5.966(44), 3.698(100), 3.638(40), 2.989(35), 2.614(86) Å.

TABLE I Lattice Constants of A_2RECl_5-Type Compounds[9] (orthorhombic, Pbnm, Z = 4)

	a/pm	b/pm	c/pm
Cs_2DyCl_5	1523.3(3)	954.9(3)	749.7(3)
Cs_2YCl_5	1522.6(2)	953.3(1)	746.9(1)
Cs_2HoCl_5	1520.2(2)	951.6(2)	745.4(1)
Cs_2ErCl_5	1519.1(2)	949.9(2)	744.2(1)
Cs_2TmCl_5	1517.7(2)	948.1(2)	741.8(1)
Cs_2YbCl_5	1514.7(5)	945.6(3)	740.8(2)
Cs_2LuCl_5	1514.2(2)	944.8(2)	738.5(1)
Rb_2ErCl_5	1466.6(3)	951.3(2)	727.4(2)
Rb_2TmCl_5	1462.1(3)	946.5(4)	727.1(3)
Rb_2LuCl_5	1460.9(2)	939.8(2)	724.6(1)

8   Solid State

Investigations of the alkali metal halide/rare earth metal halide phase diagrams are still fragmentary, as may be seen from Fig. 2.

Compounds of the A_2REX_5 type, indicated in Fig. 2 by filled circles, together with their lattice constants as determined from Guinier powder patterns are listed in Table I, and those of $A_3RE_2X_9$ type in Table II.

Like the elpasolites $A_2^IB^IRE^{III}X_6$ (see Section 2) A_3REX_6-type compounds contain "isolated" [REX_6] "octahedra," presumably of low symmetry in a low-symmetry crystal structure. The compounds K_3ScCl_6 and K_3LuCl_6 [15] crystallize with the K_3MoCl_6 type[16] (monoclinic, $P2_1/a$, $Z = 4$). Lattice constants, characteristic d values in Å, and relative intensities on a 1-100 scale are: for K_3ScCl_6, $a = 1226.5(6)$, $b = 758.0(1)$, $c = 1282.1(5)$ pm, $\beta = 109.03(2)°$; $d(I)$, 6.085(50), 5.250(34), 3.723(27), 2.968(39), 2.745(55), 2.625(100) Å; for K_3LuCl_6, $a = 1251.5(4)$, $b = 767.6(2)$, $c = 1301.1(5)$ pm, $\beta = 109.79(3)°$; $d(I)$, 6.183(100), 6.121(38), 5.888(30), 5.304(70), 3.776(40), 3.665(50), 2.652(56). The characteristic feature of Cs_2RECl_5-type compounds is an extended, infinite zweier-single zigzag chain of octahedra, *cis*-connected via common corners, while $A_3RE_2X_9$-type compounds contain "isolated" confacial bioctahedra [RE_2X_9]

TABLE II   Lattice Constants of $A_3RE_2X_9$-Type Compounds[a]
(mostly $Cs_3Tl_2Cl_9$ structure type, trigonal, $R\bar{3}c$, $Z = 6$)

	Chlorides, X = Cl		Bromides, X = Br	
	a/pm	c/pm	a/pm	c/pm
$Cs_3Sm_2X_9$			1379.7	1937
$Cs_3Gd_2X_9$			1372.1	1939
$Cs_3Tb_2X_9$			1367.2	1938
$Cs_3Dy_2X_9$			1363.8	1935
$Cs_3Y_2X_9$	1306.9	1831		
$Cs_3Ho_2X_9$	1305.3	1832	1360.3	1936
$Cs_3Er_2X_9$	1301.5	1829	1357.7	1937
$Cs_3Tm_2X_9$	1299.3	1828	1355.1	1934
$Cs_3Yb_2X_9$	1296.9	1829	1352.0	1932
$Cs_3Lu_2X_9$	1293.9	1829	1350.0	1931
$Cs_3Sc_2X_9$	1270.4	1810.9	768.0[b]	1926
$Rb_3Er_2X_9$			1340.2(4)	1916 (1)
$Rb_3Tm_2X_9$			1338.3(2)	1913.4(3)
$Rb_3Yb_2X_9$			1334.3(2)	1914.1(3)
$Rb_3Lu_2X_9$			1331.8(1)	1916.3(2)
$Rb_3Sc_2X_9$	1239.8(1)	1789.3(2)	756.8(1)[b]	1914.9(8)

[a] Data on Cs compounds were taken from Reference 6, Rb compounds from Reference 15.
[b] $Cs_3Cr_2Cl_9$ structure type, hexagonal, $P6_3/mmc$, $Z = 2$.

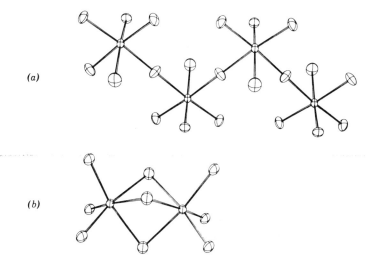

Fig. 4. Increasing octahedral "condensation" with decreasing alkali metal halide content. (a) A section of the infinite cis-connected zig-zag chain of octahedra in Cs_2RECl_5; (b) a confacial bioctahedron as the characteristic structural feature of $A_3^I RE_2^{III} X_9$ type compounds in both the $Cs_3Tl_2Cl_9$ and $Cs_3Cr_2Cl_9$ structure types.

(Fig. 4). Two structure types are found: $Cs_3Tl_2Cl_9$ [11-13] (stacking sequence of $CsCl_3$ layers: ABABAB...) and $Cs_3Cr_2Cl_9$ [17] (ABACBC...).

References

1. *Gmelins Handbuch der Anorganischen Chemie*, 8th edition, system no. 39, part C 5, Springer Verlag, New York, 1977.
2. J. Kutscher and A. Schneider, *Z. Anorg. Allgem. Chem.*, **408**, 135 (1974).
3. R. Blachnik and D. Selle, *Z. Anorg. Allgem. Chem.*, **454**, 90 (1979).
4. R. Blachnik and A. Jaeger-Kasper, *Z. Anorg. Allgem. Chem.*, **461**, 74 (1980).
5. K. R. Poeppelmeier, J. D. Corbett, T. P. McMullen, D. R. Torgeson, and R. G. Barnes, *Inorg. Chem.*, **19**, 129 (1980).
6. G. Meyer and A. Schoenemund, *Mat. Res. Bull.*, **15**, 89 (1980).
7. K. R. Poeppelmeier and J. D. Corbett, *Inorg. Syn.*, this volume, Chap. 1, Sec. 4.
8. G. Meyer, *Z. Anorg. Allgem. Chem.*, **491**, 217 (1982).
9. G. Meyer, J. Soose, and A. Moritz, unpublished work (1980/81).
10. G. Meyer, *Z. Anorg. Allgem. Chem.*, **469**, 149 (1980).
11. J. L. Hoard and L. Goldstein, *J. Chem. Phys.*, **3**, 199 (1935).
12. H. M. Powell and A. F. Wells, *J. Chem. Soc.*, 1008 (1935).
13. G. Meyer, *Z. Anorg. Allgem. Chem.*, **445**, 140 (1978).
14. A. Bellanca, *Period. Mineral.*, **17**, 59 (1948); see also: Ch. O'Connor, B. S. Deaver, Jr., and E. Sinn, *J. Chem. Phys.*, **70**, 5161 (1979).

15. G. Meyer and U. Strack, unpublished results (1981).
16. Z. Amilius, B. van Laar, and H. M. Rietveld, *Acta Cryst.*, **B 25**, 400 (1969).
17. G. J. Wessel and D. J. W. Ijdo, *Acta Cryst.*, **10**, 466 (1957).

2. QUATERNARY CHLORIDES AND BROMIDES OF THE RARE EARTH ELEMENTS: ELPASOLITES $A_2^I B^I RE^{III} X_6$ $(r(A^I) > r(B^I))$

Submitted by GERD MEYER*
Checked by S.-J. HWU† and J. D. CORBETT†

A. CESIUM LITHIUM THULIUM CHLORIDE, $Cs_2LiTmCl_6$

$$4CsCl + Li_2CO_3 + Tm_2O_3 \xrightarrow{HCl(aq)} (2Cs_2LiTmCl_6)(hyd)$$

$$\xrightarrow[-H_2O]{HCl(g)\ 500°} 2Cs_2LiTmCl_6$$

Quaternary chlorides and bromides of the rare earth elements (RE = Sc, Y, La, lanthanides), the so-called elpasolites, $A_2^I B^I RE^{III} X_6$ $(r(A^I) > r(B^I))$, (A^I = Cs, Rb, Tl, K; B^I = Li, Ag, Na, K; X = Cl, Br) are known in many of the theoretically possible cases.[1-7] They have been proved to be useful tools in investigating the properties of the trivalent rare earth ions because of the often ideal octahedral surrounding of RE^{3+} by X^- (site symmetry O_h). Elpasolites have been and will be widely used for optical measurements and Raman and ESR spectra. Susceptibility data have also been collected and interpreted to evaluate the behavior of ions with 4f electrons in ligand fields of different strength. Distances between RE^{3+} ions are relatively large; the $[REX_6]$ octahedra are "isolated" so that no magnetic interactions between the paramagnetic ions occur.

Procedure

Cesium chloride, 673.4 mg (4 mmole), 73.9 mg (1 mmole) Li_2CO_3, and 385.9 mg (1 mmole) Tm_2O_3, all available as reagent grade or ultrapure from various

*Institut für Anorganische und Analytische Chemie I, Justus-Liebig-Universitaet Giessen, Heinrich-Buff-Ring 58, 6300 Giessen, West Germany.
†Department of Chemistry and Ames Laboratory, Iowa State University, Ames, IA 50011. The Ames Laboratory is operated for the U.S. Department of Energy by Iowa State University under Contract No. W-7405-ENG-82.

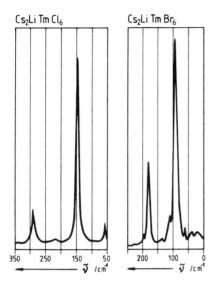

Fig. 1. Raman spectra of $Cs_2LiTmCl_6$ and $Cs_2LiTmBr_6$.

sources, are used as starting materials. Approximately 200 mL concentrated hydrochloric acid are added to the reactants as in Section 1-A, and the same further procedure followed as described there.

$Cs_2LiTmCl_6$ is obtained as a slightly sintered powder. Because of its sensitivity to moisture, it should be stored and handled in a dry box and for longer periods of time sealed under inert gas (N_2 or Ar) in Pyrex tubes.

The yield is approaching the theoretical value (2 mmole $Cs_2LiTmCl_6$: 1.309 g).

Properties

$Cs_2LiTmCl_6$ is colorless and moisture sensitive. For its characterization, common X-ray (Debye-Scherrer or, better, Guinier patterns) and/or spectroscopic techniques are used. The first nine d values of cubic-face-centered $Cs_2LiTmCl_6$ (a = 1043.9 pm) are: 6.022(54), 3.688(100), 3.145(25), 3.011(30), 2.608(64), 2.393(9), 2.129(43), 2.007(11), 1.844(39) Å; numbers in parentheses are calculated intensities on a 1-100 scale.

Fig. 1 shows the Raman spectra of $Cs_2LiTmCl_6$ and $Cs_2LiTmBr_6$.

B. OTHER CHLORO AND BROMO ELPASOLITES

Other chloro elpasolites are obtained by the same preparative route by selecting the appropriate starting materials (carbonates and chlorides as well; Ag_2O is also useful). For the preparation of bromo elpasolites, chlorides as starting materials of course have to be replaced by bromides, hydrochloric acid by hydrobromic acid, and HCl gas by HBr gas, which is also obtainable in steel cylinders or lecture bottles.

When Pr or Tb compounds are to be prepared, the use of "Pr_6O_{11}" or "Tb_4O_7" (the sesquioxides are not available commercially) requires more time to obtain a clear solution (maybe 15 min or more), and additional HCl solution may be needed, for example, to dissolve precipitated binary halides such as AgCl and TlCl.

Larger-scale syntheses should be possible with appropriate sizes of the crucible, quartz tube, and tubular furnace.

In some cases (elpasolites $Cs_2NaRECl_6$ [1]) the desired products crystallize from aqueous HCl solution so that drying the residue in the HCl gas stream is not necessary. However, to make sure that the expected compound forms, the drying step should be carried out. In general, melting in the open flow system described above should be avoided to prevent undesirable effects such as incongruent melting, sublimation of binary components, or creeping of the material over the edges of the crucible. Therefore, the temperature should be maintained at 400-450° in the case of the rubidium chlorides and compounds of the $Cs_2LiREBr_6$ type.

Large single crystals of $Cs_2NaRECl_6$- and $Cs_2NaREBr_6$-type compounds can be grown by a Bridgman technique.[8-11]

Four different crystal structure types are known for elpasolites depending on the composition and the intensive variables (temperature, pressure). K is a cubic face-centered K_2NaAlF_6 type,[12] *the* elpasolite (there is also a lower symmetry variety, T, which can be indexed tetragonally[5]) with [BX_6] and [REX_6] octahedra sharing common corners, and B^I and RE^{III} occupying the octahedral interstices alternately between layers of composition AX_3 with the stacking sequence ABCABC....

The *2L* form has the trigonal Cs_2LiGaF_6 structure type,[13] with a stacking sequence AB.... The [BX_6] and [REX_6] octahedra all share common faces. Both *6L* (HT-K_2LiAlF_6-type,[14] stacking sequence: ABCBAC...) and *12L* (rhombohedral Cs_2NaCrF_6 type,[13] stacking sequence ABABCACABCBC...) are polytypes between the polymorphs K and *2L* with confacial bioctahedra (*6L*) and triple octahedra (*12L*), respectively, connected via common corners with single octahedra. Fig. 2 shows the characteristic structural features.

Thermally induced phase transitions are found for $Rb_2NaTmCl_6$ ($T \rightarrow K$)[5] and $Cs_2LiLuCl_6$ ($6L \rightarrow K \rightarrow 6L$)[15] as well as for others. In general, higher temperatures favor structures with more face-sharing octahedra, i.e., with larger molar volumes.

Known chloro and bromo elpasolites are summarized in Tables I and II to-

TABLE I Lattice Constants (in pm) of Cubic and Tetragonal Chloro and Bromo elpasolites

RE	$Rb_2LiRECl_6$	$Rb_2NaRECl_6$	$Cs_2LiRECl_6$	$Cs_2AgRECl_6$	$Cs_2NaRECl_6$	Cs_2KRECl_6	$Cs_2LiREBr_6$	$Cs_2NaREBr_6$
La			1072.4		1099.2	(1137.9)[a]	1127.9	1151.9[b] 1158.5 1150.3
Ce			1067.5	1090.7	1094.6	1128.1 1135.1	1123.6	1146.8
Pr			1065.1	1087.8	1091.2	1130.9	1120.7	1144.6
Nd			1061.9	1085.2	1088.9	1126.8	1118.2	1139.3
Sm	1042.7[b] 1050.0		1057.7	1080.0	1083.4	1121.3	1113.4	
Eu	1040.4 1048.2	1064.1 1069.9	1055.4	1078.0	1081.0	1118.0	1111.4	1137.8
Gd	1038.8 1043.8	1061.3 1067.6	1053.4	1075.8	1079.2	1116.4	1109.7	1136.1
Tb	1037.8		1050.4	1073.5	1076.7	1116.0	1107.3	1133.4
Dy	1035.4	1057.6 1065.2	1048.7	1071.2	1074.3	1115.9	1105.7	1131.4
Y	1034.4	1056.6 1063.3	1047.9	1069.3	1073.2	1112.8	1104.7	1130.1
Ho	1033.9	1056.2 1062.5	1047.0	1069.5	1072.9	1111.6	1103.4	1129.3
Er	1030.7	1053.9 1060.5	1045.1	1068.2	1070.4	1109.9	1102.1	1127.9
Tm	1029.6	1053.9 1059.9	1043.9	1065.4	1068.6	1107.2		1125.3
Yb	1028.3	1050.9 1057.6	1041.8	1064.4	1067.7	1107.4		1124.5
Lu	1026.9	1049.4 1057.0		1062.6	1065.5	1102.7		1123.3
Sc	1010.7	1036.4		1045.2	1048.8	1088.0		1107.0
Ref.	5	5	4	3	1	2[c]	7	6, 7

[a] Presumably tetragonal. [b] Tetragonal with a above and c below. [c] Ce, Lu, and Sc compounds are own observations.

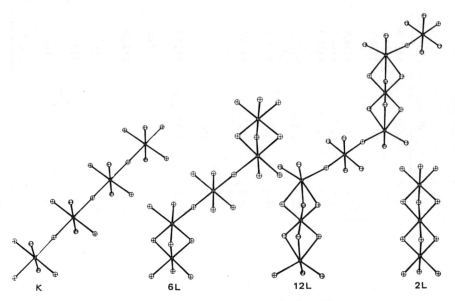

Fig. 2. Characteristic structural features in $A_2^I B^I RE^{III} X_6$-type compounds ("elpasolites"). Sections in $[00.1]$ direction show the principal connection of $[BX_6]$ and $[REX_6]$ octahedra (see text).

gether with their lattice constants. Mixed crystals with their particular properties have also been prepared.[5]

All these compounds are most conveniently characterized by powder X-ray diffraction methods. To calculate line intensities, computer programs such as LAZY-PULVERIX[16] are readily available; the required crystallographic data may be found in References 12–14.

TABLE II Trigonal and Hexagonal Chloro and Bromo elpasolites: Crystal Structures and Lattice Constants[4,7]

	Structure Type	a/pm	c/pm
$Cs_2LiLuCl_6$	$6L$	738.6	1821
$Cs_2LiScCl_6$	$12L$	730.0	3601
$Cs_2LiTmBr_6$	$6L$	777.4	1925.2
$Cs_2LiYbBr_6$	$6L$	777.0	1924.1
$Cs_2LiLuBr_6$	$6L$	775.9	1919.6
$Cs_2LiScBr_6$	$2L$	761.8	651.3

References

1. L. R. Morss, M. Siegal, L. Stenger, and N. Edelstein, *Inorg. Chem.*, **9**, 1771 (1970).
2. G. Baud, L. Baraduc, P. Gaille, and J.-C. Cousseins, *C. R. Hebd. Seances Acad. Sci.*, C **272**, 1328 (1971).
3. G. Meyer and P. Linzmeier, *Rev. Chim. Miner.*, **14**, 52 (1977).
4. G. Meyer and H.-C. Gaebell, *Z. Anorg. Allgem. Chem.*, **445**, 147 (1978).
5. G. Meyer and E. Dietzel, *Rev. Chim. Miner.*, **16**, 189 (1979).
6. G. Meyer and P. Linzmeier, *Z. Naturforsch.*, **32b**, 594 (1977).
7. G. Meyer and H.-C. Gaebell, *Z. Naturforsch.*, **33b**, 1476 (1978).
8. R. W. Schwartz, T. R. Faulkner, and F. S. Richardson, *Mol. Phys.*, **38**, 1767 (1979).
9. R. W. Schwartz and P. N. Schatz, *Phys. Rev.*, B **8**, 3229 (1973).
10. G. Mermant and J. Primot, *Mat. Res. Bull.*, **14**, 45 (1979).
11. H.-D. Amberger, *Z. Naturforsch.*, **35b**, 507 (1980).
12. L. R. Morss, *J. Inorg. Nucl. Chem.*, **36**, 3876 (1974).
13. D. Babel and R. Haegele, *J. Solid State Chem.*, **18**, 39 (1976).
14. H. G. F. Winkler, *Acta Cryst.*, **7**, 33 (1954).
15. G. Meyer and W. Duesmann, *Z. Anorg. Allgem. Chem.*, **485**, 133 (1982).
16. K. Yvon, E. Parthe, and W. Jeitschko, *J. Appl. Cryst.*, **10**, 73 (1977).

3. TANTALUM AS A HIGH-TEMPERATURE CONTAINER MATERIAL FOR REDUCED HALIDES

Submitted by JOHN D. CORBETT[*]
Checked by ARNDT SIMON[†]

A refractory metal container is virtually mandatory for syntheses and reaction studies at high temperatures ($>$500-600°) involving the more active transition and inner transition (rare earth) elements and their reduced phases. This requirement arises from the fact that these elements generally react with glass or ceramic containers to form very stable oxides, silicides, borides, carbides, nitrides, etc. Considerations of strength, inertness, tightness, ease of fabrication, and availability rapidly limit these container possibilities to tantalum, niobium, molybdenum, and tungsten. Although the two group-VI elements generally provide the greater inertness, they are used only when necessary because of greater difficulty in their fabrication. Tungsten can generally be formed only with sintering techniques, whereas for molybdenum, heat treatment and special care to avoid surface

---

[*]Ames Laboratory and Department of Chemistry, Iowa State University, Ames, IA 50011. The Ames Laboratory is operated for the U.S. Department of Energy by Iowa State University under contract No. W-7405-Eng-82.

[†]Max-Planck Institut für Festkörperforschung, Heisenbergstrasse 1, 7000 Stuttgart 80, West Germany.

and welding atmosphere contamination are necessary. Furthermore, cold working often leads to brittleness and cracking because of a low solubility of nonmetallic impurities.[1,2] On the other hand, tantalum (and niobium) are ductile and easily worked at room temperatures, since the nonmetallic impurities remain in solution. The service limits for these containers are extraordinary relative to silica; for example, >2000° or ≈30 atm internal pressure in standard 6-9-mm diameter Ta tubing, and even then reactions of many reduced compounds and metals with the container wall may be virtually undetectable.

Applications

We have found tantalum to be especially suitable for synthetic reactions and equilibrations involving elements and their reduced halides which lie to the left of group V in the periodic table, namely: the alkali metals and alkaline earth metals; scandium, yttrium, and the lanthanides; titanium,* zirconium, hafnium, thorium, and uranium. Tantalum and niobium are also uniquely suitable containers for the syntheses of their own lowest halides, for example, Ta_6Br_{14}[3] and $CsNb_6I_{11}$.[4] Tantalum containers have been extensively employed for the synthesis of halides, but reduced compounds of some other nonmetals, some oxides, for example, perhaps can be handled as well.

Other useful applications of tantalum are:

1. For equilibration of molten salts with metals or intermetallics which would normally reduce a Pyrex or silica container (e.g., Al, Na_3Bi) as well as for syntheses of the higher-melting and reactive intermetallic phases such as NaSb.[5]
2. As a condenser or container for sublimation or fusion of some halides, the rare earth metal triiodides, for example (these undergo serious metathetical reactions with glass at elevated temperatures to yield SiX_4 and the corresponding oxide or silicate). This pertains especially to the liquid and gaseous halides.[6]
3. In reductions of rare earth metal fluorides and chlorides either with an active metal (Ca, Mg, Li) or by electrochemical means.[7]
4. In many metallurgical applications where tantalum proves to be inert to a large variety of metals.

On the other hand, one should not attempt reactions in tantalum that involve either nonmetallic elements (Si, P, As, Se, Te) or relatively oxidized compounds of nonmetals (e.g., ZrS_2, Ta_2O_5) which form very stable compounds with Ta.

The usefulness of tantalum (and niobium) as container material derives from two physical properties of these elements. First, their ductility remains high even in the presence of significant nonmetallic impurities, as noted above. Sec-

*Titanium may be borderline; there is some evidence that Ta promotes the disproportionation of TiI_2 through dissolution of the Ti product.

ond, because of the high melting point of the metals (Ta, 2996°; Nb, 2468°), eutectic, eutectoid, etc., points with most other metals also lie at very high temperatures, whereas stable intermetallic phases with them are not common. Niobium may be the metal of choice when a very high melting point and strength are not critical because of its substantial cost advantage (see below). Though niobium has not been investigated as thoroughly, the enthalpies of formation of its lowest halides are generally close to those of tantalum, and so a similar inertness is expected. The high melting points of these materials also mean that interdiffusion of other metals into tantalum or niobium at moderate temperatures will be very slow and solid solution limits, low. For example, zirconium or rare earth elements generally cannot be detected on emission spectroscopic examinations of the inner walls of Ta containers which have been used for metal–metal halide reactions with these elements for weeks at 1000-1100°, and vice versa, though liquid rare earth metals dissolve appreciable Ta.[7] Nonmetallic impurities, especially hydrogen and oxygen, may be mobile in tantalum.

Metal Sources

Both tantalum and niobium are available* in foil, sheet, and deep-drawn tubing prepared from both welded and seamless stock. Tubing with a wall thickness of from 10-30 mil (0.25-0.76 mm) is most useful. Seamless tantalum tubing sells in quantity for about $75/m in 6-9-m diameter sizes, whereas an equal amount (volume) of niobium tubing costs only about 20% as much. The materials now available have been electron-beam melted under vacuum, a process which gives a substantial decrease in impurities through volatilization of most of the O, N, H, and C (as CO). Principal impurities in Ta are thus the heavy metals (e.g., 0.01-0.05% each of W and Nb), and guaranteed purities are >99.85% total. Niobium runs 99.7-99.8% total and contains principally small amounts of W and Ta.

Welding

Tantalum and niobium (as well as Mo, W, Zr and other metals) are readily sealed by arc welding in helium or argon as well as by electron-beam welding in high vacuum. Only the former technique is feasible for many laboratories. Cold welding with commercial pinch-off tools has not proved reliable even on smaller-diameter tubing.

An arc welding apparatus which has received extensive use has been described by Miller et al.[8] This description should be consulted for details. The schematic of Fig. 1 includes some recent changes. It will aid in understanding the principles

*Fansteel Metals, North Chicago, IL; Kawecki Berylco Industries, Reading, PA; Aeromet, Inc., Englewood, NJ.

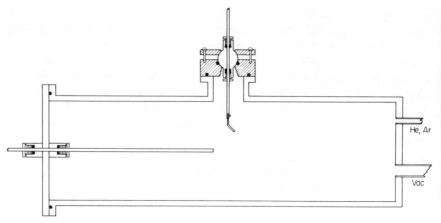

Fig. 1. Cross section of an arc-welder modified after that described in Reference 8. (Fastenings not shown.)

and techniques involved. A "stinger" cathode of Ta (or W) is mounted on an insulated and water-cooled rod projecting through a Bakelite cap on the side of a stainless steel chamber approximately 66 cm long by 20 cm dia. This electrode can be moved from side to side as well as in and out with the aid of a ball joint and compressed slip ring, both sealed with O rings. The normal anode is coaxial to the welding chamber and enters through another slip seal on the entry cover on the end. Materials to be welded can be clamped to this. However, tantalum containers loaded with materials that might melt or volatilize if heated strongly are better welded while resting upright on the bottom of the welder below the stinger with the sample container(s) held in a split metal block, pipe section, or collar with a hose clamp or set screw. Copper or brass materials are better here because of their higher thermal conductivities. The body of the welder is now the anode, and sheets of rubber are laid across the top of the chamber to prevent accidental (minor) shocks to the operator manipulating the cathode. For eye protection during welding, a sheet of welding glass must be flipped over the fused silica sight port that is located on the side of the welder below and $90°$ from the cathode. Although the top of a Ta container necessarily reaches a temperature above $3000°$ during the operation, the mass of the mounting block and the water cooling of the welder chamber allow the successful enclosure of compounds which develop appreciable vapor pressures by $150-200°$ without loss. With care, even K (mp $62°$) can be sealed in Ta without melting.

The welding current is best controlled through foot switches, one for off-on and the other for the necessary variation of the current according to the material, size, and progress of the job. The original design utilized a dc generator and achieved current control through variation in the depth of immersion of two electrodes in a (hard) water bath. A solid-state rectifying and control circuit has been designed and put into use more recently.[9] A welding supply would also be a suitable source, perhaps with some resistors in series to help stabilize the arc, and an arc-melting furnace of the sort illustrated in Volume XIV (p. 132) of *Inorganic Syntheses* would also work but only for pieces less than about 7.6 cm high unless the device is modified. Welding within the dry box where the samples are loaded would also be feasible if trimming, etc. are not necessary (see below) and the higher gas pressure left within the container is of no consequence. Ordinarily, noble gas pressures used during welding range from about 100 torr (Ar) to approximately 400-500 torr (He), the actual room temperature pressure left within the heated container being somewhat less. The absence of air leakage into the welder, especially in the slip seals, is very important and can be ascertained with a vacuum gage after pump down (below).

Container Fabrication

Commercially drawn tubing with 10-20 mil (0.25-0.51 mm) wall thickness is the easiest to use for small containers, although less expensive items as well as those of more diverse shapes can be fabricated from sheet stock with considerably more skill and a greater investment in time.[8,10]

A good snug fit is essential for successful welding, no matter what the size or shape. Some container seals and designs which have been found useful are shown in Fig. 2. The simplest seal is the crimped and welded type A, while the capped type B closure serves for larger containers or higher pressures. Caps of the type used in B have a height of about 3 mm and can be prepared with deep-drawn dies or spun on a mandril. More specialized container designs include C for isopiestic equilibration between adjoining phases, D for thermal analysis, and E for sealing tantalum under vacuum in the absence of an electron beam welder. Finally, design F illustrates a two-piece sublimation container and condenser which has been found convenient for purification of volatile halides when either the crude material or the condensate reacts with silica. The design also facilitates recovery of the product in the dry box. (Flaring to increase the diameter at the joint can also be used rather than swaging as shown.) Of course, some systems involve sufficiently low vapor pressures of any component that open, rather than sealed, tantalum crucibles of type A or B are sufficient, aided perhaps by a blanket of noble gas to retard evaporation.

Operationally, to avoid sealing difficulties or contamination of the reactants, the tube is first cleaned with a solution consisting of 50% (by vol)·conc. H_2SO_4, 25% conc. HNO_3, and 25% HF (48% strength).

■ **Caution.** *HF causes severe chemical burns–see Vol. III (p. 112) of Inorganic Syntheses regarding HF precautions. Work in the hood and wear gloves. Rinse Ta articles well with distilled water and dry thoroughly. A wash bottle containing aq $NaHCO_3$ is useful for identifying and neutralizing spills.*

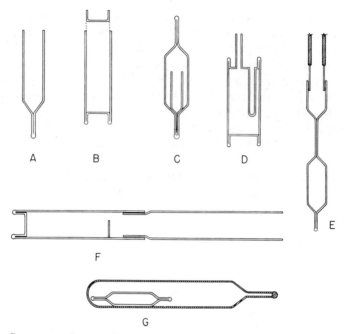

Fig. 2. Some tantalum seals and apparatus. (A) *Crimped and welded;* (B) *capped and welded;* (C) *apparatus for biphasic, isopiestic equilibrations;* (D) *thermal analysis container;* (E) *tantalum tube to be welded under vacuum [tube is brazed to metal-to-glass seal and closed with joint and stopcock at top) not shown)];* (F) *sublimation container with reservoir or condenser swaged or flared to fit;* (G) *reaction container sealed in silica jacket under vacuum.*

This mixture can be quite reactive with tantalum. Therefore, it is preferable to add HF to the first two components, as required, for mild etching. The used solution can be stored in polyethylene and reactivated with HF.

After cleaning, the lower end of the tube, A or B, is welded closed, and cleaned again, if necessary. After filling, the second end of A is flattened over at least a 4-5 mm length in the dry box with a small vise, taking care that the added reactants do not remain on the inside walls at the crimp. The sample container is carried to the welder within a polyethylene or glass container for protection of the contents until transfer to the welder. A further flattening of the crimp with a shop-weight vise (perhaps while enclosed in polyethylene for protection) is often helpful, as are trimming the closed end with a cutter to give a smooth, even end and ultimate buffing on a wire wheel. A close fit of the two walls across the crimp is essential. Spot welding of the crimp in larger tubing and careful control of current can be helpful in preventing any warping during welding. Sealing of tubes larger than 13 mm in diameter is usually easier with caps (style B). Contamination of the contents in A during the short periods of handling in air necessary for the above steps does not seem to be a problem with well-crimped tubes except for the most susceptible systems; end-capped (B) tubes have a shorter leakage path, but these can be rapidly and directly transferred to the welder which is then evacuated to at least 10^{-3} torr before back-filling with noble gas. Nitrogen within the tube from a dry-box atmosphere is readily removed during the evacuation and does not produce the nitride during the welding.

Use of the evacuated apparatus (style E) may improve some transport reaction rates. The tantalum portion is attached to a glass joint and stopcock by a brazed metal-to-glass seal. The flat part can be produced by a pair of vise blocks which are driven together by tightening bolts through them while the apparatus is attached to the vacuum line. The evacuated and closed container is then welded from the side at the flat portion and cut off above the weld. The same blocks may be utilized within the dry box instead of a vise.

Finally, the welded containers must be placed in sealed jackets or maintained under active vacuum to protect them from attack by the atmosphere at high temperatures unless vacuum or noble gas-filled furnaces are used. Either silica (Fig. 2G) or stainless steel is suitable. Careful cleaning of the former (dilute HF) prior to the final evacuation, outgassing (heating with torch), and sealing will reduce Ta volatilization (transport) and SiO_2 devitrification at the higher temperature. Ordinarily, holes in unsatisfactory welds can be detected by eye, perhaps with low magnification. Loss of volatile components through pinhole leaks which develop in unsatisfactory welds can easily be detected if a portion of the glass jacket is left projecting from the furnace during the initial heat up, thereby preventing dangerous buildup of volatile materials within the jacket. Clearly, leakage of a volatile component into a uniformly heated jacket before it can be reduced by metal to a less volatile lower halide within the Ta tube can present a hazard. Failures of gas-tight containers because of excessive pressures therein very rarely, if ever, occur by sudden rupture but rather through small holes which develop at welds after considerable distention. The capped style B is considerably stronger than are those with crimped ends (A) in this respect. For example, 13-19-mm dia. vessels with 0.40-.50-mm-thick walls and end caps constructed from 0.7-mm-thick sheets have allowed the retention of autogenous pressures of 30-40 atm within them on a regular basis. Niobium has a somewhat lower tensile strength at high temperatures than tantalum, but this is important only in extreme cases. The effective upper limit for a silica jacket in air is about 1100°, while that for steel is determined by its melting point. The use limit for the tantalum or niobium generally appears to be determined only by its melting point or by the formation of a liquid eutectic or intermetallic phase with its contents. A serviceability limit for Ta based on its recrystallization, which was cited earlier,[8] does not pertain to the more nearly pure material currently available that has been electron beam-melted and thus does not develop intergranular segregation of nonmetallic impurities unless contaminated.

Note should be made of the fact that tantalum is permeable to some gases at high temperature. Hydrogen readily diffuses through the metal walls above about 600°. This can be used to advantage either to remove H_2 traces from contained phases or as a means of forming new hydrides within a sealed tube while preventing the loss of other volatile components within, for example, of $ZrCl_4(g)$ above solid ZrClH.[11] Oxygen at low pressure (and probably other nonmetals) will also effectively pass through tantalum under a concentration gradient. This has been observed as oxyhalide contamination of metal–metal halide reactions of rare earth metals, thorium, etc. *within* the container at ~800–1000° after development of very small air leaks in either the sealed or pumped glass jacket. The integrity, appearance, and ductility of the tantalum may remain quite good under such gettering conditions.

The tantalum and niobium tubes are readily trimmed and opened after use with the spin-cutter of the sort used for cutting copper pipe. A tapered flaring tool is also helpful during the construction of apparatus. The tantalum tubing can often be reused if contamination from contents or outgassing of the jacket is not serious. Ductility or softness (low yield stress) is a good indication of purity.

Acknowledgment

The major credit for the above applications belongs to a long succession of able and persistent graduate students in the author's laboratories. Also, Professors A. H. Daane and Karl A. Gschneidner, Jr. have generously accommodated us on their welders through these years.

References

1. C. A. Hampel, *Rare Metals Handbook*, 2nd Ed., Reinhold Publ. Co., New York, 1961, p. 298.
2. *Molybdenum Metal*, Climax Molybdenum Co., Climax, Colorado, 1960, p. 78–82.
3. R. E. McCarley and J. C. Boatman, *Inorg. Chem.*, 4, 1486 (1965).
4. H. Imoto and J. D. Corbett, *Inorg. Chem.*, 19, 1241 (1980).
5. M. Okada, R. A. Guidotti, and J. D. Corbett, *Inorg. Chem.*, 7, 2118 (1968).
6. J. D. Corbett, *Inorg. Nucl. Chem. Lett.*, 8, 337 (1972).
7. B. J. Beaudry and K. A. Gschneidner, Jr., *Handbook on Physics and Chemistry of the Rare Earths*, K. A. Gschneidner and L. Eyring (eds.), North-Holland Publishing Co., Amsterdam, 1978, Chapter 2.
8. A. E. Miller, A. H. Daane, C. E. Haberman, and B. J. Beaudry, *Rev. Sci. Inst.*, 34, 644 (1963).
9. U. Löchner, Ph.D. Thesis, University of Karlsruhe, West Germany, 1980 (Prof. H. Bärnighausen).
10. A. H. Daane, *Rev. Sci. Inst.*, 23, 245 (1952).
11. H. Imoto, J. D. Corbett, and A. Cisar, *Inorg. Chem.*, 20, 145 (1981).

4. CESIUM SCANDIUM(II) TRICHLORIDE

$$3CsCl + 2ScCl_3 + Sc \longrightarrow 3CsScCl_3$$
$$Cs_3Sc_2Cl_9 + Sc \longrightarrow 3CsScCl_3$$

Submitted by K. R. POEPPELMEIER* and J. D. CORBETT*
Checked by GERD MEYER†

The synthesis of the title compound[1] is a good example of a reaction that is made possible through the proper choice of the container material. In this preparation, the tantalum reaction vessel serves to contain both the potentially reactive and volatile scandium(III) chloride melt and the strongly reducing scandium metal under the conditions necessary for complete reaction. This choice of components is required since the simple binary scandium(II) compound $ScCl_2$ has not been prepared; thus, the direct reaction of $ScCl_2$ with alkali metal halide cannot be used to obtain $CsScCl_3$. The reduction starts instead with either the separate components $CsCl + ScCl_3$ (method A), or the compound $Cs_3Sc_2Cl_9$ obtained by indirect means[2] (method B). The ternary halide $CsScCl_3$ is of value since it exhibits an unusual nonstoichiometry on the metal-poor side and contains the electronically interesting d^1 ion of scandium(II).

Procedure

Standard vacuum line and dry-box techniques must be used for manipulation and storage of the starting materials and the product; both moisture and oxygen must be scrupulously excluded to avoid contamination by ScOCl.

A. SYNTHESIS WITH CsCl

The CsCl ($\geqslant$99.9% purity) is dried at 200° under high vacuum and then melted in a silica container. The latter step reduces the possibility of recontamination by surface moisture and makes weighing and complete sample transfer easier in the dry box.

Sublimed $ScCl_3$ (750°, 10^{-5} torr in a tantalum jacket) which has received

---

*Ames Laboratory and Department of Chemistry, Iowa State University, Ames, Iowa 50011. The Ames Laboratory is operated by the U.S. Department of Energy by Iowa State University under contract No. W-7405-ENG-82.
†Institut für Anorganische und Analytische Chemie I, Justus-Liebig-Universität, 6300 Giessen, West Germany.

minimum exposure to the dry-box atmosphere will give the best product, and that from the synthesis procedure in Chapter 1, Section 8 has been preferred. (Although anhydrous scandium(III) chloride is available from a few suppliers, those samples have not been sufficiently pure. Sublimation and a check for other metals, especially Fe, are advisable.) The product after sublimation will give typical recoveries for both chloride and metal of 100.0 ± 0.2 wt. % with Cl:Sc = 3.00 ± 0.01. Scandium is titrated with EDTA using Xylenol Orange as the indicator[3] and chloride is determined gravimetrically as AgCl. However, since the analysis is not nearly sensitive enough to detect deleterious amounts of moisture or hydrolysis products and the freedom of the product from ScOCl, the $ScCl_3$ melting point ($967 \pm 3°$) is probably a better measure of purity.

The next step in the preparation is to combine in the dry box the stoichiometric amounts (molar ratio 3:2:1) of CsCl, $ScCl_3$, and Sc* (powder, if available; otherwise small chunks or foil) respectively, the total quantities conveniently varying from one to several grams depending on the size of the tantalum tube and the precision of weighing possible.† The tantalum container is crimped and sealed by arc welding using the apparatus described earlier in this volume (Chapter 1, Sec. 3) and the welded tube next sealed under vacuum in a fused silica tube. The sample is now heated to 700° for 2-3 days and then cooled to 650° and annealed for at least 24 hr. The latter step is necessary since the fully reduced phase appears to melt somewhat incongruently. The reaction product should be examined closely to determine whether pieces of unreacted scandium metal can be detected. If so, the sample should be ground, resealed, and the procedure repeated. (Lines at 6.99, 4.70, 4.19, and 3.78 Å distinguish $Cs_3Sc_2Cl_9$ from the similar lattice of $CsScCl_3$,[1] but these are weak to very weak even with the pure phase and so are not very useful in checking for complete reduction.) The reaction can also be run with excess metal if it is in a form (foil, etc.) which can easily be recovered.

Either procedure when properly applied will give an essentially quantitative yield. The compound ScOCl, which haŝ a light pink tinge in reducing systems, is the most likely impurity but can be avoided by careful work. The stronger lines in the ScOCl powder pattern (with intensities in parentheses) are about: 8.2(vs),

---

*High-purity metals are available from a variety of suppliers. These inlude Atomergic-Chemicals Corp., 100 Fairchild Avenue, Plainview, NY 11803; Leico Industries, Inc., 250 West 57th Street, New York, NY 10019; Lunex Company, P.O. Box 493, Pleasant Valley, IA 52767; Molycorp, Inc., 6 Corporate Park Drive, White Plains, NY 10604; Research Chemicals Division, Nuclear Corporation of America, P.O. Box 14588, Phoenix, AZ 85031; Ronson Metals Corporation, 45-65 Manufacturers Place, Newark, NJ 07105; Rare Earth Products, Ltd., Waterloo Road, Widnes, Lancashire, England (United States representative of rare earth products is United Mineral & Chemical Corp., 129 Hudson St., New York, NY 10013; and Kolon Trading Co., Inc., 540 Madison Avenue, New York, NY 10022.

†A useful electronic weighing sensor with remote output is available from Scientech, Inc., Boulder, CO.

3.56(vs), 2.63(vs), 2.023(ms), 1.878(s), 1.585(ms) Å. It is estimated that ScOCl can be detected at ≤1% visually and ≤2% by careful Guinier work.

B. SYNTHESIS WITH $Cs_3Sc_2Cl_9$

Alternatively, $Cs_3Sc_2Cl_9$ can be used as the starting material for reduction. This is prepared by dehydration of a mixture of the hydrated chloride and $Cs_2ScCl_5 \cdot H_2O$ under flowing HCl according to the procedure of Meyer (this volume, Chapter 1, Sec. 1-B). The driving force of the formation of the anhydrous ternary compound $Cs_3Sc_2Cl_9$ avoids the production of ScOCl which would otherwise result when $ScCl_3$ preparation is attempted by this method in the absence of CsCl.

Properties

The compound $CsScCl_3$ has the $2H$ (hexagonal perovskite) structure of $CsNiCl_3$, space group $P6_3/mmc$, $a = 7.350(2)$ Å, $c = 6.045(3)$ Å, and represents the first example of a scandium(II) compound, although many more reduced metal-metal bonded phases are known.[4] The material is black in color in bulk and blue when ground. The material is relatively stable to moisture for a reduced scandium compound but still readily reacts with water with the evolution of hydrogen gas.

The stoichiometry of $CsScCl_3$ can evidently be varied in single phase between the $6R$-type $(Cs_3Tl_2Cl_9)$ structure of the scandium(III) salt $Cs_3Sc_2Cl_9$ and $Cs_3Sc_3Cl_9 (= CsScCl_3)$ according to the reaction

$$Cs_3Sc_2Cl_9 + xSc \longrightarrow Cs_3Sc_{2+x}Cl_9, \quad 0 \leqslant x \leqslant 1.0$$

without the formation of coherent superstructures according to powder X-ray diffraction (Guinier technique). The material is a semiconductor, there being evidently only a weak interaction between the Sc(II) ions even though they form an infinite chain (surrounded by confacial chloride trigonal antiprisms) with a separation of only 3.02 Å.

Analogous phases exist with $CsScBr_3$, $CsScI_3$, $RbScCl_3$, and $RbScBr_3$ but have not been obtained with potassium. But the stoichiometry range noted above exists only with $Rb_3Sc_2Cl_3$ because the other $M_3^I R_2 X_9$ phases have a different and less favorable structure ($Cs_3Cr_2Cl_9$ type).

References

1. K. R. Poeppelmeier, J. D. Corbett, T. P. McMullen, R. G. Barnes, and D. R. Torgeson, Inorg. Chem., **19**, 129 (1980).

2. G. Meyer and A. Schönemund, *Mat. Res. Bull.*, **15**, 89 (1980).
3. J. Korbl and R. Pribil, *Chemist-Analyst*, **45**, 102 (1956).
4. K. R. Poeppelmeier and J. D. Corbett, *J. Am. Chem. Soc.*, **100**, 5039 (1978).
5. G. Meyer and J. D. Corbett, *Inorg. Chem.*, **20**, 2627 (1981).

5. ZIRCONIUM MONOCHLORIDE AND MONOBROMIDE

$$3Zr \text{ (sheet, turnings)} + ZrX_4(l, g) \xrightarrow[Ta]{350, 600, 800°} ZrX(s)$$

Submitted by RICHARD L. DAAKE* and JOHN D. CORBETT†
Checked by DONALD W. MURPHY ‡

The two zirconium monohalides represent the first examples of metallic salts with double metal layers sandwiched between two halogen layers in a close-packed array.[1,2,3] At the present time, they are also the most stable of these monohalides and the easiest to prepare in microcrystalline form.

Direct high-temperature reactions of Zr and the corresponding ZrX_4 are at present the only practical route to these phases, but these processes are characteristically slow and incomplete, giving mixtures with higher halides. In one of the procedures to be described, the zirconium is finely divided. Even so, the product is impure and requires reequilibration with fresh metal. A second approach utilizes thin turnings in a stoichiometric reaction which is carried out at increasing temperatures as the reduction proceeds through the less volatile, intermediate phases. Both procedures require the use of tantalum containers, not only to contain the sizable pressures involved but also to provide the necessary inertness to the products, a property that glass does not possess. Techniques and equipment necessary for effective use of tantalum as a container material are described elsewhere in this volume (Chapter 1, Section 3).

Procedure

Because zirconium halides are air- and moisture-sensitive, handling of these compounds must be done in an inert atmosphere dry box equipped to maintain both

---

*Bartlesville Wesleyan College, Bartlesville, OK 74003.
†Ames Laboratory and Department of Chemistry, Iowa State University, Ames, IA 50011. The Ames Laboratory is operated by the U.S. Department of Energy by Iowa State University under contract No. W-7405-Eng-82.
‡Bell Laboratories, 600 Mountain Avenue, Murray Hill, NJ 07974.

oxygen and water at or below the 10-ppm level. Prepurified nitrogen low in oxygen serves as an adequate working atmosphere. The nitrogen is passed over activated Linde Molecular Sieves to maintain an acceptably low moisture level. Generally, the reactivity of the zirconium halides to air and moisture decreases with decreasing oxidation state.

The $ZrCl_4$ and $ZrBr_4$ starting materials should be of high purity, particularly with respect to other nonmetallic components, for these are apt to be carried into the final monohalide product. Commerical products* should be sublimed under high vacuum (at ca. 300°), preferably through an intervening, coarse-grade glass frit to reduce entrainment of oxyhalide, etc.

The metal for the reduction needs to be low in nonmetallic impurities in order to have adequate ductility for rolling. Reactor-grade (crystal bar) quality or equivalent is recommended.† Thin metal sheet 0.25-0.5-mm-thick or turnings are made using standard vacuum melting, cold rolling, and machining techniques. Thin uniform turnings ($\leqslant 0.1$ mm) are highly desirable for the second procedure described below.

Because of the relatively high intermediate pressures (20-40 atm) generated in the reactions, tantalum end caps with thickness 30-50% greater than that of the tubing itself are recommended.

■ **Caution.** *Crimp welds cannot dependably withstand the necessary pressures. A steel or Inconel tube furnace liner is recommended as partial protection against possible explosion hazard. It also serves to smooth out temperature gradients.*

Thermal cycling will be lessened if the control (but not the measuring) thermocouple is placed between the liner and the furnace wall. It is highly advisable that the fused silica jacket about the welded container be sufficiently long to extend from the end of the furnace to permit condensation of the tetrahalide, should a leak develop. Unless a thermal gradient is required, the tantalum tube itself should be centered in the furnace. Although the explosion hazard is very slight if these procedures are followed, the open ends of the furnace should still be shielded from workers.

A. SYNTHESIS WITH ZIRCONIUM FOIL

The original two-step method described required zirconium sheet or foil in large excess to compensate for a surface-limited reaction. The sheet form also allows the ready separation of the product from unreacted metal. Although the litera-

---

*Available from Materials Research Corp., Orangeburg, NY; Alfa Division, Ventron Corp., Danvers, MA; Cerac, Butler, WI.
†Available from Materials Research Corp., Orangeburg, NY; Goodfellow Metals, Cambridge, England; Atomergic Chemetals Corp., Plainview, NY.

ture describes this method only for ZrCl (and HfCl),[1] the technique works equally well for ZrBr.

A 3–8 g quantity of tetrahalide together with a fourfold (by weight) excess of zirconium sheet is contained in a length of 9–19-mm-diameter tantalum tubing with at least a 0.4-mm wall which is capped as described in the welding section (Chapter 1, Section 3). This is sealed in an evacuated fused silica tube and heated in a gradient with the Zr at the hot end. The gradient serves mainly to provide a cool reservoir for volatile phases (ZrX_4 and ZrX_3) while also keeping the metal foil at a sufficient temperature to maintain an acceptable reaction rate. The reaction temperature is raised slowly over 2–3 days from a 350°/450° gradient to a 500°/600° gradient where it is held for 4–6 days. (The gradient is measured with thermocouples strapped to the outside of the silica jacket and the real gradient is doubtless less.)

This first-stage reaction gives a nonequilibrium mixture of reduced phases, owing to physical blockage of the metal surface to further reaction. As expected, the somewhat volatile trichloride is found at the cooler end of the reaction tube, often as a plug adjacent to a band of well-formed $ZrCl_3$ needles up to 1 cm or more in length. The metal foil is found coated with a relatively thick (up to 1 mm) layer of strongly adhering blue-black product of variable composition ($1.2 \leqslant Cl:Zr \leqslant 1.5$) containing ZrCl and $ZrCl_2$ (according to X-ray powder pattern analysis).

In the second stage, the $ZrCl/ZrCl_2$ mixture, scraped from the foil in the dry box, is equilibrated isothermally with a 10-fold excess of *fresh* foil in a new reaction tube of the same type. The temperature is initially set at 600° and slowly raised to 800° over a period of 4–6 days, resulting in an essentially quantitative conversion to ZrCl, which can be recovered easily from the metal substrate in excellent purity.

If only a small amount (<1 g) of ZrCl or ZrBr is desired, it can be prepared directly in a single step with a 15-fold excess (by weight) of metal foil, the temperature being gradually raised from 400° to 800° over 10–14 days.

B. SYNTHESIS USING ZIRCONIUM TURNINGS

Thin turnings of metal can also be caused to react with a stoichiometric amount of tetrahalide to achieve 100% conversion to ZrCl or ZrBr. The method is also suitable for much larger quantities (50–60 g) in a single batch and is the preferred route if thin, uniform turnings can be obtained. The turning process is believed to produce cracks along grain boundaries, thereby making the bulk of the metal more accessible to the gaseous ZrX_4. (Very thin foil has not been tried.) The monohalides are essentially line compounds, and a homogeneous product is obtained without taking special precautions regarding moderate thermal gradients along the reaction tube.

The weighed tetrahalide is first transferred into the cap-welded end of a tantalum tube of size noted above (tube volume $\simeq 1.5$ cm^3/g ZrX). The appropriate weight of turnings ($\pm 0.1\%$) is then packed into the tube. Reactants are loaded in this order to prevent welding difficulties from the volatile tetrahalides.

After the second cap is welded in place, the tube is jacketed in fused silica and the assembly is placed in a tube furnace with an Inconel liner. The temperature is then slowly raised from 350° to 850° over a period of 7-10 days and maintained there for 3-5 days.

Both ZrCl and ZrBr are stoichiometric line compounds as far as can be judged. Although standard chemical analyses for metal and halogen[1] may be useful in assaying the monohalides produced from reactions using excess metal foil, X-ray diffraction patterns are suitable indicators of phase purity for most purposes. If normal-abundance zirconium is used, there will be some fractionation of the fraction of hafnium contained therein, particularly in the first procedure. This must be taken into account in gravimetric procedures, but it will affect X-ray results only slightly. The values (Å) and intensities (10 maximum) reported below were obtained with a precision flat-plate Guinier camera and NBS silicon powder as an internal standard. The platelet morphology of ZrCl and ZrBr causes preferred orientation in the flat-plate technique, so that Debye-Scherrer intensities may deviate substantially from those reported here.

For ZrCl, the 14 most intense lines in the Guinier pattern ($\theta < 45°$) are: 8.90 Å(5), 2.943(2), 2.706(10), 2.224(2), 2.215(3), 1.710(10), 1.679(6), 1.596(3), 1.481(3), 1.445(3), 1.354(3), 1.120(2), 1.118(5), 1.104(4).[3] (Areas of the pattern are usually banded together because of grinding damage.) Some characteristic lines of potential impurity phases occur at 5.53, 2.832, 2.682, and 2.212 Å for ZrCl$_3$; 6.46, 2.507, 2.338, and 1.691 Å for $3R$-ZrCl$_2$,[4] and 6.92, 6.49, 2.541, and 1.797 (doublet) Å for the Zr$_6$Cl$_{12}$ phase.[5] A line at 2.461 Å is the best measure of α-Zr in the presence of any of these halides.

For the monobromide, the 11 most intense lines in the Guinier pattern[2] are: 9.36 Å(2), 3.016(2), 2.965(2), 2.669(10), 2.419(2), 2.060(3), 1.752(5), 1.464(4), 1.402(2), 1.279(2), 1.124(3). Characteristic lines of potential impurity phases occur at 5.84, 3.157, 2.976, and 2.305 Å for ZrBr$_3$; 6.86, 3.053, 2.540, and 1.763 Å for one form of ZrBr$_2$;[6] and 7.29, 3.077, 2.669, 1.892, and 1.883 Å for Zr$_6$Br$_{12}$.[5]

Properties

Zirconium monochloride and monobromide are black powders or highly reflective microcrystals. Although the monohalides appear to be stable in air for days to weeks, they should be kept and handled under inert atmospheres if high purity is required, owing to a probable slow reduction of water vapor to form the hydrides (see below) and ZrO$_2$. These monohalides possess a rhombohedral ($R\bar{3}m$), three-slab structure in which each slab consists of four tightly bound, close-packed layers sequenced X-Zr-Zr-X.[2,3] The two monohalides actually possess slightly different structures which are interrelated through interchange of two of the four-layered slabs. Very strong Zr–Zr bonding is evident between the pairs of metal layers. In ZrCl, for example, each metal atom has six like neighbors in the same layer at 3.43 Å plus three in the adjoining layer at 3.09 Å. For comparison, the average internuclear distance in the 12-coordinate metal is 3.20 Å. Both phases are Pauli paramagnetic. As is typical of most layered compounds, the weak Van der Waals interactions between two neighboring chlorine layers in two adjacent slabs allow easy cleavage.[3] The zirconium monohalide structures, formally d^3 for zirconium, have also been found for a number of d^2 examples, namely in ScCl[7] and many of the rare earth elements.[8]

The zirconium monohalides are essentially two-dimensional metals with good delocalization and conduction within the double metal sheets. The metallic conductivity has been established by both X-ray photoelectron spectroscopy[2] and conductivity measurements where single-crystal data (obtained by an unspecified method) indicated a conductivity of 55 ohm^{-1} cm^{-1} parallel to the plates and about 10^{-3} ohm^{-1} cm^{-1} normal to them.[9]

Both compounds are sufficiently ductile that they may be pressed at room temperature to 97% of theoretical density.[2] The ZrCl particularly shows extensive damage on grinding, analogous to that in graphite, which is evident in the powder pattern through banding of certain classes of reflections. In contrast the bromide shows this only to a small extent, evidently because of different bonding interactions with second nearest neighbors.[2] It has not proved possible to intercalate these phases. However, both monohalides do take up hydrogen readily just above room temperature to form discrete hemihydride and monohydride phases.[10] The hydrogen atoms therein are mobile above ~80°, occupying primarily the tetrahedral holes.[11] The monohydride phases appear to be metastable and to disproportionate above 600° into ZrH_2 and the corresponding Zr_6X_{12} phases, thereby providing a better route to these compounds than achieved by direct reactions.[5]

The monohalides are useful reducing agents for the preparation of a number of intermediate zirconium chloride and bromide phases since their reactivities are considerably greater than those of the refractory metal. Thus, ZrX–ZrX_4 reactions have been used to obtain the trihalides in 100% yields[12] as well as $ZrCl_2$(3R–NbS_2 type)[4] and a different polytype of $ZrBr_2$.[6]

References

1. A. W. Struss and J. D. Corbett, *Inorg. Chem.*, 9, 1373 (1970).
2. R. L. Daake and J. D. Corbett, *Inorg. Chem.*, 16, 2029 (1977).
3. D. G. Adolphson and J. D. Corbett, *Inorg. Chem.*, 15, 1820 (1976).
4. A. Cisar, J. D. Corbett, and R. L. Daake, *Inorg. Chem.*, 18, 836 (1979).
5. H. Imoto, J. D. Corbett, and A. Cisar, *Inorg. Chem.*, 20, 145 (1981).

6. R. L. Daake, Ph.D. Thesis, Iowa State University, 1976.
7. K. R. Poeppelmeier and J. D. Corbett, *Inorg. Chem.*, **16**, 294 (1977).
8. Hj. Mattausch, A. Simon, N. Holzer, and R. Eger, *Z. Anorg. Allgem. Chem.*, **466**, 7 (1980).
9. S. I. Troyanov, *Vestn. Mosk. Univ., Khim.*, **28**, 369 (1973).
10. A. W. Struss and J. D. Corbett, *Inorg. Chem.*, **16**, 360 (1977).
11. T. Y. Hwang, R. G. Barnes, and D. R. Torgeson, *Phys. Lett.*, **66A**, 137 (1978).
12. R. L. Daake and J. D. Corbett, *Inorg. Chem.*, **17**, 1192 (1978).

6. LANTHANUM TRIIODIDE
(AND OTHER RARE EARTH METAL TRIIODIDES)

$$2La(s) + 3HgI_2(l) \xrightarrow{330°} 2LaI_3(s) + 3Hg(l)^1$$

$$2La(s) + 3I_2(g) \xrightarrow{800°} 2LaI_3(l)^2$$

Submitted by JOHN D. CORBETT*
Checked by ARNDT SIMON†

Lanthanum triiodide may be obtained from the oxide by a number of routes, such as dissolution of the oxide in HI(aq) followed by precipitation of the hydrated iodide and vacuum dehydration (20–350°), perhaps in the presence of excess NH_4I.[3] However, the yield will be low because of significant hydrolysis to LaOI, and one or two vacuum sublimations or distillations of the LaI_3 product will be necessary to obtain a useable material. On the other hand, the ready availablility of lanthanum (and other rare earth elements) as good-quality metal makes direct reactions considerably more attractive. A simple route suitable for small quantities is the reduction of excess liquid HgI_2 by lanthanum in a sealed Pyrex container.[1] The direct combination of the elements is less expensive and much better for larger quantities; however, some attention must be paid to the choice of container. The metal must be maintained above 734° (the LaI_2-LaI_3 eutectic[4]) so that a layer of solid iodides does not block the surface of the metal to further oxidation, but the reducing melt above that temperature acts as a flux for the spontaneous reduction of SiO_2 by La. Therefore, both the metal and the reduced melt must be kept out of contact with silica (and most other ceramic

*Ames Laboratory and Department of Chemistry, Iowa State University, Ames, IA 50011. The Ames Laboratory is operated by the U.S. Department of Energy by Iowa State University under Contract No. W-7405-Eng-82.
†Max-Planck Institut für Festkörperforschung, Heisenbergstrasse 1, 7000 Stuttgart 80, West Germany.

materials). Among the available metal containers that are suitable for the rare earth elements only tungsten is also sufficiently inert to iodine at about 0.1-1 atm and 800° to give minimal side reactions.[2] The reaction also illustrates the safe application of a hot-cold, sealed reaction tube for the reaction of materials with very different volatilities.

The reactions described below have been applied to the syntheses of the triiodides of the rare earth elements as well as those of yttrium and scandium.

A. SYNTHESIS USING HgI_2

Lanthanum metal* in the form of sheet, small lumps, or turnings should be stored under vacuum or inert gas and protected from the atmosphere as much as possible. (The reactivity of the rare earth metal decreases across the series so that the heavier members are relatively inert to air and moisture at room temperature.) These metals all form very stable carbides and hydrides, and degreasing may be necessary on the materials as received.

Preparation

The container for the HgI_2-La reaction consists of a Pyrex tube 15-25 mm in diameter and 6-10 times as long. This is closed at one end, and the other end is connected to a ground-glass joint through a section of Pyrex tubing of 1.5-2 mm wall thickness, 6-10 mm i.d., suitable for seal-off under vacuum. The main tube is either bent 20-30° near the middle to form a flat inverted V or a dike is inblown there to allow separation of the products. A weighed quantity of La (1-5 g) and at least a three-fold excess of HgI_2 are placed in the reaction container, and the tube is evacuated under high vacuum ($<10^{-5}$ torr) and sealed off at the thick-walled tubing. (The seal-off should be practiced ahead of time if this is a new technique. Go slowly and do not attempt to reheat the seal-off. Sealing off silica glass rather than Pyrex requires much more heat but is less troublesome.)

The container is placed in the furnace slightly inclined so that the melt will cover the metal in the lower closed end, and the tube is heated for 12-48 hr at 300-330°.

■ **Caution.** *The reaction should be run in a good hood in case of unexpected container failure and release of toxic mercury vapor. A reliable automatic controller should also be used to avoid accidental overheating and dangerous pressures of HgI_2 and Hg. (The total pressure is 1.95 atm at 354° neglecting dilution of HgI_2 by Hg_2I_2 and LaI_3.)*

*See Chapter 1, Section 4 for list of suppliers.

After the reaction is complete, the container is partly withdrawn from the furnace to expose the empty end, and heating is continued until the HgI_2 and Hg are distilled from the LaI_3 product. A final heating of the LaI_3 to 600° in high vacuum (preferably in a tantalum crucible) will remove the last traces of mercury compounds.

B. SYNTHESIS FROM THE ELEMENTS

Preparation

As noted above, a temperature in excess of 734° is necessary to obtain an adequate rate of reaction of the metal with I_2, either as an exposed metal surface or with reduced iodide melts. (The metal readily dissolves in liquid LaI_3, giving solid LaI_2 only at high concentrations or lower temperatures.) Iodine is highly volatile (boiling point 184°) and must be maintained at a relatively low temperature to avoid dangerous pressures. Though this difference has been accommodated through the use of rather complex flow systems,[3] reactions run in sealed hot-cold tubes are in general simple, direct, and safe if attention is paid to one condition:

■ **Caution.** *A portion of the reaction tube must be kept at low enough temperature that the volatile component will not exert an unsafe pressure. Use of an evacuated container permits rapid volatilization of that species in and out of the cooler zone.*

Two designs that can be used to accomplish this synthesis are shown in Fig. 1. Both are constructed from fused silica. One furnace is used to maintain iodine either in the bottom of design **a** or in the side arm of design **b** at a temperature T_2 sufficient to generate a pressure of 0.2-1 atm, while a second furnace is used to heat the metal and its salt products in a tungsten crucible to the higher temperature T_1 necessary to give an adequate reaction rate. The separation of the two arms in design **b** then must be sufficient to accommodate the combined wall thicknesses of two tubular furnaces.

The tungsten crucible, about 3.2 cm diameter by 5 cm high,* is sealed within the apparatus, taking care that it rests either on a fairly flat tube end or on a silica support so that greater thermal expansion of tungsten will not lead to cracking of the glass vessel during heating. The crucible is fairly porous and so should be strongly heated (>500°) in high vacuum before loading. Lanthanum chunks or turnings (not powder) (5-20 g) and reagent grade I_2 are quickly loaded in air into the crucible with the aid of a small funnel which fits through the joint and neck. The apparatus is evacuated (through a trap cooled with dry ice to protect the pumps), and then the portion containing iodine is also cooled to -80°

---

*Available from Kulite Tungsten Co., Ridgefield, NJ, and Ultramet, Pacoima, CA.

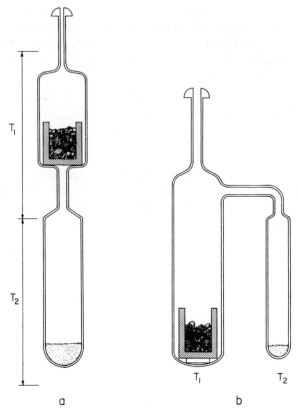

Fig. 1. Two designs for sealed hot-cold tubes.

once the ambient moisture has been pumped out. The apparatus is then evacuated below 10^{-3} torr and sealed off below the joint.

■ **Caution.** *Welder's glasses are required for eye protection.*

The iodine is then sublimed to the lower or right-hand portion of apparatus **a** or **b**, respectively, and two tubular furnaces arranged as to provide the temperature profiles shown. Insulation (Fiberfrax, Carborundum Co.) is used to cover the tops of the furnaces and especially the smaller horizontal tube in **b**. The hotter furnace is positioned so that sublimed or molten LaI_3 which may overflow the crucible will not block the narrow tubes. The temperature of the metal end is run right up to ~800°, with the I_2 reservoir held at 110-130°.

■ **Caution.** *A reliable automatic temperature controller should always be used on the lower temperature furnace to avoid dangerous I_2 pressures.*

After some salt has formed and covered the metal, the I_2 furnace temperature is increased to about 180°. The reaction will take 4-12 hr, depending on quantities and temperatures. The furnace around the I_2 can be lowered occasionally to judge progress. If excess I_2 has been used, the iodine reservoir is cooled to $-80°$ after reaction is complete while the furnace around the salt is still at 200-300° and the I_2 reservoir sealed off. If a slight deficiency of I_2 has been used (or the reaction has been incomplete) the excess metal can be separated (and intermediate iodides allowed to disproportionate) through sublimation of the crude product in high vacuum (below).

Purification

The LaI_3 product obtained from either of the above routes is sufficiently pure for some purposes. However, some impurities will have been introduced by handling of the reactants, La especially, in the air. One may also choose to use a relatively poor grade of metal (with respect to nonmetals) for the synthesis, anticipating the purification afforded by the sublimation. Also, some reaction of LaI_3 with fused silica to give SiI_4 plus LaOI or a lanthanum silicate will occur[5] in the high temperature reaction B. Although the amount would not seem serious in a sealed vessel, some SiI_4 may diffuse back to the metal, producing silicide. In any event the quality of either product, especially as judged by color and melting point, will be definitely improved by vacuum sublimation or distillation.[6] This should be carried out at about 900°, either in a sealed Ta tube designed so as to avoid refluxing or under a high vacuum using something like the two-piece apparatus shown in Fig. 2F of the earlier section on tantalum welding. A second sublimation may be necessary if entrainment is large in the first. Given adequate starting materials and the absence of adventitious contamination during the process, a sublimed product of high purity is assured, $\geqslant 99.9\%$. The yield is nearly quantative if the initial reactions go to completion but is reduced by up to 5% per sublimation through recovery losses.

The most likely impurity is LaOI which has the PbFCl structure as do many of the other rare earth metal oxyiodides.[3] The stronger 25% of the powder diffraction lines for LaOI (Å), with intensities in parentheses, are: 3.05(10), 2.92(8), 2.06(4), 1.72(5), 1.71(5). The powder pattern of LaI_3 is not especially useful for establishing its purity unless all the lines from a high-resolution (Guinier) pattern are compared.

Properties

The light yellow LaI_3, melting point 778-779°, exhibits the $PuBr_3$-type structure. The material is very sensitive to traces of both moisture and O_2. The absence of cloudy appearance on the dissolution of LaI_3 (and other rare earth metal trihalides) in absolute ethanol is *not* a good assurance of purity unless the material has been heated strongly to ensure the formation of crystalline LaOI (or other MOX phases) from absorbed moisture, hydroxide, etc. The same applies to the appearance of MOI in the powder pattern.

References

1. F. L. Carter and J. F. Murray, *Mat. Res. Bull.*, 7, 519 (1972).
2. L. F. Druding and J. D. Corbett, *J. Am. Chem. Soc.*, 83, 2462 (1961).
3. D. Brown, *Halides of the Lanthanides and Actinides*, Wiley-Interscience, New York, 1968, p. 219.
4. J. D. Corbett, L. F. Druding, W. J. Burkhard, and C. B. Lindahl, *Discuss. Farad. Soc.*, 32, 79 (1962).
5. J. D. Corbett, *Inorg. Nucl. Chem. Lett.*, 8, 337 (1972).
6. J. D. Corbett, R. A. Sallach, and D. A. Lokken, *Adv. Chem. Ser.*, 71, 56 (1967).

7. LANTHANUM DIIODIDE

$$2LaI_3(l) + La(s) \xrightarrow[Ta]{870°} 3LaI_2(l)^1$$

Submitted by JOHN D. CORBETT*
Checked by ARNDT SIMON†

This straightforward reaction provides a method of preparing a simple metallic salt, $La^{3+}(I^-)_2e^-$. The principal requirements for the synthesis are high-purity reactants, an inert container, and a good dry box (not a glove bag). A small amount of H_2O in the LaI_3 can have a large effect on the purity of the LaI_2 product owing to the formation of LaOI and LaH_2.[2]

Procedure

The synthesis involves the direct reaction of LaI_3 with a modest excess of La. A Ta (Nb or Mo) container is essential (see section on tantalum welding, Chapter 1, Sec. 3) and sublimed LaI_3 is preferable.[3] An open crucible can be used under a noble gas atmosphere with only a small amount of sublimation of LaI_3 during the reaction (and accompanying reaction with the silica walls), but a sealed container gives better control of stoichiometry and purity. Lanthanum metal is somewhat reactive toward air and the best product is obtained by minimizing atmospheric exposure of the lanthanum metal used in the reduction.

---

*Ames Laboratory and Department of Chemistry, Iowa State University, Ames, IA 50011. The Ames Laboratory is operated by the U.S. Department of Energy by Iowa State University under contract No. W-7405-Eng-82.

†Max-Planck Institut für Festkörperforschung, Heisenbergstrasse 1, 7000 Stuttgart 80, West Germany.

The LaI_3 (5-25 g) and somewhat more than the stoichiometric amount of La* (13.46% of LaI_3 by weight) are weighed in the dry box and transferred into the crucible. If an open crucible is used, this is placed in the closed end of a fused silica tube long enough so that the upper end, which is equipped with a standard taper joint, will extend 10-15 cm from the (vertical) furnace. This end is capped with a Pyrex top equipped with a standard taper joint and stopcocks to allow evacuation and introduction of a noble gas. The joint should be greased with a high-temperature grease (e.g., Apiezon T) and cooled with a small air blower. A closed crucible is crimped or capped in the dry box, welded, and jacketed as described. The reactants are heated to 840-900° (melting point LaI_2 = 830°) for 3-4 hr, after which the product is cooled by turning off power to the furnace.

According to the phase diagram,[1] LaI_2 melts congruently in equilibrium with La. The reaction can thus be carried out using excess metal if a bulk form is used so that the excess can be readily removed after grinding the product. Alternatively, the sample may be heated above the melting point of La (845°) in this system to aggregate the excess metal.

Samples of LaI_2 equilibrated for weeks near 850° with a large excess of metal, preferably with high surface area, have been found to contain also the cluster phase $La(La_7I_{12})$[4] (isostructural with Sc_7Cl_{12}[5]). The phase relationships of this with the other phases are not known, but the yield is negligible when the excess metal has limited surface area.

The surest identification of LaI_2 is accomplished by the determination of melting point and powder pattern. The latter may be calculated from the known crystal structure.[6] The stronger lines (Å) are at 6.98(m), 3.00(vs), 2.773(s), 2.275(m), 1.961(m), and 1.641(m). A search for the neighboring phases by the same means is a good check of purity; the stronger lines (in Å) are, for La_7I_{12}, 3.117(vs), 2.209(mw), and 2.197(mw),[4] and for $LaI_{2.5}$,[1,7] 3.46(vs), 3.08(vs), 3.015(s), 2.838(s), and 2.202(vs). The dissolution of LaI_2 in water for analytical purposes should be performed in a closed container to avoid loss of HI (and perhaps I_2) during the vigorous reaction. A small amount of acetic acid is added afterwards to dissolve the hydrolysis products. Standard methods for determining La^{3+} and I^- are used.[1] With sufficient care regarding purity of the reactants and execution of the reaction, a substantially quantitative yield of the single phase is obtained, and wet analyses should give $LaI_{2.00 \pm .02}$ with 100.0 ± 0.3% of recovery of La plus I.

Properties

The blue-black lanthanum diiodide is a very good reducing agent and should be stored and handled in the absence of O_2, H_2O, etc. It is one of the better char-

*For list of suppliers see Chapter 1, Section 4.

acterized members of a small group of metallic diiodides and is best formulated $La^{3+}(I^-)_2e^-$ so as to emphasize delocalization of the differentiating electron. The phase exhibits a conductivity comparable to that of lanthanum metal and a small (Pauli) paramagnetism appropriate to the metallic state and diamagnetic cores. The above representation probably overstates the ionic character of the bonding, however; covalency and iodine participation are thought to be important in the band formation. Support for this comes from the fact that none of the elements that form metallic diiodides (La, Ce, Pr, Gd, Th) yield similar chlorides or bromides in spite of the closer approach of the cations which would be possible with smaller anions.[3]

The structure of LaI_2 and of the isostructural CeI_2 and PrI_2 is of the $MoSi_2$ (or $CuTi_2$) type and gives an almost alloy-like impression with its square layers. For LaI_2 the lattice constants are $a = 3.922$, $c = 13.97$ Å, space group $I\frac{4}{m}mm$, $d = 5.46$ g/cm^3.[6]

The synthetic procedure described for LaI_2 should also work for CeI_2 and PrI_2, subject to two conditions. First, the compounds melt increasingly incongruently in the order listed, meaning that complete conversion is best accomplished by quenching the equilibrated melt (to avoid segregation) followed by equilibration with excess metal below the peritectic melting point.[1] Small amounts of the more reduced M_7I_{12} are formed only on extended reaction with excess metal.[4] In addition, PrI_2 and CeI_2, to a limited extent, have been shown to form a variety of diiodide polymorphs at the higher temperatures, namely $CdCl_2$, $2H_1$-MoS_2, $3R$-MoS_2, and a $M_4I_4I_4$ cluster type.[6] The metallic $MoSi_2$-type material appears to be more stable at lower temperatures.[7] These same three MI_3-M systems also contain an isostructural series of M_2I_5 phases of known structure, now melting congruently only for M = Pr, and techniques similar to those described here can also be used for their preparation.[1,7]

References

1. J. D. Corbett, L. F. Druding, W. J. Burkhard, and C. B. Lindahl, *Discuss. Farad. Soc.*, **32**, 79 (1962).
2. J. D. Corbett, *Prep. Inorg. Reactions*, **3**, 10 (1966).
3. J. D. Corbett, R. A. Sallach, and D. A. Lokken, *Adv. Chem. Ser.*, **71**, 56 (1967).
4. E. Warkentin and A. Simon, private communication.
5. J. D. Corbett, K. R. Poeppelmeier and R. L. Daake, *Z. Anorg. Allgem. Chem.*, **491**, 51 (1982).
6. E. Warkentin and H. Bärnighausen, *Z. Anorg. Allgem. Chem.*, **459**, 187 (1979).
7. E. Warkentin and H. Bärnighausen, private communication.

8. TRICHLORIDES OF THE RARE EARTH ELEMENTS, YTTRIUM, AND SCANDIUM

$$M(s) + 3HCl(g) \longrightarrow MCl_3 + 3/2H_2$$

Submitted by JOHN D. CORBETT*
Checked by SMITH L. HOLT†

Among the possible methods for the preparation of these trichlorides, the conversion of the oxides or the hydrated chlorides with various chlorinating agents has been utilized most frequently. Ammonium chloride is probably the most effective and inexpensive reactant,[1-3] but careful control of conditions and a rather lengthy reaction period are necessary. For example, in a recent study,[4] conditions for the production of $GdCl_3$ were optimized by this method, and it was found that periods of 20 hr or more under controlled temperature and flow rate were necessary to achieve 98-99.8% conversion of Gd_2O_3 to $GdCl_3$. The level of the unconverted oxide (presumably present as oxyhalide) is unsatisfactory for some purposes (the synthesis of the more reduced halides, for example), and one or more distillations or sublimation steps would be necessary. Now that the rare earth metals are available in good quality, their direct reaction with high-purity HCl has the advantage of both speed and a higher purity for the preparation of moderate amounts of their trichlorides. Of course, less pure starting materials can be used if additional sublimations are included.

Containment of this particular combination of metal, trichloride, and HCl at reaction temperatures presents a problem of the same character as occurs in the preparation of the triiodides of these elements (this volume, Chapter 1, Sec. 6). These metals all dissolve in (reduce) the liquid trichlorides to some degree, and this process greatly accelerates the spontaneous reduction of SiO_2 by the metal to form silicide and oxide. The same problem pertains to all known ceramic materials. Container metals which are the most inert to both the rare earth elements and HCl are limited to tungsten and molybdenum and then only if reactions such as $Mo(s) + 4HCl(g) \rightarrow MoCl_4(g) + 2H_2(g)$ are suppressed by the addition of H_2. The last is not a serious problem; for example, with $P_{H_2}/P_{HCl} \geqslant 10^{-4}$ the pressures of all gaseous molybdenum chlorides are estimated to be $<10^{-5}$ atm at 800°.[6] Molybdenum is the material of choice because of its greater ease of fabrication.[7]

---

*Ames Laboratory and Department of Chemistry, Iowa State University, Ames, IA 50011. The Ames Laboratory is operated by the U.S. Department of Energy by Iowa State University under contract No. W-7405-Eng-82.
†Department of Chemistry, Oklahoma State University, Stillwater, OK 74078.

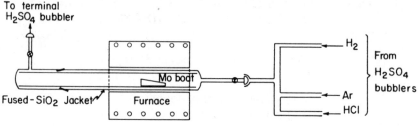

Fig. 1. Reaction apparatus.

Procedure

A convenient apparatus for the reaction is shown in Fig. 1. A molybdenum boat is folded from thin sheet or foil to contain the metal. The fused silica cylinder (~30-mm o.d.) which holds this is not necessary but does make product removal easier. The outer fused silica tube (~40-mm o.d.) is terminated on the outlet end through a standard taper joint to a (Pyrex) cap with stopcock and connection to an exit bubbler (H_2SO_4, not H_2O) which prevents back diffusion of air. The inlet end (which is slightly elevated) is connected through a stopcock and joint to a (Pyrex) mixing manifold, the three inlets of which connect through bubblers filled with H_2SO_4 for flow measurement to sources of HCl, H_2, and Ar or He (not N_2).

The usual commercial grade of HCl is a chlorination by-product, and the chlorinated hydrocarbons therein contribute a significant carbide impurity in the trichloride through their reaction with the heated metal. Subsequent removal of this appears to require one extra sublimation. However, high purity "electronic-grade" HCl (available from several gas suppliers) greatly reduces contamination from traces of the chlorocarbons, O_2, N_2, and H_2O which are present in the usual grade of HCl.

The Mo boat is loaded with the requisite metal* (5-20 g), in the dry box if a light rare earth metal is involved and maximum purity is desired. If loaded in the air, the assembled apparatus is flushed for about 15 min with a good flow of Ar (or He) (*not* H_2). Once the air has been displaced, the furnace is heated and a moderate flow of Ar and H_2 (50:50) introduced at about 200°. During the subsequent heat-up, at least some of the metal will be converted to the hydride which probably enhances its reactivity. Once the temperature reaches 600-800°, the H_2 flow is greatly reduced and a good flow of HCl + Ar (1:1) is started.

■ **Caution.** *The reaction is strongly exothermic, and chloride plus hydride formation may consume essentially all of the added HCl during active periods. A sufficient inert gas flow will not only moderate the reaction but is necessary to*

---

*See Chapter 1, Section 4 for a list of suppliers.

prevent a pressure drop in the vessel and a suck-back through the outlet. A higher HCl:Ar ratio ($\approx 2:1$) can be used once the active period is past.

The reaction temperature depends on the metal chloride involved. The minimum temperature is that necessary (see Reference 2) to melt or volatilize the trichloride which coats the metal and thereby inhibits reaction. A better rate of reaction will be achieved 50-150° higher, under which conditions the halide sublimes as well as melts and overflows the crucible. The conversion takes 0.5-4 hr depending on quantities and temperatures.

The H_2 produced in the reaction suppresses the formation of $MoCl_4$. The stage at which the reaction is complete or at least very slow is signaled by the appearance of white fumes of molybdenum chloride in the outlet tube. At this time, a small amount of H_2 should be added to the gas stream and the reaction continued for another 30 min, since some metal covered by molten chloride may still remain. A flow of argon is used to displace the HCl during cooling, and the reaction tube (closed at stopcocks) is then taken into the dry box for product removal.

If the metal has been completely oxidized, this product is suitable for many purposes. Generally, the impurities will be outside the sensitivity of the usual wet analytical techniques for metal and halide. Nonetheless, sublimation or distillation will noticeably improve the purity as judged by appearance, melting point, and especially by a diminished quantity of the ubiquitous oxyhalides formed during subsequent production of highly reduced halides.[8] Purification may be accomplished either in a sealed tantalum tube or under dynamic vacuum by using a two-piece tantalum crucible and condenser such as shown in Fig. 2F in Chapter 1, Section 3 of this volume. Contact of the molten trichlorides with silica at high temperatures under dynamic vacuum should be avoided because of the metathetical reaction forming MOCl (or silicate) plus gaseous $SiCl_4$.[9]

The yield of trichloride is largely limited by recovery losses after sublimation, being >95% after one sublimation and $\approx 80\%$ after four. The powder diffraction pattern of the trichloride as such is not a particularly useful measure of purity unless a high-resolution (Guinier) technique is used and all lines are measured. Examination of the X-ray diffraction patterns for impurities is better, especially for the oxychloride. Most of the MOCl phases have the PbFCl structure, and their patterns are listed in the ASTM file.[10] The strongest 25% of the lines for LaOCl (in Å, intensities in parentheses) are: 3.54(9), 2.914(8), 2.642(10), 2.060(4), 1.624(4).

Wet analytical methods may be used to verify the composition and, to a lesser degree, the purity of the trichlorides. Weighed samples should generally be dissolved in water in a closed container to avoid loss of HCl through hydrolysis. Titrimetric methods for the rare earth elements[7] together with gravimetric determination of chloride will typically give analytical compositions of Cl:M = 3.00 ± 0.01 with $100 \pm 0.2\%$ weight recovery as metal plus chloride.

The absence of a cloudy appearance on the dissolution of the trichlorides in absolute ethanol is not a good assurance of purity, unless the material in question has been strongly heated to ensure the formation of crystalline MOCl from absorbed moisture, hydroxide, etc. The same applies to the appearance of oxychloride in the powder pattern. For demanding synthetic applications, the trihalides are best stored in sealed, evacuated ampules.

Properties

These trichlorides exist in several different structures, the UCl_3-type for La-Gd, the YCl_3-type for Dy-Lu, the $PuBr_3$ form for $TbCl_3$ and $DyCl_3$, and the $FeCl_3$ structure for $ScCl_3$. The phases are relatively well characterized as concerns melting point, vapor pressure, and other properties.[2,3] High-purity trichlorides are starting materials for the preparation of a remarkable variety of reduced halides, such as conventional salts with oxidation states near two[5] and the clusters or extended metal–metal bonded structures exhibiting formal metal oxidation states between one and two.[8,11]

References

1. J. B. Reed, B. S. Hopkins, and L. F. Audrieth, *Inorg. Syntheses*, 1, 28 (1939).
2. D. Brown, *Halides of the Lanthanides and Actinides*, Wiley-Interscience Publishers, New York, 1968, p. 149.
3. J. M. Haschke, *Handbook on Physics and Chemistry of the Rare Earths*, Vol. 4, Chap. 32, K. A. Gschneidner and L. Eyring, (eds.), North Holland Publishing Co., Amsterdam, 1978.
4. Y. S. Kim, F. Planinsek, B. J. Beaudry, and K. A. Gschneidner, Jr., in *The Rare Earths in Modern Science and Technology*, Vol. 2, G. M. McCarthy, J. J. Rhyne, and H. B. Silber (eds.), 1980, Plenum Publishing Co., p. 53.
5. J. D. Corbett, *Rev. Chim. Minerale*, 10, 239 (1973).
6. L. Brewer, L. A. Bromley, P. W. Gilles, and N. L. Lofgren, *Chemistry and Metallurgy of Miscellaneous Materials–Thermodynamics*, L. L. Quill, (ed.), McGraw-Hill Book Co., New York, New York, 1950, paper 8.
7. L. F. Druding and J. D. Corbett, *J. Am. Chem. Soc.*, 83, 2462 (1961).
8. Hj. Mattausch, J. B. Hendricks, R. Eger, J. D. Corbett, and A. Simon, *Inorg. Chem.*, 18, 2128 (1980).
9. J. D. Corbett, *Inorg. Nucl. Chem. Lett.*, 8, 337 (1972).
10. "1980 Powder Diffraction File," JCPDS–International Center for Diffraction Data, Swarthmore, PA 19081.
11. J. D. Corbett, *Adv. Chem. Ser.*, 186, 329 (1980).

9. SINGLE CRYSTAL GROWTH OF OXIDES BY SKULL MELTING: THE CASE OF MAGNETITE (Fe_3O_4)

Submitted by H. R. HARRISON*, R. ARAGÓN,† J. E. KEEM,‡ and J. M. HONIG*
Checked by J. F. WENCKUS§

The skull-melting technique represents a versatile method for single crystal growth of a large variety of metal oxides; the methodology will be illustrated by reference to Fe_3O_4. In the past, a variety of crystal growth methods have been employed including: Verneuil,[1] Czochralski,[2] pedestal pulling,[3] arc transfer,[4,5] flux,[6] hydrothermal,[7] chemical transport,[8,9] and Bridgman growth in Pt crucibles.[10,11] Several of these methods (Verneuil, pedestal pulling, arc transfer) are crucibleless melt growth techniques in which the ambient atmosphere cannot be controlled. Others (flux, hydrothermal, chemical transport) also may be said to be free of crucible reaction (partly because of operating at lower than melt temperatures), but in these techniques the ambient atmosphere also cannot be controlled. In addition, the inclusion of impurities from the medium is to be anticipated. The remaining methods (Czochralski and Bridgman) are melt growth techniques which allow for control of the ambient atmosphere, but are not crucibleless. Bridgman growth in Pt crucibles has produced the largest (~ few centimeters) good quality crystals of all the methods mentioned. This procedure, while reliable, is quite tedious and also consumes expensive Pt crucibles, which must be cut to free the crystals within. In addition, Fe diffuses into the Pt, thus changing the stoichiometry of the crystals. The skull melting technique is superior to all the above in that it is a melt growth technique which is crucibleless while at the same time it allows the use of any desired atmosphere. It is capable of providing large single crystals of good quality.

Procedure

The technique and past history of skull melting has been described in several articles from this and other laboratories.[12-19] Basically, radio frequency power from a commercial power supply (50 KW at 3-5 MHz) is transferred to a work coil tightly wrapped about a skull crucible consisting of a set of cold fingers that

---

*Central Materials Preparation Facility and the Department of Chemistry, Purdue University, West Lafayette, IN 47906.
†Departmento de Fisica, Facultad de Ciencias Exactas, Universidad Nacional de La Plata, 1900 La Plata, Pcia de Buenos Aires, Argentina.
‡Energy Conversion Devices, 1675 W. Maple Road, Troy, MI 48084.
§Ceres Corporation, 411 Waverly Oaks Park, Waltham, MA 02154.

Fig. 1. Photograph of skull crucible.

are individually water cooled. A typical assembly is shown in Fig. 1.[20] The spacings between fingers are sufficiently small to prevent leakage of the charge and, at the same time, sufficiently wide to permit penetration of the radio frequency field into the interior, while also avoiding excessive power dissipation through the generation of large eddy currents in the copper container. A schematic of the setup is shown in Fig. 2.

Provision must also be made for the proper ambient atmosphere to obtain a single phase of appropriate stoichiometry; this matter may be understood by reference to Fig. 3, which shows the stability diagram for iron oxides in terms of the equilibrium oxygen fugacity, f_{O_2}, as a function of temperature.[21] It is clear that to produce Fe_3O_4, the oxygen fugacity must be maintained in the region bounded by the curves marked Fe_3O_4-Fe_2O_3 and $Fe_{1-x}O$-Fe_3O_4. It is seen that the requisite oxygen partial pressures may be maintained by use of CO_2/CO mixtures[22] in the ratio $100 \geqslant CO_2/(CO + CO_2) \geqslant 94\%$ over the whole temperature range until the composition is kinetically frozen in.

Depending on the size of the crucible, a total charge of ~500 g of 99.9% Fe_2O_3 powder is placed incrementally into the container, together with a graphite susceptor ring for initial coupling to the power source. As the susceptor heats up, the temperature of the surrounding hematite rises sufficiently that its conductivity increases dramatically; this portion then couples to the power source. Gradually the entire charge becomes coupled; the applied power

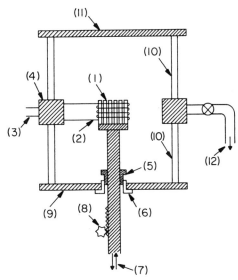

Fig. 2. A schematic diagram of the skull melting apparatus. Not shown are protective screens around the vacuum cylinders, a transparent fused silica sleeve between the skull crucible and the r.f. work coil, and a fume hood above the apparatus. (1) Skull crucible; (2) work coil; (3) to r.f. generator; (4) 12-port vacuum collar; (5) vacuum quick-connect coupling; (6) Teflon insulating flange; (7) water supply (in and out) for crucible; (8) motor-driven gear system for lowering crucible; (9) base plate and supporting frame, (10) 45.7-cm diameter × 30.5 cm high Pyrex vacuum cylinders; (11) aluminum top plate; (12) to mechanical pump.

can then be raised until virtually all of the charge is converted to liquid Fe_3O_4. One should note that this is a crucibleless process; the copper container and the molten Fe_3O_4 are isolated from each other by a thin skull of sintered material next to the cold fingers. As the powder melts, its volume shrinks considerably; hence more powder must be added to the container to achieve a satisfactory level of molten material. In the course of this operation the graphite susceptor ultimately burns off as CO or CO_2.

To achieve proper stoichiometry control, the chamber above the container is closed, evacuated with a mechanical pump, and then filled with CO_2 which is allowed to flow through the system; the boule has a sufficiently high conductivity that it can be remelted without a susceptor by reapplication of radio frequency power. The boule is kept under circulating CO_2/CO for 90 min to achieve complete equilibration. The skull crucible is finally lowered out of the stationary power coil at a rate of 15 mm/hr.

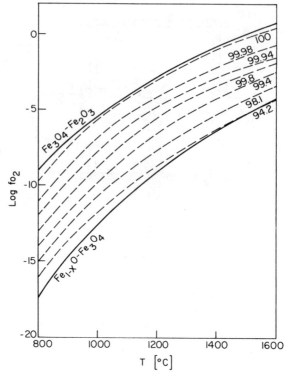

Fig. 3. Log f_{O_2} vs temperature. Fe_2O_3-Fe_3O_4 and Fe_3O_4-$Fe_{1-x}O$ buffer curves (solid lines). CO/CO_2 gas mixtures (dotted lines) (21). Numbers are percentages of CO_2, that is, $CO_2/(CO + CO_2)$ (22).

Properties

A photograph of the resulting boule is shown in Fig. 4. The individual single crystal pieces may be clearly discerned: At the bottom there exists poorly defined polycrystalline Fe_3O_4. The single crystals at the center are of high quality: X-ray Laue back-reflection photographs indicated that the center pieces were indeed single crystals, some of these weighing 5 g or more. Standard wet chemical analysis by oxidation titration[23] yielded analyses that correspond to $Fe^{3+}_{2(1+x)}Fe^{2+}_{1-3x}\square_x O_4$ with $0.008 \leqslant x \leqslant 0.012$. X-ray diffraction analysis provided no indication of the presence of any phase other than magnetite. Polarized reflected light microscopy represents perhaps the most sensitive technique for the detection of extraneous phases. With some practice, it is possible to distinguish the various iron oxides from each other, using shadings of grey as criteria, the highest oxide (Fe_2O_3 in this case) appearing lightest and the lowest oxide ($Fe_{1-x}O$ in this case) appearing darkest; free iron appears as very bright regions

Fig. 4. Interior of a boule. Note that individual crystals may be clearly seen. Scale in inches.

in the image. Examination under polarized reflected light microscopy with an oil immersion lens indicated the presence of some wüstite and free iron restricted to within a few microns of grain boundaries. A thin hematite layer was encountered at the top; the bulk of the single crystals was free of these defects.[18]

References

1. E. J. Scott, *J. Chem. Phys.*, **23**, 2459 (1955).
2. F. H. Horn, *J. Appl. Phys.*, **32**, 900 (1961).
3. R. P. Poplawsky, *J. Appl. Phys.*, **33**, 1616 (1962).
4. B. A. Smith and I. G. Austin, *J. Cryst. Growth*, **1**, 79 (1967).
5. V. S. Pavlov, M. M. Mochalov, and E. S. Vorontsov, *Izvestiya Vysshikh Uchebnykh Zaredenii, Chernaya Metallurgiya*, No. 3, 49 (1973).
6. M. Vichr and H. Makram, *J. Cryst. Growth*, **5**, 77 (1968).

48 Solid State

7. E. D. Kolb, A. J. Caporaso, and R. A. Laudise, *J. Cryst. Growth*, 19, 242 (1973).
8. Z. Hauptman, *Czech. J. Phys.*, 12B, 148 (1962).
9. P. Peshev and A. Toshev, *Mat. Res. Bull.*, 10, 1335 (1975).
10. J. Smiltens, *J. Chem. Phys.*, 20, 990 (1952).
11. Y. Syono, *Jap. J. Geophys.*, 4, 71 (1965).
12. D. Michel, *Rev. Int. Hautes Temp. Refract.*, 9, 225 (1972).
13. V. I. Aleksandrov, V. V. Osiko, A. M. Prokhorov, and V. M. Tatarintsev, *Vestn. Akad. Nauk SSSR*, No. 12, 29 (1973).
14. J. F. Wenckus, M. L. Cohen, A. G. Emslie, W. P. Menashi, and P. F. Strong, Tech. Report AFCRL-TR-75-0213 (1975).
15. V. I. Aleksandrov, V. V. Osiko, A. M. Prokhorov, and V. M. Tatarintsev, *Current Topics in Materials Science*, Vol. 1, E. Kaldis (ed.), North-Holland, Amsterdam, 1978, p. 421.
16. D. Michel, M. Perez y Jorba, and R. Collongues, *J. Cryst. Growth*, 43, 546 (1978).
17. J. E. Keem, H. R. Harrison, S. P. Faile, H. Sato, and J. M. Honig, *Am. Ceram. Soc. Bull.*, 56, 1022 (1977).
18. H. R. Harrison and R. Aragón, *Mater. Res. Bull.*, 13, 1097 (1978).
19. H. R. Harrison, R. Aragón, and C. J. Sandberg, *Mater. Res. Bull.*, 15, 571 (1980).
20. J. F. Wenckus, W. P. Menashi, and A. Castonquay, U.S. Patent 4,049,384 (1977).
21. J. S. Heubner, *Research Techniques for High Pressure and High Temperature*, G. C. Ulmer (ed.), Springer Verlag, New York, 1971, pp. 123–177.
22. P. Deines, R. H. Nafzinger, G. C. Ulmer, and E. Woerman, "Temperature-Oxygen Fugacity Tables for Selected Gas Mixtures in the System C-H-O at One Atmosphere Total Pressure," Bulletin of the Earth and Mineral Science Experiment Station No. 88, The Pennsylvania State University, University Station, Pennsylvania, 1974.
23. W. F. Hildebrand and G. E. F. Lundell, *Applied Inorganic Analysis*, John Wiley & Sons, New York, 1946, p. 781.

10. LITHIUM NITRIDE (Li$_3$N)

$$6Li + N_2 \longrightarrow 2Li_3N$$

Submitted by E. SCHÖNHERR,* A. KÖHLER,* and G. PFROMMER*
Checked by Kj. Fl. NIELSEN†

In the last several years, lithium nitride has excited much interest as a solid ionic conductor.[1-7] Lithium nitride is formed by direct reaction between lithium metal and nitrogen gas. The reaction goes to completion if the temperature of the lithium and the ambient pressure of the nitrogen are sufficiently high. The synthesis of Li$_3$N ("AzLi$_3$") was first described by Ouvrard,[8] who passed nitrogen over glowing lithium in an iron boat. Several other groups have investi-

*Max-Planck-Institut für Festkörperforschung, Heisenbergstrasse 1, 7000 Stuttgart 80, West Germany.
†Physics Laboratory III, The Technical University of Denmark, DK-2800, Lyngby, Denmark.

gated the applicability of various materials as containers for lithium during the reaction,[9,10] Additional information on this synthesis was given by Zintl and Brauer,[11] who used a LiF-coated ZrO_2 boat between 400 and 800°. A detailed procedure for Li_3N synthesis is described by Masdupuy and Gallais.[12] These authors used Fe boats for temperatures between 370 and 450°. The application of elevated N_2 pressures in order to effect the low-temperature synthesis of Li_3N has been described by Neumann, Kröger, and Haebler,[13] by Yonco, Veleckis, and Maroni,[14] and by Kutolin and Vulikh.[15]

The phase diagram of the Li–Li_3N system was first described by Bol'shakov, Fedevov, and Stepina,[16] while exact thermodynamic data for Li_3N are reported by Yonco, Veleckis, and Maroni.[14] Lithium nitride forms a simple eutectic with Li with the eutectic point near 0.05 mole % Li_3N at 180.3°. The melting point of Li_3N is 813 ± 1°.

A. POLYCRYSTALLINE MATERIAL

Procedure

The reactants used throughout this synthesis are N_2 gas (99.996%) and Li metal rods (99.9%)* which are about 12 mm in diameter and up to 200 mm in length and are packed in argon-filled cans. Opening of the cans and handling of the Li rods is carried out in a glove box which contains high-purity argon (flow rate 50 m³/hr, $H_2O \approx 1$ ppm). The impurity layer on the rods is removed with a knife. The cleaned rods are placed into a conical crucible of tungsten or molybdenum,† and the remaining rods are stored in an evacuated container of stainless steel mounted within the glove box. The crucible is about 50 mm in diameter and 70 mm in height. The crucible is filled with about 20 g Li, closed with a rubber stopper, and transferred to the external synthesis equipment.

■ **Caution.** *Lithium reacts vigorously with water, liberating explosive H_2 gas. It will cause severe burns if allowed to contact bare flesh. Appropriate precaution should be taken.*

The synthesis equipment consists of a stainless steel cylinder with an inside diameter of 100 mm and a height of 265 mm and is shown schematically in Fig. 1. The wall of the cylinder is 3 mm and the top plate 10 mm in thickness. The cylinder is mounted with 12 MX10 stainless steel screws on a base plate made of stainless steel. The base plate is furnished with a rabbet which contains the O-ring Teflon seal between the cylinder flange and base. In addition, three 6-mm o.d. stainless steel lines are welded to the base. One line leads to a vacuum pump, another line to the gas bottle, and the third to a mechanical safety valve which opens for pressure higher than 75 atm. The inlet line has connections to a

---
*Alfa Products, Danvers, MA 01923.
†Available from Metallwerk Plansee GmbH, A-6600 Reute, Austria.

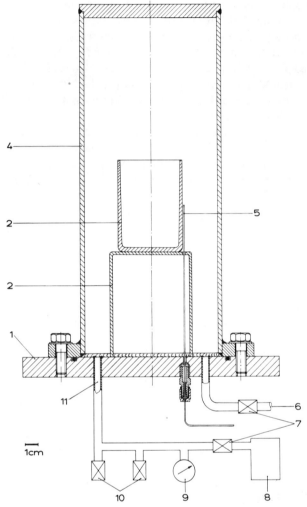

Fig. 1. Apparatus for Li_3N synthesis. (1) Base Plate; (2) pedestal; (3) crucible; (4) vessel; (5) thermocouple; (6) to vacuum pump; (7) valves; (8) N_2 gas bottle; (9) manometer; (10) safety valves; (11) N_2 gas inlet.

manometer and a second electromechanical safety valve which can be controlled by the manometer. A Pt/Pt 10% Rh, stainless steel shielded thermocouple allows the temperature to be recorded during the procedure. The rubber stopper of the crucible is removed after the crucible has been placed on the pedestal of the synthesis equipment. The equipment is then evacuated to a pressure of about 10^{-2} torr, flushed two times with N_2 at about 5 atm pressure and loaded with 10 atm of N_2.

■ **Caution.** *Since the gas pressure within the vessel (ca. 2 L in volume) is higher than 1 atm, the precautions for operating containers with compressed gases must be followed. The equipment described in this synthesis may not be safe for applied pressures higher than 90 atm.*

The crucible is heated with a heating strip, mounted at the upper half of the vessel, until the reaction starts. The temperature of the strip should be limited to 250°. The increase of N_2 pressure during heating is about 20% (i.e., from 10 to 12 atm). The reaction between Li and N_2 goes to completion spontaneously, owing to the high heat of formation, which produces a reaction temperature of approximately 800°. To initiate the reaction, the rate of heating has to be sufficiently high, that is, higher than 10°/min; otherwise the Li_3N layer that is formed near the melting point of Li can prevent further nitridation. The reaction can start at temperatures as low as 100°, however, if the lithium is contaminated with LiOH. If the nitridation has not commenced by the time the temperature has reached 230°, cooling the container to room temperature and reheating will in general initiate the reaction. When a rapid temperature increase during the synthesis is noticed, heating is stopped and the sample allowed to cool. The remaining nitrogen is vented, lowering the pressure to 1 atm before opening. The final product tends to stick to the walls of the crucible, depending on the degree of surface oxidation of the crucible. Therefore, a crucible having a conical shape is preferred. Since the crucible is in contact with Li_3N at high temperatures for only a short time, even a Ni crucible can be used without serious contamination of the product.

Properties

The Li_3N is brown red in color in reflected light and consists of thin shells and compact material. The shape of the thin shells corresponds partly with the initial nitridated surface of the Li rods. The compact material consists of agglomerated thin plates up to 3 mm in diameter. The plates are intensely red in transmitted light. The color of the surface changes to dark blue and violet on exposure to air. The compound forms NH_3 in humid air.

Lithium nitride crystallizes with a hexagonal point symmetry. The lattice constants are $a = 3.648$ Å and $c = 3.875$ Å,[11,17] and the space group is P6/mm.[11,17,18] Along the c axis, Li_3N forms an alternating structure of Li and Li_2N layers. The movement of Li^+ ions may be preferred within the Li_2N layers.[18]

B. SINGLE CRYSTALS OF Li_3N

Procedure

Small crystals in the millimeter range can be grown from solution,[11,17,19] or from a mixture of lithium and sodium as a solvent.[20] Large crystals can be grown by the Czochralski technique similar to that described in Reference 21. The growth equipment is schematically shown in Fig. 2. The vessel (1) is made from stainless steel and furnished with four slots at the top. The center slot carries the seed holder (2) which consists of a nickel clamp welded to a stainless steel tube. An inner tube supplies the clamp with water for cooling. One slot (3) is used as the inlet for N_2 gas. The crucible is illuminated with a 150-W halide bulb through a small-diameter slot (20 mm i.d.) behind the center slot (not drawn in the Fig. 2). The fourth slot (4) with an inside diameter of 32 mm is used for the observation of the crystal growth. At least three thin stainless steel tubes with an inside diameter of 8 mm and 50 mm in length are mounted

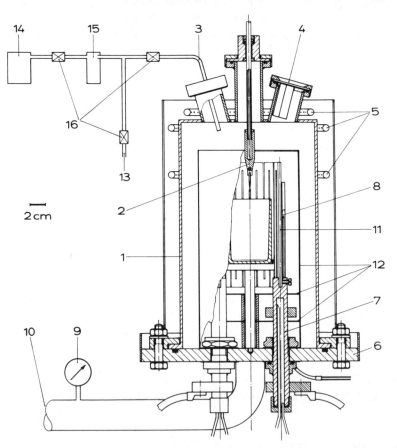

Fig. 2. Apparatus for Li_3N crystal growth. (1) Vessel (ss); (2) seed holder; (3) N_2 gas inlet; (4) observation window; (5) cooling tubes; (6) base plate (Ni); (7) electrodes (Ni); (8) thermocouple (Pt/Pt–10% Rh, inconel shielded); (9) manometer; (10) to vacuum pump; (11) furnace (ss); (12) heat shields (Mo); (13) N_2 gas outlet; (14) N_2 gas bottle; (15) flow meter; (16) valves.

along the slot to prevent the contamination of the window by Li_3N deposition. Three copper tubes (5) with an array of fine holes supply the outside of the vessel with water for cooling.

The nickel base plate (6) supports water-cooled electrodes made of nickel (7), a thermocouple (8), and a pipe leading to a manometer (9) and to a vacuum pump system (10).

A resistance furnace (11) with sufficient mechanical stability and chemical resistance is machined from stainless steel 150 mm in height, 70 mm in outside diameter, and 2 mm in wall thickness by cutting it lengthwise into 24 elements. The clearance between elements is 2 mm. A power supply with 30 V and 200 A is sufficient. The lifetime of the furnace is about 80 hr in the presence of molten Li_3N. A graphite heater cannot be used because of the formation of CO_2, which reacts with the Li_3N crystals. The temperature is controlled by a Pt/Pt 10% Rh thermocouple which is encapsulated in an Inconel tube with 1 mm outside diameter. The thermocouple is mounted on the outside of the furnace. The heat loss and material loss by convection are reduced by several heat shields made of molybdenum (12).

For crystal growth, a slightly conical crucible of tungsten with an outside diameter of 50-49 mm, a height of 70 mm, and a wall thickness of 2 mm is used. Tungsten has been found to be the most suitable material, showing less corrosion by the Li_3N melt than either tantalum or zirconium. The Li_3N melt reacts with W to form what appears to be Li_6WN_4, which is not noticeably soluble in the Li_3N melt and solid. The Li-W-N compound is formed at the crucible walls and sticks to neither the tungsten nor the Li_3N cast. The average weight loss of the crucible is 0.6 ± 0.2 g/hr. The crucibles can be used, for example, either 20 times for 11 hr or 13 times for 22 hr, before they become leaky. Molybdenum crucibles are also fairly compatible with the Li_3N melt. (Use of them however, reduces the maximum stable growth rate.) Crucibles of iron or vitreous carbon cannot be used for crystal growth.

For crystal growth the crucible can be loaded with presynthesized Li_3N material, but to prevent additional contact of Li_3N with air, the crucible can be loaded with Li metal, and the synthesis of Li_3N carried out in the growth equipment. For this purpose the crucible is filled with Li rods and heated to about 300° in the glove box (Section A) under an Ar atmosphere. The floating impurities are removed from the molten Li metal with a stainless steel spoon. When the purified Li has cooled, the crucible is closed with a rubber stopper and transferred to the growth equipment.

After loading, the growth chamber is evacuated to 10^{-5} torr. The lithium is then heated to 350-400° to remove hydrogen. After a period of 15 hr, the chamber is filled with N_2 until the pressure is approximately 600 torr. The temperature is increased until the Li metal begins to glow and the N_2 pressure drops, an indication that the reaction has started. Immediately, N_2 is added to restore the original pressure and the crucible is heated in less than 5 min to the melting point of the Li_3N. If longer times are used, the synthesized material may solidify. This will lead to bubbling when it is remelted. When the temperature of about 830° is reached, the N_2 pressure is reduced to atmospheric pressure by venting any excess N_2 through the outlet (13).

The seeding can be controlled either with a small Li_3N crystal with dimensions of 4 × 4 × 25 mm or more conveniently with a tungsten pin imbedded in a copper rod. Details of the seeding device are shown in Fig. 3. If the tapered end is not inserted too far into the melt, only a few seeds are formed, fewer than in the case when a Li_3N seed crystal is used. Material is deposited from the vapor phase at the initial grains so that the grains grow towards the seed holder. After about an hour, the deposited material reaches the copper rod which then serves for removing the heat of fusion.

The growth of the boules is achieved with a pulling rate of 0.7–3 mm/hr and a rotation rate of 30 rpm. The resulting growth rate is 3.3 ± 0.8 mm/hr. The effective time of growth is 5–20 hr. The growth is terminated by withdrawing the boules from the melt and the cooling to room temperature at a rate of 40°/hr.

■ **Caution.** *Finely crystalline Li_3N is deposited at all inner parts of the equipment which have been cooler than the furnace. When the vessel is opened, the deposited Li_3N can ignite. Burning melt droplets can be formed which drop from the walls of the vessel or seed holder. The finely divided Li_3N also ignites on contact with H_2O.*

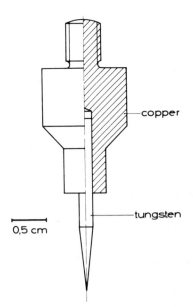

Fig. 3. Schematic sketch of the seeding tip.

■ **Caution.** *The equipment should be opened in a hood since NH_3 gas and Li_2O smog may be formed.*

The equipment is cleaned in a hood with water and then evacuated until a pressure of 10^{-5} torr is reached with the furnace at 600°.

Properties

The shiny red-brown boules are up to 50 mm in length and 35 mm in diameter. When the boules are stored in air, only the surface becomes dark. The boules contain along the long axis single crystalline grains up to 30 mm in length and up to 15 mm in diameter. Those grains are dark red in transmission. The density is determined to be 1.294 ± 0.006 g/cm^3. Although the impurities in the Li metal are Na ≈ 70, Mg ≈ 10, and Cu ≈ 20 ppm, the main impurities in the Li_3N are Na ≈ 50, K ≈ 50, Ca ≈ 25, Mg ≈ 50, and Cu ≈ 100 ppm.

References

1. F. Gallais and E. Masdupuy, *C. R. Hebd. Seances Acad. Sci.*, **227**, 635 (1948).
2. B. A. Boukamp and R. A. Huggins, *Phys. Lett.*, **58A**, 231 (1976).
3. U. v. Alpen, A. Rabenau, and G. H. Talat, *Appl. Phys. Lett.*, **30**, 621 (1977).
4. B. A. Boukamp and R. A. Huggins, *Mater. Res. Bull.*, **13**, 23 (1978).
5. S. G. Bishop, P. J. Ring, and P. J. Bray, *J. Chem. Phys.*, **45**, 1525 (1966).
6. U. v. Alpen, *J. Solid State Chem.*, **29**, 379 (1979).
7. A. Rabenau, in *Festkörperprobleme* (Adv. in Solid State Phys.) Vol. 18, J. Treusch (ed.), Vieweg, Braunschweig, 1978, p. 77.
8. L. Ouvrard, *C. R. Hebd. Seances Acad. Sci.*, **114**, 120 (1892).
9. M. Guntz, *C. R. Hebd. Seances Acad. Sci.*, **123**, 995 (1896).
10. W. Lenz-Steglitz, *Ber. Deut. Pharm. Ges.*, **20**, 227 (1910).
11. E. Zintl and G. Brauer, *Z. Elektrochem.*, **41**, 102 (1935).
12. E. Masdupuy and F. Gallais, *Inorg. Syntheses*, **4**, 1 (1953).
13. B. Neumann, C. Kröger, and H. Haebler, *Z. Anorg. Allgem. Chem.*, **204**, 81 (1932).
14. R. M. Yonco, E. Veleckis, and V. A. Maroni, *J. Nuclear Mater.*, **57**, 317 (1975).
15. S. A. Kutolin and A. I. Vulikh, *Zh. Prikl. Khim*, **41**, 2529 (1968).
16. K. A. Bol'shakov, P. I. Fedevov, and L. A. Stepina, *Izv. Vysshikh. Ucheb. Zaved, Tsve t. Met.*, **2**, 52 (1959).
17. A. Rabenau and H. Schulz, *J. Less Common Metals*, **50**, 155 (1976).
18. H. Schulz and K. Schwarz, *Acta Cryst.*, **A34**, 999 (1978).
19. M. D. Lyutaya and T. S. Bartnitskaya, *Inorg. Mater.*, **6**, 1544 (1970).
20. M. G. Down and R. J. Pulham, *J. Crystal Growth*, **47**, 133 (1979).
21. E. Schönherr, G. Müller, and E. Winckler, *J. Crystal Growth*, **43**, 469 (1978).

11. THE ALKALI TERNARY OXIDES A_xCoO_2 AND A_xCrO_2 (A = Na, K)

Submitted by C. DELMAS,* C. FOUASSIER,* and P. HAGENMULLER*
Checked by J. F. ACKERMAN†

Several ternary oxides of formula A_xMO_2 (A = Na, K; M = Co, Cr) have been synthesized by solid-state reactions. Depending on the stability and the reducing character of the materials obtained, various synthesis methods may be used. They are described here from the simplest to the most sophisticated.

A. SODIUM COBALT OXIDES: Na_xCoO_2 ($x \leqslant 1$)

$$3xNa_2O_2 + 2Co_3O_4 + (2 - 3x)O_2 \longrightarrow 6Na_xCoO_2$$

($x = 0.60; 0.64 \leqslant x \leqslant 0.74; 0.77; 1$).

Procedure

All manipulations are carried out in a dry box. The peroxide, Na_2O_2, can be replaced by the hydroxide, NaOH, or the oxide, Na_2O. The reaction temperatures are the same as for Na_2O_2. Nevertheless, it is advantageous to start the reaction with Na_2O_2 because it is the purest of the three sodium compounds.

Stoichiometric amounts of the two starting oxides (1 g of Co_3O_4, 4.15 mmole, and the corresponding Na_2O_2 amount) are intimately mixed by grinding in an agate mortar. The mixture is introduced into an alumina crucible and then heated for 15 hr in an oxygen stream.

As the resulting phases are not thermally very stable the reaction temperature required is dependent on the value of x. To obtain phases which have x equal to 0.60, 0.77, or 1, the reaction temperature is 550° while for pure phases in the range $0.64 \leqslant x \leqslant 0.74$ the reaction temperature is 750°.

Properties

The Na_xCoO_2 phases are obtained in the form of black powders which are very sensitive to atmospheric moisture.

The structure is derived from $(CoO_2)_n$ sheets of edge-sharing octahedra. The

---

*Laboratoire de Chimie du Solide du C.N.R.S., Université de Bordeaux 1, 351 Cours de la Libération, 33405 Talence Cedex, France.
†General Electric Company, P.O. Box 8, Schenectady, NY 12301.

TABLE I Crystallographic Data of the Na_xCoO_2 Phases

x	Symmetry	Cell Parameters	Oxygen Packing
0.60	Monoclinic	$a = 4.839$ Å $b = 2.831$ Å $c = 5.71$ Å $\beta = 106.3°$	AABBCC
0.64, 0.74	Hexagonal	$a = 2.833$ Å $c = 10.82$ Å	AABB
0.77	Monoclinic	$a = 4.880$ Å $b = 2.866$ Å $c = 5.77$ Å $\beta = 111.3°$	ABCABC
1	Rhombohedral	$a = 2.880$ Å $c = 15.58$ Å	ABCABC

Na^+ ions are inserted between the slabs in a trigonal prismatic ($x = 0.60$ and $0.64 \leq x \leq 0.74$) or octahedral ($x = 0.77$ and $x = 1$) environment.[1]

The crystallographic data are summarized in Table I.

Although the nonstoichiometric compounds have a metallic character, $NaCoO_2$ is a semiconductor.

B. POTASSIUM COBALT OXIDE BRONZES: $K_{0.50}CoO_2$, $K_{0.67}CoO_2$

$$6xKOH + 2Co_3O_4 + \frac{4 - 3x}{2} O_2 \longrightarrow 6K_xCoO_2 + 3xH_2O$$

Procedure

As commerical potassium hydroxide always contains a few per cent of potassium carbonate and water, it is necessary to determine (by acidimetric titration) the potassium content of the starting material.

The procedure is similar to that used for preparing the Na_xCoO_2 phases (part A). The reaction temperature for preparing either phase is 500°.

Properties

Both phases are obtained in the form of dark blue powders, extremely sensitive to atmospheric moisture.

The K^+ ions are inserted in a trigonal prismatic environment between $(CoO_2)_n$ sheets. The two structural types differ in the oxygen packing: AABBCC for $x = 0.50$ and AABB for $x = 0.67$.[2] The $K_{0.50}CoO_2$ phase crystallizes in the rhombohedral system. The hexagonal parameters are: a = 2.829 Å, c = 18.46 Å. The $K_{0.67}CoO_2$ phase crystallizes in the hexagonal system with the cell parameters: a = 2.837 Å, c = 12.26 Å.

They are both metallic conductors.

C. POTASSIUM COBALT OXIDE: $KCoO_2$

$$4K_2O + KO_2 + 3Co_3O_4 \longrightarrow 9KCoO_2$$

Procedure

The stoichiometric mixture of the starting materials is ground in a dry box in an argon or nitrogen atmosphere whose O_2 and H_2O content is less than 4 ppm. (K_2O is not stable in the presence of oxygen and leads to peroxide, K_2O_2.) As the potssium oxide K_2O is not a commercial product, it is prepared by controlled oxidation of liquid potassium by oxygen diluted with argon according to Rengade.[3] The resulting mixture (K_2O:0.1609 g, 1.708 mmole, KO_2:0.0303 g, 0.427 mmole, and Co_3O_4:0.3083 g, 1.281 mmole) is introduced into a gold tube (4 mm diameter, 0.4 mm thickness, 60 mm length). This tube is sealed, weighed, and heated for 15 hr.

Depending upon the thermal treatment, two allotropic varieties of $KCoO_2$ can be obtained. A synthesis temperature of 500° leads to the low-temperature α form, while the high-temperature β form, which is metastable at room temperature, is obtained by quenching in air from 900°.

As a precautionary measure, the tube is weighed again and opened in a dry box. (A difference in the two weights indicates a crack in the gold tube with resultant reaction with water vapor or oxygen or K_2O volatilization.)

Properties

Whereas α-$KCoO_2$ is a brown powder, β-$KCoO_2$ is black. Both phases are very sensitive to moisture. The structural transition between the α and β forms is reversible; the transition temperature is around 650°. As previously mentioned, the β form can be maintained at room temperature, but it is metastable and leads to α-$KCoO_2$ when heated above about 250°.

The α variety crystallizes in the tetragonal system with the cell parameters: a = 3.797 Å, c = 7.87 Å. Its structure is unknown.

The product β-$KCoO_2$ also crystallizes in the tetragonal system. The cell parameters are: a = 5.72 Å, c = 7.40 Å.[2] The structure of β-$KCoO_2$ is related to that of high-temperature cristobalite, the K^+ ion occupying the twelvefold coordinate site which is empty in β-SiO_2.

D. POTASSIUM CHROMIUM OXIDE: $KCrO_2$

$$K_2O + Cr_2O_3 \longrightarrow 2KCrO_2$$

The compound, $KCrO_2$, is used as starting material for the preparation of the K_xCrO_2 phases (part E).

Procedure

A small amount of potassium is added to a stoichiometric mixture of potassium and chromium oxides (K:5×10^{-3} g, 0.127 mmole, K_2O:0.382 g, 4.06 mmole, Cr_2O_3:0.617 g, 4.06 mmole). After grinding, the powder is introduced into a silver crucible which is placed in a Pyrex tube. The tube is sealed under vacuum and heated for 15 hr at 450°. Theoretically, potassium metal does not participate in the reaction, but it creates a strongly reducing atmosphere in the sealed tube that is necessary for the synthesis of $KCrO_2$. During the reaction, the potassium reacts with the glass and is thus eliminated. After 15 hr at 450°, pure $KCrO_2$ is obtained. All manipulations are carried out in a dry box in an oxygen-free argon or nitrogen atmosphere (O_2 and H_2O content less than 5 ppm).

Properties

The product is a green powder, extremely sensitive to moisture or oxygen. At room temperature in air, it is immediately oxidized giving a mixture of $K_{0.50}CrO_2$ and K_2CrO_4. It crystallizes in the rhombohedral system. The hexagonal parameters are: $a = 3.022 \pm 0.004$ Å, $c = 17.76 \pm 0.03$ Å. Its structure consists of $(CrO_2)_n$ layers between which the K^+ ions are inserted in an octahedral environment.[4]

E. POTASSIUM CHROMIUM OXIDE BRONZES: K_xCrO_2

$$0.70 \leqslant x \leqslant 0.77$$
$$0.50 \leqslant x \leqslant 0.60$$
$$KCrO_2 \longrightarrow K_xCrO_2 + (1-x)K$$

Procedure

One gram of $KCrO_2$ is put in a gold crucible which is introduced into the apparatus shown in Fig. 1. The temperature is increased while holding the reaction mixture under a vaccum of 10^{-2} torr. At 700° $KCrO_2$ evolves potassium metal which distills and deposits on the cooled part of the Vycor reaction vessel.

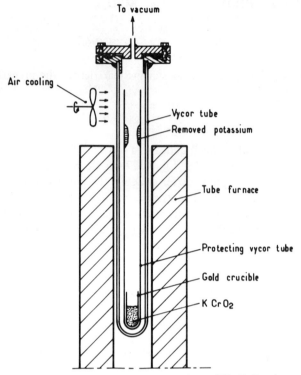

Fig. 1. Apparatus for the preparation of K_xCrO_2 phase.

After 3 hr at 700° a potassium-deficient phase whose formula is K_xCrO_2 ($0.70 \leqslant x \leqslant 0.77$) is obtained. If the temperature is raised to 950°, more potassium volatilizes and a phase more deficient in potassium ($0.50 \leqslant x \leqslant 0.60$) is obtained. All materials must be manipulated in an inert atmosphere.

Direct syntheses of K_xCrO_2 from the oxides K_2O, KO_2, and Cr_2O_3 or K_2O, Cr_2O_3, and CrO_2 cannot be realized by this technique. These reactions give only mixtures of Cr_2O_3 and K_2CrO_4, K_3CrO_4 or K_4CrO_4, depending on the values of x.

Properties

The two phases obtained are dark brown powders, very sensitive to atmospheric moisture. The phase with the higher potassium content is also sensitive to oxygen.

Fig. 2 shows the variation of x (as determined by thermogravimetric analysis and X-ray diffraction) versus temperature. The parts I, III, and V of the curve correspond respectively to $x = 1$, $0.70 \leqslant x \leqslant 0.77$, and $0.50 \leqslant x \leqslant 0.60$.

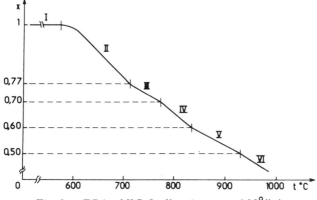

Fig. 2. TGA of $KCrO_2$ (heating rate: $100°/hr$).

The structures of the bronzes are built up of $(CrO_2)_n$ sheets with K^+ ions inserted in distorted octahedral sites for $0.70 \leqslant x \leqslant 0.77$ and trigonal prismatic sites when $0.50 < x < 0.60$.[4] K_xCrO_2 ($0.70 \leqslant x \leqslant 0.77$) crystallizes in a monoclinic system ($a = 5.062$ Å, $b = 2.986$ Å, $c = 6.581$ Å, $\beta = 112.61°$) and K_xCrO_2 ($0.50 \leqslant x \leqslant 0.60$) in a rhombohedral system. Its hexagonal parameters are: $a = 2.918$ Å, $c = 18.44$ Å.

References

1. C. Fouassier, G. Matejka, J. M. Réau, and P. Hagenmuller, *J. Solid State Chem.*, **6**, 532 (1973).
2. C. Delmas, C. Fouassier and P. Hagenmuller, *J. Solid State Chem.*, **13**, 165 (1975).
3. E. Rengade, *C. R. Hebd. Seances Acad. Sci.*, **114**, 754 (1907).
4. C. Delmas, M. Devalette, C. Fouassier, and P. Hagenmuller, *Mater. Res. Bull.*, **10**, 393 (1975).

12. ZEOLITE MOLECULAR SIEVES

Submitted by L. D. ROLLMANN* and E. W. VALYOCSIK*
Checked by R. D. SHANNON†

Zeolites are three-dimensional, crystalline networks of AlO_4 and SiO_4 tetrahedra, a unit negative charge being associated with each AlO_4 tetrahedron in the framework. Their crystallization is often a nucleation-controlled process occurring

---

*Mobil Research and Development Corp., Central Research Division, P.O. Box 1025, Princeton, NJ 08540.
†Central Research and Development Dept., E. I. du Pont, Wilmington, DE 19898.

from molecularly inhomogeneous, aqueous gels, and the particular framework structure which crystallizes can be strongly dependent on the cations present in these gels.[1]

Synthesis methods are described below for four very different but important zeolite structures, A,[2] Y,[3] tetramethylammonium (TMA) offretite[4] and tetrapropylammonium (TPA) ZSM-5.[5,6] These four were selected because they span the composition range from 1:1 Si:Al to a potentially aluminum-free zeolite structure (A to ZSM-5). In addition, these syntheses provide examples of fundamental concepts in crystallization such as templating (TMA offretite and ZSM-5), low-temperature nucleation (Y), and variable reactant (silica) sources.

The chemical description of a zeolite synthesis mixture requires special comment. It is conventional to present reaction mixtures as mole ratios of added ingredients:

$$\frac{SiO_2}{Al_2O_3}, \frac{H_2O}{SiO_2}, \frac{OH^-}{SiO_2}, \frac{Na^+}{SiO_2}, \frac{R}{SiO_2},$$

wherein R (if present) may be a component such as a quaternary ammonium cation, potassium ion, etc. By convention, moles of hydroxide are calculated by assuming, for example, that sodium silicate is a mixture of silica, sodium hydroxide, and water; that sodium aluminate is an analogous mixture but with alumina in place of silica; and that aluminum sulfate is a mixture of alumina, water, and sulfuric acid. In the descriptions given below, it is also recognized that alumina consumes 2 moles of hydroxide (i.e., acts as 2 moles of acid) on its incorporation into a zeolite framework as aluminate ion. Hydroxide:silica ratios are then calculated by subtracting moles of acid added from moles of hydroxide and then dividing by moles of silica present. Organics such as amines are never included in calculating $OH:SiO_2$ ratios.

Determination of purity in zeolites is a second area of concern. Elemental analysis is generally not a satisfactory criterion since almost all zeolite structures can exist in a range of compositions (i.e., of $SiO_2:Al_2O_3$ ratio). For example, the A structure has been crystallized with a $SiO_2:Al_2O_3$ ratio from 2 to 6;[7] at the other extreme, ZSM-5 has even been synthesized with essentially no aluminum.[6]

X-ray powder diffraction patterns are the most common measure of purity in zeolite samples. If the diffraction pattern shows no evidence for crystalline (or amorphous) contaminants, purity is then estimated by comparing intensities of reflections (at d spacings smaller than about 6 Å) with those of an authentic sample of the same composition and crystal size. Except for such large-scale commercial products as NaA and NaY, "authentic" samples are normally obtained by repeated and varied crystallization experiments.

A. ZEOLITE A[2]

$$2NaAlO_2 + 2(Na_2SiO_3 \cdot 9H_2O) \longrightarrow$$
$$Na_2O \cdot Al_2O_3 \cdot 2SiO_2 \cdot 4.5H_2O + 4NaOH + 11.5H_2O$$

Procedure

Sodium aluminate (13.5 g, approximately 0.05 mole alumina and 0.07 mole Na_2O; commercial sodium aluminate contains about 40% Al_2O_3, 33% Na_2O and 27% H_2O) and sodium hydroxide (25 g, 0.62 mole) are dissolved in 300 mL of water in a magnetically stirred 600-mL beaker and brought to a boil. The aluminate solution is added, with vigorous stirring, to a hot solution of sodium metasilicate, $Na_2SiO_3 \cdot 9H_2O$ (14.2 g, 0.05 mole), in 200 mL water in a 1-L beaker, also equipped with a Teflon-coated magnetic stirrer. The entire mixture is heated with stirring at about 90° until the suspension will settle quickly when stirring is stopped (2-5 hr). The suspension is then filtered hot and the solid washed repeatedly with water (four 100-mL portions) and dried in an oven at 110° to give about 7-8 g (80-90% yield based on SiO_2) of $Na_2O \cdot Al_2O_3 \cdot 2SiO_2 \cdot 4.1H_2O$.

Anal. Calcd.: Na_2O, 17.3; Al_2O_3, 28.5; SiO_2, 33.5. Found: Na_2O, 16.1; Al_2O_3, 28.8; SiO_2, 34.2. The purity of the sample is determined by inspection of its X-ray diffraction pattern.

Properties[7]

The product, NaA, is a white crystalline solid with a crystal density of 1.27 g/cm^3. Its crystals are normally 1-2 μm in diameter and have cubic symmetry. A typical unit cell formula is $Na_{12}[(AlO_2)_{12}(SiO_2)_{12}] \cdot 27H_2O$, the sodium ions being readily exchanged in aqueous solution by cations such as calcium or potassium. In the sodium form, after dehydration at 350-400° in vacuum, zeolite A will sorb about 0.25 g H_2O per gram of ash (at room temperature, 4 torr), but it will not sorb hexane. In the dehydrated calcium form, 0.27 g H_2O per gram (4 torr) and 0.145 g hexane (10 torr) per gram are sorbed, but benzene is not.

The X-ray diffraction pattern of NaA is as follows:

hkl	d (Å)	I/I(0)
100	12.29	100
110	8.71	69
111	7.11	35
210	5.51	35

hkl	d (Å)	I/I(0)
211	5.03	2
220	4.36	6
221, 300	4.107	36
311	3.714	53
320	3.417	16
321	3.293	47
410	2.987	55

B. ZEOLITE Y[3]

$$2NaAlO_2 + 5SiO_2 + xH_2O \longrightarrow Na_2O \cdot Al_2O_3 \cdot 5SiO_2 \cdot xH_2O$$

Procedure

A solution of 13.5 g sodium aluminate (0.05 mole alumina and 0.07 mole Na_2O) and 10 g sodium hydroxide (0.25 mole) in 70 g water is prepared in a 200-mL beaker and added, with vigorous magnetic stirring, to 100 g 30% silica sol (0.5 mole SiO_2; commercial colloidal silica suspensions typically contain 30% SiO_2 together with 0.1%–0.5% Na_2O, as stabilizer) in a 250-mL polypropylene bottle. The reaction mixture, which is defined by the following mole ratios of components:

$$\frac{SiO_2}{Al_2O_3} = 10 \quad \frac{H_2O}{SiO_2} = 16 \quad \frac{OH^-}{SiO_2} = 0.6 \quad \frac{Na^+}{SiO_2} = 0.8$$

is then set aside to age. After 24–48 hr at room temperature, the bottle is placed in a steam chest at about 95°. After 48–72 hr, daily samples of the solid in the bottom of the bottle are taken, filtered, washed, dried, and analyzed by X-ray diffraction for crystallinity. Special care should be taken during sampling not to mix the sample or inadvertently to seed the mixture. When the diffraction pattern reaches a limiting intensity, the hot mixture is removed from the steam chest and filtered and the solid washed with water (four times 100 mL) and dried at 110°. About 30 g of NaY (50–60% based on SiO_2) is obtained, with an approximate molar composition of $Na_2O \cdot Al_2O_3 \cdot 5.3SiO_2 \cdot 5H_2O$.

Anal. Calcd.: Na_2O, 10.5; Al_2O_3, 17.3; SiO_2, 53.9. Found: Na_2O, 10.9; Al_2O_3, 17.2; SiO_2, 53.4. A final X-ray diffraction pattern is taken for inspection and for comparison with that of an authentic sample of NaY.

Properties[7]

The NaY produced is a white crystalline solid with a crystal density of 1.27 g/cm^3. Its crystals are usually smaller than 1 μm and have cubic symmetry. A typical unit cell formula would be $Na_{56}[(AlO_2)_{56}(SiO_2)_{136}] \cdot 250H_2O$, about 70% of the sodium ions being readily exchanged by cations such as ammonium ion. The remainder can be exchanged with persistence. Ammonia can be removed from the resultant NH_4Y by heating to about 450°. In the sodium form, after dehydration at 350–400° in vacuum, zeolite Y will sorb about 25% of its weight in water (1 torr, 25°), about 19% in hexane (10 torr), and about 25% in benzene (10 torr). The X-ray diffraction pattern of NaY is as follows:

hkl	d (Å)	I/I(0)
111	14.29	100
220	8.75	9
311	7.46	24
331	5.68	44
333, 511	4.76	23
440	4.38	35
620	3.91	12
533	3.775	47
444	3.573	4
711, 551	3.466	9
642	3.308	37
731, 553	3.222	8
733	3.024	16

C. TMA OFFRETITE[4]

Procedure

A solution of 5.2 g sodium aluminate (0.02 mole alumina and 0.03 mole Na_2O), 14.6 g NaOH (0.36 mole), and 8.2 g KOH (0.15 mole) in 76 g water is prepared in a 200-mL beaker equipped with a magnetic stirrer. To this is added 11.0 g 50% tetramethylammonium chloride (0.05 mole), and the resultant solution is poured quickly into 112 g 30% silica sol (0.56 mole SiO_2) in a 250-mL polypropylene bottle. The mixture, which has the following composition:

$$\frac{SiO_2}{Al_2O_3} = 27 \quad \frac{H_2O}{SiO_2} = 16 \quad \frac{OH^-}{SiO_2} = 1.0 \quad \frac{Na^+}{SiO_2} = 0.75$$

$$\frac{K^+}{SiO_2} = 0.27 \quad \frac{TMA^+}{SiO_2} = 0.09,$$

is vigorously shaken and placed in a steam chest at 95° to crystallize. After 48–72 hr, daily samples of the solid in the bottom of the bottle are taken, filtered out, washed, dried, and analyzed by X-ray diffraction for crystallinity. When the diffraction pattern reaches a limiting intensity, the mixture is removed from the steam chest and filtered, and the solid washed with water (four times 100 mL) and dried at 110°. About 16 g of TMA offretite (25–30% based on SiO_2) is obtained, with an approximate molar composition of 0.3 $(TMA)_2O$ · 0.5K_2O · 0.4Na_2O · Al_2O_3 · 7.7SiO_2 · 7.1H_2O.

Anal. Calcd.: Na_2O, 3.1; N, 1.0; C, 3.5; K_2O, 5.8; Al_2O_3, 12.5; SiO_2, 56.9. Found: Na_2O, 2.9; N, 1.0; C, 3.6; K_2O, 5.9; Al_2O_3, 12.6; SiO_2, 56.9. A final X-ray diffraction pattern is taken for inspection and for comparison with that of an authentic sample of TMA offretite.

Properties[7]

Tetramethylammonium offretite has a crystal density of 1.55 g/cm^3. Its crystals are usually oval agglomerates, about 0.3 × 1.5 μm in size, with hexagonal symmetry. A typical unit cell formula would be TMA · 2K · Na[$(AlO_2)_4(SiO_2)_{14}$] · 7H_2O, the TMA (and a portion of the potassium ions) being trapped within gmelinite- and ε-cages, respectively, and therefore not readily exchanged by sodium or ammonium ions. In the as-synthesized form, after calcination to 500° in air, this zeolite will sorb about 12% of its weight in butane and 7% isobutane (100 torr). Its X-ray diffraction pattern is as follows:

hkl	d (Å)	I/I(0)
100	11.45	100
001	7.54	16.5
110	6.63	55.2
101	6.30	9.9
200	5.74	15.0
201	4.57	26.5
210	4.34	43.3
211, 002	3.76	89.2
102	3.59	43.0
220	3.31	18.6
202	3.15	17.4

In particular, a pure sample of TMA offretite will not show reflections at d = 9.2, 5.34, and 4.16 Å, which are the "odd-l" lines of erionite.

D. ZSM-5

Procedure

A solution of 0.9 g NaAlO$_2$ (0.0035 mole alumina, 0.01 mole NaOH) and 5.9 g NaOH (0.15 mole) in 50 g water is prepared in a 200-mL beaker equipped with magnetic stirrer and labeled "solution A". In a second beaker, a solution ("B") is prepared by adding 8.0 g tetrapropylammonium bromide (0.03 mole) to a stirred mixture of 6.2 g of 96% H$_2$SO$_4$ (0.12 mole) and 100 g H$_2$O. Solutions A and B are poured simultaneously into a solution of 60 g 30% silica sol (0.3 mole SiO$_2$, 0.003 mole Na$_2$O, and 50 g H$_2$O in a 250-mL polypropylene bottle). The bottle is immediately capped and vigorously shaken to form a gel with the composition:

$$\frac{SiO_2}{Al_2O_3} = 85 \quad \frac{H_2O}{SiO_2} = 45 \quad \frac{OH^-}{SiO_2} = 0.1 \quad \frac{Na^+}{SiO_2} = 0.5 \quad \frac{TPA^+}{SiO_2} = 0.1.$$

It is then placed in a steam chest at 95° to crystallize. After 10–14 days, periodic samples of the solid in the bottom of the bottle are taken, filtered out, washed, dried, and analyzed by X-ray diffraction for crystallinity. When the diffraction pattern reaches a limiting intensity, the mixture is removed from the steam chest and filtered, and the solid washed with water (four times 100 mL) and dried at 110°. (Crystallization times can be reduced to 1 day or less when conducted in a stirred autoclave at higher temperatures; at 140–180°, for example.) About 19 g of ZSM-5 (85% yield based on SiO$_2$) is obtained with a molar composition of 1.8 (TPA)$_2$O · 1.2Na$_2$O · 1.3Al$_2$O$_3$ · 100SiO$_2$ · 7H$_2$O.

Anal. Calcd.: Na$_2$O, 1.1; N, 0.7; C, 7.4; Al$_2$O$_3$, 1.9; SiO$_2$, 85.3. Found: Na$_2$O, 1.1; N, 0.7; C, 8.0; Al$_2$O$_3$, 1.9; SiO$_2$, 84.6. A final X-ray diffraction pattern is taken for inspection and for comparison with that of an authentic sample of ZSM-5.

Properties[5,6,8,9]

ZSM-5 has a crystal density of 1.77 g/cm^3. Its crystals have orthorhombic symmetry, as synthesized, and can vary widely in size. Compositionally ZSM-5 is unusual, in comparison with the preceding examples given above, in that it can be prepared in the absence of aluminum. Organics can be removed from ZSM-5 samples by careful oxidative calcination at about 500°. Alkali metal cations, if present, can be exchanged by ammonium ion, for example, to produce NH$_4$ZSM-5. Calcined samples of NH$_4$ZSM-5 will sorb about 11% hexane (25°, 20 torr). The X-ray diffraction pattern of ZSM-5 is characterized by the following significant lines:

d (Å)	Relative Intensity
11.1	S
10.0	S
7.4	W
7.1	W
6.3	W
6.04 ⎤ 5.97 ⎦	W
5.56	W
5.01	W
4.60	W
4.25	W
3.85	VS
3.71	S
3.04	W
2.99	W

References

1. L. D. Rollmann, *Adv. Chem. Series*, **173**, 387 (1979).
2. J. Ciric, *J. Colloid Interface Sci.*, **28**, 315 (1968).
3. D. W. Breck, U.S. Patent 3,130,007 to Union Carbide Corp., 1964.
4. E. E. Jenkins, U.S. Patent 3,578,398 to Mobil Oil Corp., 1971.
5. R. J. Argauer and G. R. Landolt, U.S. Patent 3,702,886 to Mobil Oil Corp., 1972.
6. F. G. Dwyer and E. E. Jenkins, U.S. Patent 3,941,871 to Mobil Oil Corp., 1976.
7. D. W. Breck, *Zeolite Molecular Sieves*, John Wiley & Sons, Inc., New York, 1974.
8. G. T. Kokotailo, S. L. Lawton, D. H. Olson, and W. M. Meier, *Nature*, **272**, 437 (1978).
9. E. L. Wu, S. L. Lawton, D. H. Olson, A. C. Rohrman, Jr., and G. T. Kokotailo, *J. Phys. Chem.*, **83**, 2777 (1979).

13. LEAD RUTHENIUM OXIDE, $Pb_2[Ru_{2-x}Pb_x^{4+}]O_{6.5}$

$$(2+x)\,Pb(NO_3)_2 + (2-x)\,Ru(NO_3)_3 + (3/4 + x/4)\,O_2 + \left(\frac{10-x}{2}\right)H_2O$$

$$\xrightarrow[75°]{pH\ 13} Pb_2[Ru_{2-x}Pb_x^{4+}]O_{6.5} + (10-x)\,HNO_3$$

Submitted by H. S. HOROWITZ,* J. M. LONGO,* and J. T. LEWANDOWSKI*
Checked by J. MURPHY†

High electronic conductivity oxides with the pyrochlore structure have generated significant interest for both scientific and technological reasons. The existence of the pyrochlores $Pb_2Ru_2O_{7-y}$ and $Pb_2Ir_2O_{7-y}$ ($0 \leqslant y \leqslant 1$) was first suggested by Randall and Ward.[1] These two compounds were later isolated and more fully characterized by Longo et al.[2] Sleight[3] has reported the synthesis of $Pb_2Ru_2O_{6.5}$ at elevated pressures. Bouchard and Gillson[4] first prepared the pyrochlores $Bi_2Ru_2O_7$ and $Bi_2Ir_2O_7$.

All the cited literature references to the above compounds have described solid-state syntheses at temperatures of 700-1200°. Such synthesis conditions will always lead to pyrochlore structure compounds in which all of the octahedrally coordinated sites are occupied by the noble metal cation, thus requiring the post-transition metal to noble metal molar ratio always to be 1.0. This paper focuses on solution medium syntheses at quite low temperatures ($\leqslant 75°$), thereby stabilizing a new class of pyrochlore compounds in which a variable fraction of the octahedrally coordinated sites are occupied by post-transition element cations.[5,6] The specific example here involves the $Pb_2[Ru_{2-x}Pb_x^{4+}]O_{6.5}$ series. The synthesis conditions may be simply adapted, however, to accommodate preparation of a wider range of pyrochlores which can be described by the formula $A_2[B_{2-x}A_x]O_{7-y}$ where A is typically Pb or Bi, B is typically Ru or Ir and $0 \leqslant x \leqslant 1$, and $0 \leqslant y \leqslant 1$.

The synthesis method involves reacting the appropriate metal ions to yield a pyrochlore oxide by precipitation and subsequent crystallization of the precipitate in a liquid alkaline medium in the presence of oxygen.[7] The alkaline solution serves both as a precipitating agent and as a reaction medium for crystallizing the pyrochlore, thus eliminating the need for subsequent heat treatment.

---

*Corporate Research, Exxon Research and Engineering Company, Linden, NJ 07036.
†Department of Chemistry, Oklahoma State University, Stillwater, OK 74078.

Procedure

A pyrochlore of the approximate composition $Pb_2[Ru_{1.33}Pb^{4+}_{0.67}]O_{6.5}$ may be synthesized by first preparing an aqueous solution source of lead and ruthenium cations in a 2:1 lead-to-ruthenium ratio as follows: 5.0 g of commercially available* $Ru(NO_3)_3$ solution, 10% (wt.) ruthenium metal (0.005 mole ruthenium metal) is diluted with 25 mL of distilled water. Reagent grade $Pb(NO_3)_2$ (3.277 g, 0.010 mole) is dissolved in 100 mL of distilled water and added to the ruthenium aqueous solution. This aqueous solution of lead and ruthenium is then stirred for approximately 10 min.

The aqueous solution of lead and ruthenium is then added with stirring to approximately 400 mL of 0.3 M KOH contained in a polyolefin beaker (600-mL capacity). A black precipitate immediately forms. The pH and volume of this reaction medium are then adjusted to 13.0 and 600 mL, respectively, by adding appropriate amounts of KOH and distilled water. The reaction medium is kept agitated with a magnetic stirring bar for the duration of this synthesis. The beaker is then heated so that the reaction medium attains a temperature of 75°, and oxygen is bubbled through it at ~1.0 SCFH for the duration of the synthesis by means of a plastic gas bubbling tube beneath the surface of the liquid. The exact flow rate of oxygen is not considered critical; it is desirable merely to maintain a positive flow of oxygen through the system. The most important requirement of this synthesis procedure is that the reaction medium be kept at an oxidizing potential sufficient to stabilize the tetravalent lead which partially occupies the octahedrally coordinated site. More specifically, it is found that an oxidizing potential within the range of 1.0-1.1 V versus a reversible hydrogen electrode must be maintained, and this can easily be accomplished by keeping the above described reaction medium sparged with oxygen. The beaker is equipped with a cover and any large leaks are sealed with parafilm. A leak-tight system is not necessary. The essential requirement is that the pH be maintained at a roughly constant level, and for this reason excessive evaporation losses should be minimized. If large evaporative losses are observed, the pH can still be maintained at an acceptably constant level by replacing the lost volume with water at any time necessary during the synthesis. The reaction medium is maintained in this condition for 4-5 days.

The precipitate is then separated from the still-hot reaction medium by vacuum filtration using a fritted glass filter funnel (350-mL Pyrex, fine frit). At the point when the precipitate is still submerged under a minimal volume of alkaline liquid, distilled water (at ≥75°) is added to the filtration funnel and allowed to wash the precipitate, always taking care to keep the precipitate submerged under liquid until the filtration is complete. This procedure is carried out in the manner specified until all KOH is removed from the precipitate. (Phenolphthalein

*Engelhard Industries, 429 Delancy Street, Newark, NJ.

can be used to test the filtrate.) The precipitate is then dried at 100° for approximately 18 hr. To insure that any residual water or hydroxide is removed, it is recommended that the solid be fired for 2 hr at 400° in air. X-ray diffraction confirms that the solid is a single-phase pyrochlore. The yield is approximately 2.9 g.

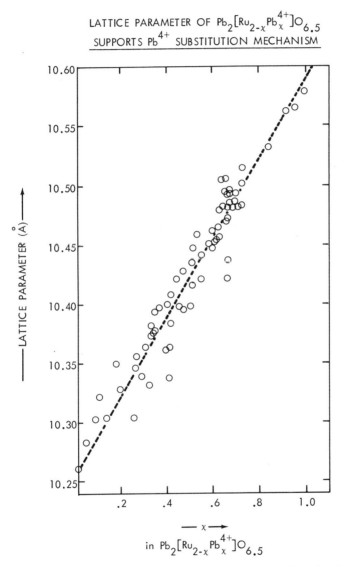

Fig. 1. Correlation between lattice parameter and composition for the series, $Pb_2[Ru_{2-x}Pb_x^{4+}]O_{6.5}$.

Properties

The resulting solid is black in color. Energy-dispersive X-ray fluorescence indicates a lead:ruthenium ratio of 2.05 ± 0.05:1. If a 2.05:1 lead:ruthenium ratio is assumed, thermogravimetric reduction in hydrogen (after in situ thermogravimetric drying at 400° in oxygen) gives an oxygen content consistent with the formula $Pb_2[Ru_{1.31}Pb^{4+}_{0.69}]O_{6.5\pm0.1}$. X-ray diffraction confirms that the material is a single-phase pyrochlore having a cubic unit cell parameter of 10.478 Å. The material is stable in oxygen to approximately 460°.

It should be noted that the exact cation stoichiometry of the product is highly sensitive to the exact metal concentration of the ruthenium source solution and temperature and pH of the reaction medium (inadvertent increases in both of these parameters lead to increased solubility of lead in the alkaline reaction medium and consequently yield solid products of lower lead:ruthenium ratios). While synthesis of a pure lead ruthenium oxide pyrochlore is relatively easy, the precise cation stoichiometry of the product is a property that is not always easy to control. A relatively quick check on the cation stoichiometry of the lead ruthenium oxide product can be obtained, however, by using the correlation between lattice parameter and composition that is displayed in Fig. 1. When lattice parameter and cation stoichiometry are independently determined, the relationship shown in Fig. 1 also provides an assessment of product purity since data points that show significant departures from the displayed linear correlation indicate the presence of impurity phases. The thermal stability of the lead ruthenium oxides decreases with increasing occupancy of tetravalent lead on the octahedrally coordinated site, but all of the ruthenium oxide pyrochlores described are stable to at least 350° in oxygen.

References

1. J. J. Randall and R. Ward, *J. Am. Chem. Soc.*, **81**, 2629 (1959).
2. J. M. Longo, P. M. Raccah, and J. B. Goodenough, *Mater. Res. Bull.*, **4**, 191 (1969).
3. A. W. Sleight, *Mater. Res. Bull.*, **6**, 775 (1971).
4. R. J. Bouchard and J. L. Gillson, *Mater. Res. Bull.*, **6**, 669 (1971).
5. H. S. Horowitz, J. M. Longo, and J. I. Haberman, U.S. Patent 4,124,539 to Exxon Research and Engineering Co., 1978.
6. H. S. Horowitz, J. M. Longo, and J. T. Lewandowski, U.S. Patent 4,163,706, to Exxon Research and Engineering Co., 1979.
7. H. S. Horowitz, J. M. Longo, and J. T. Lewandowski, U.S. Patent 4,129,525 to Exxon Research and Engineering Co., 1978.

14. CALCIUM MANGANESE OXIDE, $Ca_2Mn_3O_8$

$$2Ca^{2+} + 3Mn^{2+} + 5CO_3^{2-} \longrightarrow Ca_2Mn_3(CO_3)_5$$

$$Ca_2Mn_3(CO_3)_5 + \tfrac{3}{2}O_2 \longrightarrow Ca_2Mn_3O_8 + 5CO_2$$

Submitted by HAROLD S. HOROWITZ* and JOHN M. LONGO*
Checked by CARLYE BOOTH† and CHRISTOPHER CASE‡

Historically, the research done on the Ca-Mn-O system has been performed at high ($\geq 1000°$) temperatures.[1-9] The traditional ceramic synthesis approach to these complex oxides involves repeated high-temperature firing of the component oxides with frequent regrindings. These severe reaction conditions are necessary to obtain a single-phase product because of the diffusional limitations of solid state reactions. Such high-temperature syntheses naturally lead to crystalline, low-surface-area materials and often preclude the preparation of mixed metal oxides that are stable only at relatively low temperatures.

Achieving complete solid state reaction at low temperatures in a refractory oxide system, such as the one under investigation, is a formidable problem. Even fine reagent powders of approximately 10-μm particle size still represent diffusion distances on the order of approximately 10^4 unit cell dimensions. The use of techniques such as freeze drying[10,11] or coprecipitation[12,13] improves reactivity of the component oxides or salts because these methods can give initial particles of about several hundred Å in diameter. But this still means that diffusion must occur across 10 to 50 unit cells. To achieve complete solid-state reaction in the shortest time and at the lowest possible temperatures, one would prefer to have homogeneous mixing of the component cations on an atomic scale. Compound precursors[14,15] do achieve this objective. Unfortunately, the stoichiometry of the precursors often does not coincide with the stoichiometry of the desired product. The use of solid-solution precursors[16,17,18] provides all of the advantages of compound precursors but avoids the stoichiometry limitations.

Because both $CaCO_3$ and $MnCO_3$ have the calcite crystal structure, it is possible to prepare a complete series of Ca-Mn carbonate solid solutions. These solid solutions are then used as precursors for subsequent reaction to Ca-Mn oxides. In this way, precursors are obtained in which the reactant cations are on the order of 10 Å apart, regardless of the precursor particle size. Since solid

---

*Corporate Research, Exxon Research and Engineering Company, Linden, NJ 07036.
†Department of Chemistry, University of Georgia, Athens, GA 30602.
‡Department of Engineering, Brown University, Providence, RI 02912.

solution in this system is complete, it is possible continuously to vary the cation composition in the precursor structure without restriction. While the specific example described here involves the synthesis of $Ca_2Mn_3O_8$, the compounds Ca_2MnO_4, $CaMnO_3$, $CaMn_3O_6$, $CaMn_4O_8$, and $CaMn_7O_{12}$ may also be prepared in a similar manner.[17,18] Not only does this synthesis technique give a route to higher-surface-area complex oxides, but it also provides an approach to the preparation of several pure mixed metal oxides that are not stable at the high temperatures usually necessary for conventional solid-state reactions between small particles.

Procedure

A solid-solution precursor is first synthesized by preparing an aqueous solution containing a 2:3 molar ratio of calcium to manganese cations. Specifically, 7.006 g (0.07 mole) $CaCO_3$ and 12.069 g (0.105 mole) $MnCO_3$ are dissolved in 100 mL of distilled water plus sufficient nitric acid to effect complete solution (pH of this solution should be 1-5). The $CaCO_3$ used is reagent grade and should be well dried. The $MnCO_3$ used is freshly precipitated from a manganese nitrate solution with a large excess of ammonium carbonate, dried at 100° in a vacuum oven and stored in a container sealed under inert gas until used. Commercially available reagent grade $MnCO_3$ is unacceptable since it usually contains significant amounts of oxidized manganese products as evidenced by its brown color. The aqueous solution of calcium and manganese cations (described above) is then added, with stirring, to 102.936 g (1.072 mole) $(NH_4)_2CO_3$ dissolved in 500 mL of distilled water. The resulting precipitate is then separated from the aqueous phase by vacuum filtration. It is dried in a vacuum oven, an inert atmosphere drying oven, or for short periods of time in a microwave oven, taking care that the precipitate is not subjected to high temperatures in the presence of oxygen, which could cause divalent manganese to be oxidized. To further prevent premature oxidation, the precipitates are usually stored in an inert atmosphere. X-ray diffraction indicates that the precipitate is a single-phase solid-solution Ca-Mn carbonate with the calcite structure.

Approximately 3-5 g of the $Ca_2Mn_3(CO_3)_5$ solid structure is fired at 700° under flowing oxygen in a tube furnace for 1-48 hr to yield the pure product, $Ca_2Mn_3O_8$.

Properties

$Ca_2Mn_3O_8$, as prepared above, is a dark brown powder. The structure is monoclinic (C2/m, a = 10.02 Å, b = 5.848 Å, c = 4.942 Å, β = 109.80°) and is isostructural with Mn_5O_8 and $Cd_2Mn_3O_8$.[19] The cation stoichiometry of $Ca_2Mn_3O_8$ is determined by reducing the compound for about 2 hr in hydrogen at 1000°. The cubic lattice parameter of the resulting single-phase Ca-Mn rock salt is refined to ±0.001 Å. Since the lattice parameter variation in the Ca-Mn-O system is linear,[20] it is possible to determine the atom % of calcium and manganese present, respectively, to ±0.3%. The cation stoichiometry, experimentally determined in this way, is 40.66% Ca, 59.34% Mn as compared to a calculated stoichiometry of 40.0% Ca, 60.0% Mn. Thermogravimetric reduction in hydrogen gives a formula of $Ca_2Mn_3O_{8.0\pm0.1}$ on the basis of a 2:3 ratio of calcium to manganese. The average manganese valence determination, obtained by measuring the reducible cation content using the oxalate method,[21] is in agreement giving a formula of $Ca_2Mn_3O_{8.02\pm.01}$. The decomposition temperature at $P_{O_2} = 1.0$ atm is 890°. The measured resistivity of pressed powder samples of $Ca_2Mn_3O_8$ is 10^5 Ω-cm, and the Mn(IV) moments are found to order antiferromagnetically near 60 K. Figure 1 shows the placement of $Ca_2Mn_3O_8$ in the subsolidus system along with the thermal stabilities of the other low-temperature Ca-Mn-O phases accessible by the solid solution precursor route.[17]

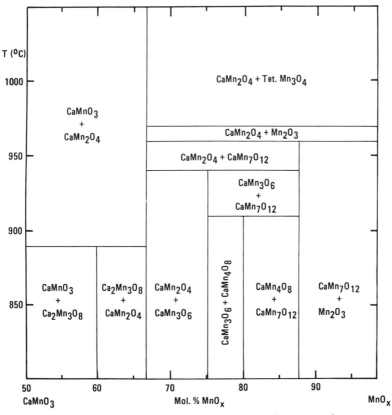

Fig. 1. Isobaric (P_{O_2} = 1.0 atm) subsolidus phase relations in the manganese-rich portion of the Ca-Mn-O system.

References

1. P. V. Ribaud and A. Maun, *J. Am. Ceram. Soc.*, **46**, 33 (1963).
2. C. Brisi and M. Lucco-Borlera, *Atti Acad. Sci. Torino*, **96**, 805 (1962).
3. C. Brisi and M. Lucco-Borlera, *J. Inorg. Nucl. Chem.*, **27**, 2129 (1965).
4. G. H. Jonker and J. H. Van Santen, *Physica*, **16**, 337 (1950).
5. H. L. Yakel, *Acta Cryst.*, **8**, 394 (1955).
6. E. O. Wollan and Koehler, *Phys. Rev.*, **100**, 545 (1955).
7. S. N. Ruddlesden and P. Popper, *Acta Cryst.*, **10**, 538 (1957).
8. C. Brisi, *Ann. Chim.*, (Rome) **51**, 1399 (1961).
9. B. Bochu, J. Chenevas, J. C. Joubert, and M. Marezio, *J. Solid State Chem.*, **11**, 88 (1974).
10. F. J. Schnettler, F. R. Monforte, and W. W. Rhodes, *Science of Ceramics*, Vol. 4, G. H. Stewart (ed.), The British Ceramic Soc., 1968, p. 79.
11. Y. S. Kim and F. R. Monforte, *Am. Ceram. Soc. Bull.*, **50**, 532 (1971).
12. A. L. Stuijts, *Science of Ceramics*, Vol. 5, C. Brosset and E. Knapp (eds.), Swedish Institute of Silicate Research, 1970, p. 335.
13. T. Sato, C. Kuroda, and M. Saito, *Ferrites*, Proc. Int. Conf., G. Hoshino, S. Iida, and M. Sugimoto (eds.), Univ. Park Press, Baltimore, 1970, p. 72.
14. W. S. Calbaugh, E. M. Swiggard, and R. Gilchrist, *J. Res. Natl. Bur Standards*, **56**, 289 (1956).
15. P. K. Gallagher and D. W. Johnson, Jr., *Thermochin. Acta.*, **4**, 283 (1972).
16. L. R. Clavenna, J. M. Longo, and H. S. Horowitz, U.S. Patent 4,060,500 to Exxon Research and Engineering Co., November 29, 1977.
17. H. S. Horowitz and J. M. Longo, *Mater. Res. Bull.*, **13**, 1359 (1978).
18. J. M. Longo, H. S. Horowitz, and L. R. Clavenna, *Solid State Chemistry: A Contemporary Overview*, Adv. Chem. Series No. 186, S. L. Holt, J. B. Milstein, and M. Robbins (eds.), American Chemical Society, Washington, D.C., 1980.
19. G. B. Ansell, H. S. Horowitz, and J. M. Longo, *Acta Cryst.*, **34A**, S157 (abstract) (1978).
20. A. H. Jay and K. W. Andrews, *J. Iron Steel Inst.*, **152**, 15 (1946).
21. D. A. Pantony and A. Siddiqui, *Talanta*, **9**, 811 (1962).

15. SYNTHESIS OF SILVER TETRATUNGSTATE

$$8AgNO_3(aq) + 4Na_2WO_4(aq) \longrightarrow Ag_8[W_4O_{16}] \downarrow + 8NaNO_3(aq)$$

Submitted by S. A. WILBER,* S. GELLER,* and G. F. RUSE*
Checked by G. LUND†

It has been shown[1] that the anion in at least the high-temperature form of silver tungstate is $[W_4O_{16}]^{8-}$; therefore, its formula should be written $Ag_8[W_4O_{16}]$ (not Ag_2WO_4). Although the crystal structure of the low-temperature form[1] has

*Department of Electrical Engineering, University of Colorado, Boulder, CO 80302.
†Dow Chemical Company, P. O. Box 150, Plaquemine, LA 70764.

not been determined, there is good reason to believe that the same tetramer exists in it. This anion is apparently present in the melt;[2] also it is the basis of the crystal structure[3] of the solid electrolyte, $Ag_{26}I_{18}[W_4O_{16}]$. It is specifically for the preparation of this solid electrolyte that the $Ag_8[W_4O_{16}]$ was synthesized. Syntheses of "Ag_2WO_4" have been reported elsewhere.[4-6] However, even in a very recently published paper[6] dealing with the solubility of this material, the authors did not realize that two different phases may be obtained depending on the pH of the reaction mixture and that the anionic unit in the solid phase is the tetramer.

Procedure

Starting materials are $AgNO_3$ and $Na_2WO_4 \cdot 2H_2O$ obtained commercially. To produce 200 g (0.1076 mole) of $Ag_8[W_4O_{16}]$ two solutions are prepared: Silver nitrate (148 g, 0.8628 mole + 1% excess) in 350 mL distilled water at 25°, solution A; sodium tungstate, $Na_2WO_4 \cdot 2H_2O$, (142.3 g, 0.4313 mole) in 350 mL distilled water at 25°, solution B.

Adjustment of pH is important. Silver tungstate is also photosensitive to some extent; thus to obtain a white product, the procedure is carried out under safe-light.

Solution A is slowly added to solution B in a 3-L beaker with vigorous stirring. A heavy, pale yellow precipitate is formed. One liter of distilled water is added with stirring. The beaker and its contents are heated on a hot plate for about 20 min until the temperature reaches 40-50°; the contents are stirred occasionally during the heating.

Dilute (1 N) NH_4OH is added until the pH is 5.5-6.0. If pH 6 is exceeded, dilute (1 N) HNO_3 is added to bring it back to between 5.5 and 6.0. **WARNING:** If the pH exceeds 8.0, insoluble compounds may form.

During the heating or after the pH is adjusted, the color of the precipitate changes from yellow to white.

After the heating and pH adjustment, the precipitate is washed by decantation with 1-L portions of distilled water, adjusted to pH 5.5-6.0 with HNO_3, until the decantate does not show the presence of Ag^+ ion. Usually seven washings suffice. Then the precipitate is washed two or three times with distilled water, or until a test for NO_3^- ion is negative.

A Büchner funnel with medium-fast pore-size filter paper is used to remove the precipitate, which is then dried in a vacuum oven at 30-40° for 24 hr, or until completely dry. Higher temperature treatment in air or vacuum causes slight loss of oxygen and color change. A yield of 99% is obtained. The product is stored in an amber bottle in a cool place.

It should be noted that if the pH is initially adjusted to 6.8-7.5 before the washing operation, a different phase of silver tungstate is obtained. It is actually the low-temperature phase. The actual transition temperature has not been determined.

Growth of Single Crystals

Single crystals of $Ag_8[W_4O_{16}]$ have been obtained by the Czochralski technique. Starting material was prepared as described above. A 2.5-cm dia by

TABLE I Powder Diffraction Data for α-$Ag_8[W_4O_{16}]$, Cu$K\alpha$ Radiation

I_{obs}	d_{obs}(Å)	d_{calc}(Å)	hkl
W-M	8.04	8.07	110
M	5.27	5.31	011
VW	3.77	3.80	211
S	2.95	2.96	002
VVS	2.83	2.84	231
MS	2.71	2.72	400
VW	2.61	2.63	240
VW	2.42	2.42	411
S	2.00	2.01	060
		2.01	350
		2.00	402
VW	1.96	1.97	242
S	1.677	1.684	233
		1.684	522
		1.682	361
M	1.655	1.660	062
		1.660	352
		1.650	071
		1.650	043
W-M	1.614	1.614	460
		1.614	550
M-S	1.591	1.593	631
		1.592	333
		1.588	262
W	1.473	1.480	004
		1.473	172
		1.469	253
		1.469	014
		1.467	104
M	1.415	1.417	462
W	1.295	1.296	073
		1.290	244
M	1.265	1.268	291
		1.267	633
W	1.236	1.237	802

2.5-cm high iridium crucible containing a 180 g charge of $Ag_8[W_8O_{16}]$ was used. (The use of an Ir crucible is not mandatory.) A 1-mm Ir wire was used to start the "seed" crystal growth. The pulling rate was 10 mm/hr, and the rotation rate was 20 rpm. The melt temperature was approximately 650°. The crystal grew along a diagonal of the B face. Small crystal specimens were cut from the boule produced for electrical[1,2] and Raman spectroscopic measurements.[2] (For further details see the latter references.)

Properties

The pure polycrystalline material is intrinsically white, and the large crystals are intrinsically water-white. The material is photosensitive, and crystals must be handled with nonmetallic-tipped forceps or with clean dry gloves. Heating the crystals in air appears to cause a slight loss in oxygen and increased conductivity.[2]

The melting point of $Ag_8[W_4O_{16}]$ has been reported[3] to be 620 ± 5°.

Crystals of the high-temperature form of $Ag_8[W_4O_{16}]$ are orthorhombic, with lattice constants a = 10.89, b = 12.03, c = 5.92 Å (all ±0.02 Å), contain two $Ag_8[W_4O_{16}]$ per unit cell, and belong to space group $Pn2n$ (C_{2v}^{10}). X-ray powder diffraction data needed for identification of the high-temperature phase are given in Table I. These data are from a powder photograph taken with CuKα radiation. Corrections were not made for absorption; therefore, the discrepancies for high d values are rather large.

To aid in identification of the low-temperature modification, the ten strongest lines in order of decreasing intensity have d spacings: 2.95, 2.77, 2.02, 1.63, 1.596, 1.252, 1.384, 1.216, 1.308 Å. Again these data are from a powder photograph taken with CuKα radiation.

References

1. P. M. Skarstad and S. Geller, *Mater. Res. Bull.*, **10**, 791 (1975); *Cryst. Struct. Commun.*, **8**, 543 (1979).
2. A Turkovič, D. L. Fox, J. F. Scott, S. Geller, and G. F. Ruse, *Mater. Res. Bull.*, **12**, 189 (1977).
3. L. Y. Y. Chan and S. Geller, *J. Solid State Chem.*, **21**, 331 (1977).
4. T. Takahashi, S. Ikeda, and O. Yamamoto, *J. Electrochem. Soc.*, **119**, 477 (1972).
5. P. H. Bottelberghs, "Phase Diagrams and Solid State Electrochemical Properties of some M_xXO_4 Compounds," Thesis, Utrecht, The Netherlands, 1976, Chap. 4.
6. J. B. Jensen and J. S. R. Buch, *Acta Chem. Scand.*, **A34**, 99 (1980), and references therein.

16. CADMIUM MIXED CHALCOGENIDES AND LAYERS OF CADMIUM (MIXED) CHALCOGENIDES ON METALLIC SUBSTRATES

Submitted by G. HODES,* J. MANASSEN,* and D. CAHEN*
Checked by ROLAND BEAULIEU† and AARON WOLD†

Cadmium chalcogenides (S, Se, Te) are widely used semiconducting materials, mainly as photoconductors, but with increasing interest focused on their use in photovoltaic[1] and photoelectrochemical[2] cells for the conversion of sunlight into electricity. For these applications, thin layers of the Cd chalcogenide on electrically conducting substrates are normally required. There are various methods of preparing such layers, such as by evaporation of either the Cd chalcogenide or the individual elements, sputtering, chemical vapor deposition, spraying, silk-screen printing, and electrochemical deposition.

For certain purposes, mixed chalcogenides are desirable, for example, Cd(S, Se) for photoconductors sensitive to a range of visible wavelengths of light, and Cd(Se, Te) which has been proposed as photoelectrode material in photoelectrochemical cells, with an extended spectral response compared with CdSe, and greater stability than CdTe.[3]

The procedures below describe two separate operations. First is the preparation of a Cd-mixed chalcogenide powder by sintering the individual Cd chalcogenides together with a flux. Any combination of the chalcogenides can be prepared in the final solid solution. Thus, quaternary Cd chalcogenides, containing S, Se, and Te, may likewise be prepared. In addition, other metal chalcogenides which form solutions with Cd chalcogenides, for example, those of Zn and Hg, may be incorporated into the compounds. The example given below may be directly extended to these possibilities.

Second, preparation of layers of a binary and a ternary Cd chalcogenide is described. An important factor here is the choice of conducting substrate. For most uses, a substrate that forms a good electrical (ohmic) contact with the Cd chalcogenide is required. For photoelectrodes, an additional factor is the desirability of a metal that is electrochemically relatively inert in the electrolyte used, to prevent local shorting of the photoelectrode due to pinholes and other exposed substrate areas. Since the layers described here are particularly suited for use as photoelectrodes, this criterion is reflected in the choice of substrate suggested here. The transition metals, Mo, Ti, and Cr are all suitable as substrates and meet these criteria. Cd chalcogenide layers on Ti, and to a greater extent,

*The Weizmann Institute of Science, Rehovot, Israel.
†Department of Chemistry, Brown University, Providence, RI 02912.

on Cr, have a tendency to peel off from the substrate if precautions are not taken; on Mo, this is not a problem. However, optimum performance of the layers as photoelectrodes is obtained on Ti, compared with either Cr or Mo.

■ *Caution. Cadmium as well as Se and Te (and to a lesser extent, S) and their compounds are very poisonous materials and care should be taken at each stage of the preparation. As far as possible, operations should be carried out in a well-ventilated hood.*

Starting Materials

High-purity semiconductors are generally used for photovoltaic purposes. In contrast, in the present procedure Analar-grade or 99% purity materials give layers with only a small reduction in performance (as photoelectrodes in photoelectrochemical cells) compared with higher purity (99.99% or 99.999%) materials. However, where available, it is safer to use high-purity materials to obtain maximal photovoltaic performance.

A. CADMIUM SELENIDE TELLURIDE ($CdSe_{0.65}Te_{0.35}$)

$$0.65 CdSe + 0.35 CdTe \xrightarrow[650°]{CdCl_2} CdSe_{0.65}Te_{0.35}$$

Procedure

Cadmium selenide (CdSe)* (248.8 mg, 1.3 mmole) and CdTe† (168.0 mg, 0.7 mmole) are ground together in a mortar with pestle until finely powdered, and well mixed. Analar-grade $CdCl_2$ (46.0 mg, 0.25 mmole) is added to the mixed powder, followed by absolute ethanol (10 drops, 0.35 mL). (The purpose of the ethanol is to disperse the $CdCl_2$ evenly on the powder, and also to facilitate grinding). This mixture is ground again in the mortar. It is then transferred to a porcelain or alumina boat of 1.5 mL capacity (or any convenient size for the above quantity of the mixture) and packed into a more or less compact mass in the boat. The mixture in the boat is dried by heating in either argon or air to ~80° (to prevent spitting when fired at the higher temperature), and then heated in a stream or argon at 600° for 30 min. The mixture is then allowed to cool to room temperature, all the time remaining in the argon atmosphere.

The mixture, which at this state is usually in the form of a sintered lump, is ground thoroughly in a mortar. The resulting powder is washed with deionized

---

*From Alfa Ventron 99.999% 3 μm.
†From Alfa Ventron 99.99% lump.

water (to remove any residual $CdCl_2$), isolated by filtration or centrifugation, and dried. The resulting powder, which is obtained in essentially 100% yield, is shown by X-ray diffraction to be a single phase consisting of hexagonal $CdSe_{0.65}Te_{0.35}$. The grain size is typically 10–20 μm. If a smaller grain size is required, grinding in a ball mill or other mechanical grinder will give an average grain size of a few micrometers.

Properties of $CdSe_{0.65}Te_{0.35}$

X-ray powder diffraction pattern (Cu$K\alpha$ radiation, Ni filter) (compare CdSe, hexagonal, File S-459 of JCPDS; wurtzite structure; space group $P6_3mc$)

d (Å)	I/I_0	(hkl)
3.81	100	100
3.59	41	002
3.37	59	101
2.61	24	102
2.20	64	110
2.03	66	103
1.90	10	200
1.88	45	112
1.84	10	201
1.68	7	202
a = 4.40 Å		c = 7.18 Å

Minor (<5%) quantities of cubic phase (a_0 = 6.24 Å, zincblende sphalerite structure, space group $F\bar{4}3m$) may sometimes be present. Further annealing of the sample will cause transformation of all the material into the hexagonal phase.

Spectral response measurements of the photocurrent obtained from photoelectrodes made from this material show an effective band gap of 1.45–1.50 eV, in agreement with that expected from this composition.[4]

B. CADMIUM SELENIDE THIN LAYERS ON TITANIUM

Procedure

Titanium metal substrates (50 × 10 × 0.6 mm) are cleaned thoroughly with 200-grade emery paper on about half the length of one side, followed by wiping with a damp cloth or tissue (to remove particles of Ti and of the emery abrasive), and rinsing in pure water. They are then preoxidized by heating in an argon atmosphere containing ~0.1% oxygen at 640° for about 30 sec. When Ti is oxidized, the buildup of oxide can be followed by the color of the oxide film, which proceeds from the clean metal through gold, purple, blue, and finally gray. The optimum thickness is when the film is a deep golden color; this is the best criterion for optimum preoxidation of the Ti. The Ti may also be heated in air (at a lower temperature) or electrochemically anodized, with the same result. Since the conditions first mentioned (640° in Ar + ~0.1% O_2) are those used for the annealing of the CdSe layer, it is convenient to carry out the preoxidation under these conditions. Such a mixture of Ar + O_2 is conveniently obtained by feeding the Ar into the furnace through silicone rubber tubing (Dow Corning SILASTIC), which is much more porous to O_2 than Tygon or ordinary rubber tubing. The set-up consists of Ar flowing at a rate of 50-100 mL/min through an 80-cm length of silicone rubber tubing (7.0 mm i.d., 1.5-mm-thick walls) into a quartz tube (60 cm × 2 cm i.d.) which is heated in a tube furnace. If a greater length of tubing must be used, then the 80-cm silicone tubing can be extended with Tygon tubing, which has little effect on the O_2 concentration in the gas entering the quartz tube.

■ **Caution.** *The exhaust gas from the furnace should be passed through a wash bottle of water (which also serves to monitor the flow rate), and vented outside. In addition, it is advisable to place the gas inlet at the same side of the quartz tube as that used for the introduction of the specimen. This prevents gas which has passed through the tube and may contain Cd and/or Se (or their compounds) from venting into the laboratory.*

Cadmium selenide* (100 mg, 0.52 mmole) is mixed with Analar-grade $ZnCl_2$ (8 mg, 0.059 mmole), and the mixture is ground in a mortar together with a 5% aqueous solution of "Nonic"† detergent (3 drops, 0.1 mL) to give a smooth thin paint. (If the paint dries out, or is too thick, more water can be added.) Using a fine paint brush, some of this mixture is painted onto a 1-cm length of the cleaned and preoxidized Ti, and allowed to dry in air at room temperature. It is then connected to a thin quartz tube sealed at one end, and with a thermocouple junction at that end (where the CdSe/Ti is placed) to monitor the temperature. The thin quartz tube can be moved in and out of the large quartz tube without allowing air to enter or Ar to escape, by use of a rubber seal. The electrode is then pushed into the large quartz tube (after it is filled with the Ar/O_2 mixture) until it is at a temperature of 200-250° (near the mouth of the furnace), and left there for 2 min. It is then introduced further into the center of furnace and left for 12 min at 640 ± 10°. Following this annealing step, it is pulled out again and

---

*As in part A.
† "Nonic" 218, Sharples, or any similar nonionic detergent, such as the Triton-X series.

left near the mouth of the furnace (200–250°) for 2 min and then cooled to room temperature in the Ar/O$_2$ atmosphere.

The resulting CdSe layer is typically 3–6 μm thick. It may be used for some purposes as it is. However, for photovoltaic and photoelectrochemical uses, a chemical etching to remove damaged surface layers is usually beneficial. A suitable etchant is 3% HNO$_3$ in concentrated HCl, in which the CdSe layer is immersed at room temperature for 2–5 sec. (A freshly made solution reaches a maximum in etching strength within ~1 hr and then gradually weakens with age over a period of days, so that the optimum etching time depends on the age of the etchant. Ideally, the CdSe should effervesce weakly in the etchant, but not too vigorously.) This etching results in a very thin layer of Se on the CdSe which may be removed by dissolution in aqueous Na$_2$S or KCN solutions.

Since ZnCl$_2$ is used as a flux it is quite possible that a Zn-rich phase exists, particularly at the surface of each CdSe crystallite. After surface etching, however, it is not likely that an appreciable amount of Zn remains in the layer.

Chromium may also be used as a substrate. In this case, chromium-plated steel is suitable, but it is preferable to modify normal chromium plating procedures from a CrO$_3$ bath[5] by plating at room temperature (instead of from a heated bath) and at somewhat higher than recommended current densities to obtain a dull, rough Cr plate, onto which the CdSe adheres better than on a bright, smooth surface. The Cr-plated steel is subjected to the same preoxidizing step as the Ti substrate. It should be noted, however, that the CdSe layers on Cr are generally somewhat poorer (judged as photoelectrodes) than those on Ti, and more importantly, their tendency to peel off the substrate is greater.

Spectral response of the CsSe, measured as for CdSe$_{0.65}$Te$_{0.35}$ in Section A, shows an effective band gap of ~1.75 eV.

C. CADMIUM SELENIDE TELLURIDE LAYERS ON MOLYBDENUM

Procedure

A CdSe$_x$Te$_{1-x}$ powder of any desired composition is prepared according to Section A. Molybdenum substrates of any suitable size for painting of 1 cm^2 with CdSe$_x$Te$_{1-x}$ are cleaned as described for Ti in Section B. The preoxidizing step required for Ti (and Cr) is unnecessary here, although also not detrimental.

A paint of the CdSe$_x$Te$_{1-x}$ is made up as described for CdSe in Section B, but substituting CdCl$_2$ for ZnCl$_2$ (in the same molar ratio), and painted onto 1 cm^2 of the cleaned Mo. Drying and annealing are carried out as described for CdSe on Ti in Section B. The O$_2$ content of the Ar gas stream is less critical when Mo is used as a substrate and, in fact, pure Ar may be used. The resulting layer may be etched, if required, in the same manner as described for CdSe.

It should be repeated that the best layers for Cd chalcogenides (measured as efficiency of photoelectrodes) are obtained on Ti. The advantage of Mo is that the annealing conditions are not as stringent as those for Ti, and the preoxidizing step can be omitted.

Properties of CdSe and $CdSe_xTe_{1-x}$

The chemical reactivity of CdSe and $CdSe_xTe_{1-x}$ (particularly with $x > 0.5$) towards common etchants is in many cases very similar. Thus, concentrated hydrochloric acid is a mild etchant for these materials, while aqua regia, Br_2/CH_3OH, and CrO_3 are strong etchants. It should be noted that these latter three, while suitable for single-crystal etching, where several microns of material may be removed, are too concentrated in their usual form for thin layers, which may be completely etched in seconds. For this reason, we generally use a dilute form of aqua regia (3% HNO_3 in HCl as previously noted), which, while a faster etchant than pure HCl, has been found suitable for these layers.

It is also worth noting the stability of these layers to air at elevated temperatures. This can be of importance if the electrodes are exposed to air after annealing, but before they have completely cooled down to room temperature. In the case of CdSe, heating in air to ~200° for ~1 min generally does not have any deleterious effect on the performance of the CdSe layers as photoelectrodes. Cadmium selenide telluride ($CdSe_xTe_{1-x}$), however, will generally lose some activity if subjected to the same treatment. The reason for this is not clear, but it seems to be a bulk effect, since mild etching to remove the surface layer does not generally reactivate these layers, as it would be expected to if the deactivation was due, for example, to an oxidized surface layer.

References

1. (a) A. G. Stanley, *Appl. Solid State Sci.*, 5, 251 (1975). (b) F. Buch, A. L. Fahrenbuch, and R. H. Bube, *J. Appl. Phys.*, 48, 1596 (1977). (c) K. Yamaguchi, H. Matsumoto, N. Nakayama, and S. Ikegami, *Jpn. J. Appl. Phys.*, 15, 1575 (1976).
2. A. Heller, ed., Semiconductor Liquid-Junction Solar Cells. *Electrochem. Soc. Proc.*, 77-3 (1977).
3. G. Hodes, *Nature*, 285, 29 (1980).
4. A. D. Stuckes and G. Farrell, *J. Phys. Chem. Solids*, 25, 477 (1964).
5. F. A. Lowenheim, ed., *Modern Electroplating*, 3rd ed., Wiley-Interscience, 1974, p. 95.

17. LAYERED INTERCALATION COMPOUNDS

Submitted by S. KIKKAWA,* F. KANAMARU,* and M. KOIZUMI*
Checked by SUZANNE M. RICH† and ALLAN JACOBSON†

The layered compound iron(III) chloride oxide FeOCl absorbs pyridine derivative molecules into its van der Waals gap forming the intercalation compounds FeOCl(pyridine derivative)$_{1/n}$. Iron(III) chloride oxide and the pyridine derivatives act, respectively, as a Lewis acid and base in the reaction with a partial transfer of the pyridine derivative electrons to the host FeOCl layer. Electrical resistivity of the host FeOCl decreases from 10^7 Ω-cm to 10^3 Ω-cm with intercalation, and the interlayer distance expands to almost twice the original value.

Methanol and ethylene glycol, which are not directly intercalated into FeOCl itself, will enter the expanded interlayer region of FeOCl(4-amino pyridine)$_{1/4}$. These molecules attack the chloride ions which are weakly bound to the previously intercalated 4-aminopyridine (APy) and cause the elimination of the 4-aminopyridine. They substitute for Cl in a FeOCl layer and are grafted to the FeO layer without the reconstruction of the host FeO layer.

The layered oxide KTiNbO$_5$ will form intercalation compounds, although it does not intercalate organic amines directly. If its interlayer potassium is replaced by protons through treatment with HCl, the product, HTiNbO$_5$, will intercalate organic amine molecules.

Herein are described the preparations of the charge-transfer-type intercalation compound FeOCl (pyridine derivative)$_{1/n}$; of grafted-type intercalation compounds FeO(OCH$_3$) and FeO(O$_2$C$_2$H$_4$)$_{1/2}$; and of some organic intercalates of HTiNbO$_5$.

A. CHARGE-TRANSFER-TYPE INTERCALATION COMPOUNDS: FeOCl(PYRIDINE DERIVATIVE)$_{1/n}$

FeOCl + 1/n (pyridine derivative) ⟶ FeOCl (pyridine derivative)$_{1/n}$

Procedure

Iron(III) chloride oxide is prepared by sealing a mixture of 100 mg α-Fe$_2$O$_3$ (0.63 mmole) and 220 mg FeCl$_3$ (1.36 mmole) in a Pyrex glass tube 20 cm long

---

*The Institute of Scientific and Industrial Research, Osaka University, Osaka 567, Japan.
†Exxon Research and Engineering Company, Corporate Research, P.O. Box 45, Linden, NJ 07036.

and 2.0 cm in diameter with a wall thickness of 2.0 mm and heating at 370° for 2-7 days.[1] The product is washed with water to remove excess $FeCl_3$. The red-violet blade-like crystals of FeOCl which are obtained are then washed with acetone, dried, and stored in a desiccator with silica gel. Prolonged exposure to moist air will cause hydrolysis of the FeOCl and render it incapable of forming intercalation compounds.

Iron(III) chloride oxide (100 mg) is introduced into a Pyrex tube ≈10 cm long, 0.8 cm in diameter, 2.0 mm wall thickness, which has been sealed at one end. To this is added 4 mL of reagent-grade pyridine (Py) or 2,4,6-trimethylpyridine (TMPy) which has been previoulsy dried over 3A molecular sieves. The tube is sealed and heated at 100° for 1 week; then the black product is collected by filtration and washed several times with dry acetone.[2] (To prepare the Apy derivative 4 mL of a 1 M acetone solution of APy, dried as above, is used and the reaction temperature is maintained at 40°).

Anal. Calcd. for $FeOCl(py)_{1/4}$: C, 11.80; N, 2.75; H, 1.00. Found: C. 13.25; N, 3.09; H, 1.11. *Anal.* Calcd. for $FeOCl(tmpy)_{1/6}$: C, 12.55; N, 1.83; H, 1.45. Found: C, 12.26; N, 1.81; H, 1.63. *Anal.* Calcd. for $FeOCl(apy)_{1/4}$: C, 11.47; N, 5.35; H, 1.16. Found: C, 12.67; N, 5.32; H, 1.94.

Properties

The d values corresponding to the interlayer distances are 7.92 Å for FeOCl, and 13.27(P), 13.57(APy), and 11.9(TMPy), for the intercalated FeOCl. Iron(III) oxide chloride is a semiconductor with a resistivity of 10^7 Ω-cm at room temperature. The intercalated complexes are still semiconductive but exhibit improved electrical conductivities along their c axes. The electrical resistivities at room temperature are 10, 10^3, and 10^3 Ω-cm, respectively, for Py, APy, and TMPy intercalated FeOCl.

B. GRAFTED-TYPE COMPOUND FROM FeOCl

1. Methanol-Grafted Compound $FeO(OCH_3)$

$$FeOCl(apy)_{1/4} + CH_3OH \longrightarrow FeO(OCH_3) + HCl + 1/4 AP$$

Procedure

Direct reaction of FeOCl with methanol does not occur up to a temperature of 100°. In contrast, the reaction of $FeOCl(apy)_{1/4}$ is a facile one due primarily to the expanded interlayer distance which permits penetration of the methanol molecules.

The intercalation compound FeOCl(apy)$_{1/4}$, 80 mg, is soaked in 2 mL of methanol (previously dried over 3A molecular sieves for 3 days) in a sealed Pyrex glass tube, which is typically 10 cm long and 0.8 cm in diameter, with a wall thickness of 2.0 mm.[3] The tube is maintained at 100° and shaken twice daily for 10 days. (If difficulty is encountered in obtaining intercalation, it may be helpful to add a small amount of APy.) The resulting brown solid is collected by filtration, washed with several volumes of dry methanol, then stored in a desiccator with silica gel.

Anal. Calcd. for FeOOCH$_3$: Fe, 54.3; C, 11.7; H, 2.94; N, 0.00; Cl, 0.00. Found: Fe, 55.9; C, 9.29; H, 2.43; N, 0.33; Cl, 4.20.

Properties

The interlayer distance in FeO(OCH$_3$) is 9.97 Å. The infrared spectrum shows a C-O stretching vibration around 1050 cm^{-1}, but no O-H stretching vibration. A study of the Mossbauer effect at room temperature indicates that the isomer shift and the quadrupole splitting are 0.37 and 0.60 mm/s, respectively.

2. Ethylene Glycol-Grafted Compound FeO(O$_2$C$_2$H$_4$)$_{1/2}$

FeOCl(apy)$_{1/4}$ + 1/2C$_2$H$_4$(OH)$_2$ ⟶ FeO(O$_2$C$_2$H$_4$)$_{1/2}$ + HCl + 1/4(apy)

Procedure

Eighty milligrams of FeOCl(apy)$_{1/4}$ is well ground and placed in a sealed Pyrex tube, 10 cm long and 0.8 cm in diameter, with a wall thickness of 2.0 mm,[4] which contains 2 mL ethylene glycol and approximately 20 mg of APy. (The addition of APy is important to drive the reaction to completion.) The reaction mixture is heated at 110° for 1 week. This produces a brown crystalline material which is collected by filtration, then washed with dry acetone until the silver chloride test is negative.

Anal. Calcd. for FeO(O$_2$C$_2$H$_4$)$_{1/2}$: Fe, 54.8; C, 11.8; H, 1.98. Found: Fe, 54.2; C, 11.6; H, 2.21.

Properties

The basal spacing of FeO(O$_2$C$_2$H$_4$) is 14.5 Å when the product is still in ethylene glycol, but it shrinks to 10.98 Å on washing with acetone. Mossbauer isomer shift and quadrupole splitting are respectively 0.38 and 0.59 mm/s at room temperature.

C. THE ORGANIC INTERCALATES OF HTiNbO$_5$

$$KTiNbO_5 + HCl \longrightarrow HTiNbO_5 + KCl$$

$$HTiNbO_5 + b \longrightarrow H_{1-1/m}(bH)_{1/m}TiNbO_5$$

where b is an amine

Procedure

White solid KTiNbO$_5$ is prepared by heating an intimate mixture of K$_2$CO$_3$, TiO$_2$, and Nb$_2$O$_5$ in the molar ratio 1:2:1 at 1100° overnight.[5] The reaction product ($\sim$3 g) is treated with 2 M HCl (200 mL) at 60° for 1 hr to produce HTiNbO$_5$.[6] The sample is washed with distilled water several times, until a flame test shows no detectable potassium. Approximately 20 mg of the sample is sealed in a Pyrex tube 15 cm long and 1.5 cm in diameter with a wall thickness of 2.0 mm which contains 3 mL of aliphatic amines CH$_3$(CH$_2$)$_n$NH$_2$ and the mixture warmed at 60° for 5 hr.[7] The products are filtered out and washed with acetone three times. The products have the composition [H$^+$]$_{1/2}$ [CH$_3$(CH$_2$)$_n$NH$_3^+$]$_{1/2}$ [TiNbO$_5^-$].

Properties

Both parent compounds (KTiNbO$_5$ and HTiNbO$_5$) and amine-intercalated HTiNbO$_5$ belong to the orthorhombic system; a = 6.44, b = 3.78, c = 17.52 Å for HTiNbO$_5$. The (TiNbO$_5$)$^-$ layer lies in the ab plane and intercalates amine in its protonated form. The interlayer distance $c/2$ expands on intercalation: c = 9.02 Å for ammonia, 22.96 Å for methanamine, 26.78 Å for ethanamine, 35.34 Å for 1-propanamine, 36.82 Å for 1-butanamine.

References

1. M. D. Lind, *Acta Cryst.*, **B26**, 1058 (1970).
2. S. Kikkawa, F. Kanamaru, and M. Koizumi, *Bull. Chem. Soc. Jpn.*, **52**, 963 (1979).
3. S. Kikkawa, F. Kanamaru, and M. Koizumi, *Inorg. Chem.*, **15**, 2195 (1976).
4. S. Kikkawa, F. Kanamaru, and M. Koizumi, *Inorg. Chem.*, **19**, 259 (1980).
5. A. D. Wadsley, *Acta Cryst.*, **17**, 623 (1964).
6. H. Rebbash, G. Desgardin, and B. Raveau, *Mater. Res. Bull.*, **14**, 1125 (1979).
7. S. Kikkawa and M. Koizumi, *Mater. Res. Bull.*, **15**, 533 (1980).

18. IRON TITANIUM HYDRIDE (FeTiH$_{1.94}$)

$$\text{FeTi} + 1/2\text{H}_2 \rightleftharpoons \text{FeTiH}_{1.0}$$
$$\text{FeTiH}_{1.0} + 0.47\text{H}_2 \rightleftharpoons \text{FeTiH}_{1.94}$$

Submitted by J. J. REILLY*
Checked by G. SANDROCK†

A number of metals and alloys react reversibly with hydrogen to form metal hydrides. Recently, interest in such compounds has blossomed because of the possibility of using those with appropriate properties in energy storage applications.[1] One material that has received much attention in this regard is iron titanium hydride. This compound was first prepared by Reilly and Wiswall, through the direct reaction of hydrogen with the alloy.[2] On the basis of pressure-composition data, a two-step reaction is involved as shown above. As indicated, the reactions are reversible, their direction depending upon the hydrogen pressure. Both hydride phases are nonstoichiometric and have a wide region of homogeneity. By convention, the monohydride is referred to as the β phase and the dihydride as the γ phase. Deuterium will react similarly to form the corresponding β and γ deuterides. The structures of the deuteride phases have been determined by neutron diffraction.[3-5] Recent work has indicated the existence of an additional deuteride[6] phase of intermediate composition. Its structure is very similar to that of the β phase and has been designated β_2. An analogous hydride phase also exists.[7,8] However, the appearance of slight modifications of parent structures is quite common in metal-hydrogen systems at low temperatures and for the present purpose the reactions as written above may be adopted.

Procedure

The apparatus for the preparation of unstable metal hydrides such as FeTiH$_x$ is shown in Figs. 1 and 2. It consists of a reactor and a pressure-vacuum manifold which is rated for pressures up to 68 atm and is constructed of stainless steel (300 series). The reactor is constructed of $\frac{1}{2}''$ tubing welded to a metal plug which forms the bottom closure. The top of the tube is welded to a $\frac{1}{2}''$ O-ring connector (VCO Type, Cajon Co. 32550, Old South Rd., Cleveland, OH) which in turn is socket-welded to a $\frac{1}{2} \times \frac{1}{4}''$ tube reducer. A sintered metal disc ($\sim$5 μm

*Brookhaven National Laboratory, Associated Universities Inc., Upton, NY 11973.
†Inco R & D Center, Suffern, NY 10901.

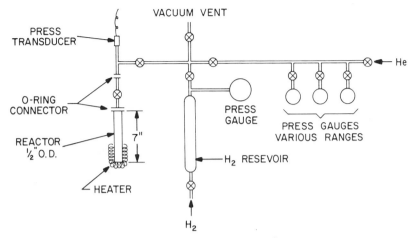

Fig. 1. Schematic of experimental apparatus.

pore size) is situated between the socket shoulder and the tube end of the O-ring fitting. The disc serves to prevent entrainment of metal particles in the gas stream when venting the reactor.

Commerical-grade FeTi is suitable for most purposes and is readily available from Ergenics Division of MPD Technology Corp., 681 Lawlins Rd., Wyckoff, NJ. Approximately 10 g of FeTi is pulverized with a hardened steel mortar and pestle of a type used to crush ore samples. FeTi is very brittle and can readily be crushed to ~40 mesh (U.S. sieve) particles, which is the recommended size range. Larger particles may be used but will require a longer activation period. Sample preparation can be carried out in air. The pulverized alloy is introduced into the reactor which is sealed and attached to the pressure-vacuum manifold. The particle bed should not be tightly packed since the alloy volume will increase almost 20% upon formation of the higher hydride. Under such circumstances, it is possible to distort the reactor wall by expansion of the alloy reactant. Such distortion can be avoided by maintaining low length-to-diameter ratios for the bed, occasional agitation, or positioning the reactor horizontally while maintaining some free volume above the entire length of the bed. The latter alternative is particularly recommended if larger than gram quantities of material are to be prepared. The reactor is evacuated and the sample is outgassed and heated to ~400°. Only the bottom portion of the reactor is heated, and care should be taken to keep the O-ring connector cool. Occasional pressurizing with low-pressure (~1 atm) helium or hydrogen will promote heat transfer. The sample should be maintained at ~400° under vacuum for ~15 min and then cooled to room temperature. At this point hydrogen is added slowly until a pressure of ~68 atm is reached. The hydrogen-alloy reaction can be monitored

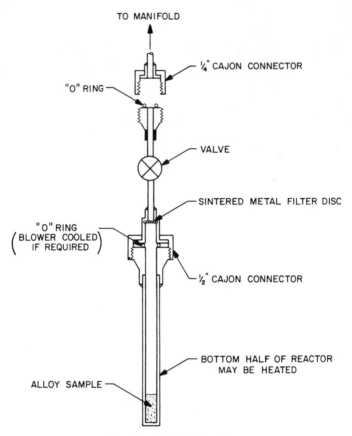

Fig. 2. Detail of reactor.

by noting the drop in pressure as a function of time. If the alloy is of high purity, a period of up to 2 weeks may be required to completely convert the metal; commercial-grade alloy will react somewhat faster. The time for completion may be shortened slightly by alternate decomposition (by gentle heating and outgassing) and rehydriding of the partially reacted alloy. When the hydrogen content corresponds to $FeTiH_{1.9}$ or greater, the reaction may be considered complete (the highest reported H content obtained is $FeTiH_{1.95}$). The first complete hydriding reaction is usually referred to as "activation". After the "activation" step the surface area: weight ratio is increased markedly, typically to ~ 0.5 m^2/g. This is a result of the numerous cracks and fissures which occur as a consequence of the large volume expansion which the alloy undergoes upon hydriding. In the "activated" state, subsequent dehydriding or hydriding reactions can be carried out extremely rapidly (a few mintues) at 25° and pressures of 1 atm

TABLE I Properties of FeTiH$_x$

<table>
<thead>
<tr><th rowspan="2">Composition</th><th colspan="6">Thermodynamic Data for the Formation of FeTiH$_x$
Relative Partial Molal Quantities, 298 K</th></tr>
<tr><th>$(\bar{H}_H - 1/2 H^0_{H_2})$,
(kcal/mole)</th><th>$(\bar{S}_H - 1/2 S^0_{H_2})$,
(e.u.)</th><th>$(\bar{G}_H - 1/2 G^0_{H_2})$,
(kcal/mole)</th><th>A^a</th><th>B^a</th><th>Reference</th></tr>
</thead>
<tbody>
<tr><td>FeTiH$_{0.1}$-FeTiH$_{1.04}$</td><td>-3.36</td><td>-12.7</td><td>0.42</td><td>-3383</td><td>12.7612</td><td>2</td></tr>
<tr><td>FeTiH$_{1.20}$</td><td>-3.70</td><td>-14.4</td><td>0.57</td><td>-3728</td><td>14.4327</td><td>2</td></tr>
<tr><td>FeTiH$_{1.40}$</td><td>-3.98</td><td>-15.6</td><td>0.65</td><td>-4020</td><td>15.6610</td><td>2</td></tr>
<tr><td>FeTiH$_{1.60}$</td><td>-4.03</td><td>-15.8</td><td>0.68</td><td>-4057</td><td>15.9165</td><td>2</td></tr>
<tr><td colspan="6">Density, 298 K FeTiH$_{1.0}$ = 5.88 g/mL; FeTiH$_{1.93}$ = 5.47 g/mL</td><td>2</td></tr>
</tbody>
</table>

<table>
<thead>
<tr><th rowspan="2">FeTi</th><th colspan="4">Ignition Temperature of Powder (°C)</th></tr>
<tr><th>Particle Size (μm)</th><th>Bulk Powder</th><th>Dust Cloud</th><th>Reference</th></tr>
</thead>
<tbody>
<tr><td>FeTi</td><td>74-149</td><td>No ignition</td><td>—</td><td>11</td></tr>
<tr><td>FeTiH$_x$</td><td>74-149</td><td>188</td><td>—</td><td>11</td></tr>
<tr><td>FeTiH$_x$</td><td><74</td><td>—</td><td>420</td><td>11</td></tr>
</tbody>
</table>

^{a}Constants in $\ln P = (A/T) + B$.

TABLE II Crystal Structure

Phase	Composition	Lattice Parameters				Strongest Three Lines, 2θ degrees (X-Rays, CuKα)			Reference
		a	b	c	β				
	Deuterides								
β_1(ortho)	FeTiD$_{1.0}$	2.956	4.543	4.388		41.8	39.7	58.5	3,4
β_2(ortho)	FeTiD$_{1.37}$	3.088	4.515	4.391		40.9	39.9	58.6	6
γ(monoc)	FeTiD$_{1.9}$	4.704	2.830	4.704	96.97°	38.5	40.9	43.1	5
	Hydrides								
β_1(ortho)	FeTiH$_{0.94}$	2.954	4.538	4.381		41.9	39.7	58.5	8
β_2(ortho)	FeTiH$_{1.40}$	3.094	4.513	4.391		40.9	39.9	58.6	8
γ(monoc)	FeTiH$_{1.9}$	4.713	2.834	4.731	97.1°	38.5	40.9	43.1	8

and 60 atm of H_2 respectively. If the hydride is removed from the reactor, some hydrogen will be evolved but upon contact with air the surface of the alloy quickly becomes poisoned and decomposition ceases. The dehydrided alloy will also be poisoned upon contact with air. Similarly, impurities such as CO and H_2O [9] in the hydrogen gas will poison the alloy surface, and in order to retain an active alloy through many hydriding-dehydriding cycles, high-purity hydrogen should be used. In the event of poisoning, in either the hydrided or dehydrided state, the alloy can be readily reactivated by heating and outgassing at 200-300° for a short time; 15 min is usually sufficient.

■ *Caution. Care should be exercised during the reactivation step if the alloy was poisoned in the hydrided state, as the hydride can decompose rapidly once the surface becomes reactivated, thus liberating large quantities of hydrogen gas in a short time even though the decomposition is endothermic.*

Properties

The hydride phases of FeTi are gray metal-like solids having essentially the same appearance as the dehydrided alloy. They are not pyrophoric at room temperature but ignition can occur at higher temperature. Thermodynamic and other properties of interest are summarized in Tables I and II.

It should also be pointed out that the behavior of the FeTi-H system can be altered in a controlled manner by the partial substitution of another transition metal for Fe. A particularly interesting alloy in this respect is $TiFe_{1-x}Mn_x$ [10] (also available from Ergenics Corp.). For example, $TiFe_{0.85}Mn_{0.15}$ has thermodynamic properties very similar to those of FeTi, but it can be activated (and reactivated) without a high-temperature outgassing step in a relatively short time. The procedure is the same as outlined above except that the initial outgassing step can be carried out at room temperature. The reaction rate of the initial hydriding (activation) step is greatly increased; the time required to reach completion ($TiFe_{0.85}Mn_{0.15}H_{1.9}$) is about 24 hr. In practical applications, the elimination of the high-temperature requirement in the initial activation (and any subsequent reactivation) step greatly simplifies reactor design.

References

1. J. J. Reilly, *Z. Physik. Chem., Neue Folge*, **117**, 155 (1979).
2. J. J. Reilly and R. H. Wiswall, Jr., *Inorg. Chem.*, **13**, 218 (1974).
3. P. Thompson, M. A. Pick, F. Reidinger, L. M. Corliss, J. M. Hastings, and J. J. Reilly, *J. Phys. F: Metal Phys.*, **8**, L75 (1978).
4. P. Fisher, W. Hälg, L. Schlapbach, F. Stucki, and A. F. Andresen, *Mater. Res. Bull.*, **13**, 931 (1978).
5. P. Thompson, J. J. Reilly, F. Reidinger, J. M. Hastings, and L. M. Corliss, *J. Phys. F: Metal Phys.*, **9**, L61 (1979).

6. J. Schefer, P. Fisher, W. Hälg, F. Stucki, L. Schlapbach and A. F. Andresen, *Mater. Res. Bull.*, **14**, 1281 (1979).
7. T. Schober and W. Schäfer, *J. Less Common Metals*, **74**, 23 (1980).
8. F. Reidinger, J. F. Lynch, and J. J. Reilly, An X-Ray Diffraction Examination of the FeTi/H$_2$ System, *J. Phys. F: Metal Phys.*, **12**, L49 (1982).
9. G. D. Sandrock and P. D. Goodell, *J. Less Common Metals*, **73**, 161 (1980).
10. J. R. Johnson and J. J. Reilly, "The Use of Manganese Substituted Ferrotitanium Alloys for Energy Storage," Proc. Int. Conf. on Alternative Energy Sources, Miami Beach, Florida, 1977.
11. C. E. Lundin and F. E. Lynch, Safety Characteristics of FeTi Hydride, Proc. Intersoc. Energy Conversion and Eng. Conf., Univ. of Delaware, Inst. of Electrical and Electronic Engineers, New York, 1975.

19. ALUMINUM LANTHANUM NICKEL HYDRIDE

$$AlLaNi_4 + 2H_2 \longrightarrow AlLaNi_4H_4$$

Submitted by M. H. MENDELSOHN,* D. M. GRUEN,* and A. E. DWIGHT†
Checked by G. D. SANDROCK ‡

Aluminum lanthanum nickel hydride is but one example of a whole series of hydrides which may be prepared in a common fashion. A more general reaction would be as follows:

$$LaNi_{5-x}M_x + yH_2 \longrightarrow LaNi_{5-x}M_xH_{2y}$$

TABLE I Illustrates Values of x and y for Various Elements (M)

Element (M)	Approx. max. solubility (x)	H$_2$ absorption for $x = 0.4$ (y)
Al	1.5	2.9
Ga	0.6	2.8
In	0.4	2.8
Si	0.5	2.2
Ge	0.6	2.5
Sn	0.65	2.7

*Chemistry Division, Argonne National Laboratory, Argonne, IL 60439.
†Department of Physics, Northern Illinois University, DeKalb, IL 60115.
‡Inco R & D Center, Suffern, NY 10901.

By varying the amount of the substituted element (M), one can vary the hydrogen dissociation pressure of the corresponding hydride over several orders of magnitude.[1] These materials are useful in hydrogen storage applications, particularly in stationary uses where weight is not the prime consideration. Also, alloys of this type have been demonstrated to be effective in a chemical heat pump system.[2]

Procedure

The ternary metal alloys are prepared by standard arc melting techniques[3] using the pure metals. For example, 3.294 g of 99.9% La (0.0237 mole), 5.573 g of 99.9% Ni (0.0949 mole), and 0.64 g of 99.9% Al (0.0237 mole) metals were melted and remelted several times in an arc furnace under an atmosphere of argon, to yield 9.469 g of $AlLaNi_4$ (0.0236 mole). The alloy button can be easily broken into smaller pieces with a hammer, and a small piece can be further ground, using a mortar and pestle, into a powder for analysis by X-ray diffraction. Four of the strongest X-ray powder diffraction lines observed for the alloy $LaNi_4Al$ are as follows: 2.97 Å (medium), 2.51 Å (medium), 2.19 Å (medium), and 2.14 Å (strong). Extraneous phases may also be detected by metallographic examination of the alloy.

After preparation of the alloy and determination of its phase composition, the material may next be hydrided. To accomplish this, a high-pressure manifold needs to be constructed. The materials of construction should be typically 304 or 316 stainless steel. Stainless steel tubing having an outside diameter of $\frac{3}{8}''$ and a wall thickness of $0.035''$ is convenient to use, but other sizes are acceptable if care is taken that their allowable working pressure loads are at least 2500 psi (the maximum pressure generally available in a high-pressure cylinder). Miniature needle valves with a maximum operating pressure of 3000–5000 psi are available from a number of manufacturers. Finally, the high-pressure system may be put together with Swagelok fittings for both easy assembly and disassembly, or for a more permanent installation, socket-weld fittings may be used. A schematic of a simple hydriding manifold is illustrated in Fig. 1. The hydriding line should be placed in a ventilated hood for the safe disposal of hydrogen gas. Air flow through the hood should be high enough to ensure that any hydrogen released will be quickly diluted below its lower limit of flammability in air (4.0 vol %).

Besides the hydriding manifold, a reactor to contain the alloy or hydride is also needed. One simple design we have used consists of a $4''$ length of stainless steel tubing having an outside diameter of $\frac{3}{8}''$ welded to a $3\frac{1}{2}''$ length of tubing having an outside diameter of $\frac{3}{4}''$. The $\frac{3}{4}''$ tube contains a stainless steel filter disc which has been beam welded to the bottom of the $\frac{3}{4}''$ tube. Finally, a Hoke miniature needle valve is attached to the top of the $\frac{3}{8}''$ tubing for connection to the hydriding manifold.

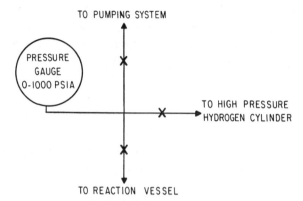

Fig. 1. *Schematic drawing of a basic hydriding apparatus.*

After the alloy has been weighed and sealed in the reactor, the reactor is connected to the hydriding manifold. The manifold and reactor are evacuated to a pressure of 1×10^{-2} torr or better. After being pumped for a few minutes, the reactor and manifold are isolated from the pump. An amount of hydrogen ranging from 100 psia to 1000 psia may then be admitted to the manifold. The higher the pressure admitted, the faster the hydriding reaction. After the amount of hydrogen added has been noted, the reactor valve is opened. The absorption of hydrogen may be followed by noting the drop in pressure. After a period of time ranging from 30 min to several hours, the reaction is complete as determined by the invariant hydrogen pressure over the solid phase. The amount of hydrogen in the solid phase may be calculated by subtracting the amount of hydrogen in the gas phase at the end of the reaction (from pressure-volume-temperature measurements) from the total amount of hydrogen added to the system initially (also determined from P-V-T measurements).

Properties

The hydride $AlLaNi_4H_4$ has X-ray powder diffraction lines similar to those of the alloy, only shifted to larger d spacings. For $LaNi_4AlH_4$, $a = 5.307 \pm 0.005$ Å and $c = 4.228 \pm 0.004$ Å. Five of the strongest X-ray powder diffraction lines are as follows: 3.10 Å (medium), 2.64 Å (medium), 2.29 Å (medium), 2.24 Å (strong), and 1.438 Å (medium). The most likely impurity is unhydrided $LaNi_4Al$, whose strongest X-ray powder diffraction lines were listed earlier. If prepared in a glass tube, one may note the disintegration of the alloy chunks into a fine powder (1-100 μm particle size).

■ **Caution.** *Care must be exercised in this case since a pressure lower than the burst pressure of the glass tube must be used.*

There is no noticeable change in the color of the materials, both the alloy and the hydride being light gray to dark gray powders. Most of the hydrides rapidly lose hydrogen if exposed to the air. The hydride can be stored in sealed tubes, either glass or metal.

■ **Caution.** *When stored in sealed containers, the hydrides must not be heated to a temperature at which their hydrogen equilibrium pressures exceed the rupture strength of the container.*

If a hydride is placed in a bottle, moisture will many times form in the bottle and sometimes an odor of ammonia may be discerned after the material has been contained for some time.

■ **Caution.** *Dehydrided alloy powders have been found, in a few cases, to burst into flames when placed on a combustible surface (for example on a piece of weighing paper). As a consequence, when exposed to the air, both the hydrided and dehydrided alloy powders should be handled as potentially pyrophoric materials.*

References

1. (a) M. H. Mendelsohn, D. M. Gruen, and A. E. Dwight, *J. Less Common Metals*, **64**, 193 (1979). (b) M. Mendelsohn, D. Gruen, and A. Dwight, *Inorg. Chem.*, **18**, 3343 (1979).
2. D. M. Gruen, M. H. Mendelsohn, and I. Sheft, *Solar Energy*, **21**, 153 (1978).
3. See, for example, T. B. Reed and E. R. Pollard, *Inorg. Synth.*, **14**, 131 (1973).

Chapter Two

TRANSITION METAL COMPLEXES AND COMPOUNDS

20. PURIFICATION OF COPPER(I) IODIDE

$$CuI + KI \longrightarrow K[CuI_2]*$$

$$K[CuI_2] \xrightarrow{H_2O} CuI \downarrow + KI$$

Submitted by GEORGE B. KAUFFMAN[†] and LAWRENCE Y. FANG[†]
Checked by N. VISWANATHAN[‡] and G. TOWNSEND[‡]

Although freshly prepared copper(I) iodide is white, the commercial product as well as samples that have been allowed to stand are usually a light tan because of the presence of iodine produced by decomposition. The simple method of purification described below makes use of the fact that copper(I) iodide is readily soluble in concentrated solutions of iodide ion[4-8] but reprecipitates when the iodide concentration is lowered. With appropriate modifications, the method may be applied to the purification of other insoluble halides that reversibly form soluble halide complexes.

*Smaller amounts of complex anions containing more iodine may also be found in solution.[1-3]
[†] California State University, Fresno, CA 93740.
[‡] The Pennsylvania State University, Fayette Campus, Uniontown, PA 15401.

Procedure

Commercial copper(I) iodide or a sample that has become discolored[9,10] (3.80 g, 0.02 mole) placed in a 250-mL Erlenmeyer flask is dissolved with magnetic stirring and moderate heating by the addition of a minimum volume (about 125 mL) of 3.5 M potassium iodide (about $\frac{1}{2}$ hr is required). Activated charcoal (e.g., Matheson, Coleman, and Bell "Darco" G-60) (about 5 g) is added to the resulting solution, which varies in color from light tan to dark brown, depending upon the extent of contamination. The mixture is stirred magnetically with moderate heating until the supernatant solution becomes colorless, at which time stirring is stopped and the charcoal is allowed to settle, indicating that the contaminating iodine has been completely adsorbed (about 10 min). The mixture is suction-filtered through a 150-mL (60-mm i.d.) sintered-glass funnel (medium porosity), and the filtrate is diluted with sufficient water (about 500 mL) to cause reprecipitation of the copper(I) iodide upon standing for about one hr.

After precipitation commences, the dense, white solid is allowed to settle for about 15 min, the supernatant liquid is decanted, and the residue is then collected on a small, sintered-glass funnel (medium porosity) and washed with several 20-mL portions of water, ethanol, and finally diethyl ether. The product is powdered and dried in a vacuum over sulfuric acid for several days. (Copper(I) iodide retains moisture tenaciously. The air-dried product contains about 4% water,[11] while the drying procedure recommended reduces this to ca. 0.2%.[12,13] The product may be dried overnight in a phosphorus(V) oxide drying pistol at 100° in the vacuum. When copper(I) iodide is heated in air below 200°, oxygen displaces iodine and copper(II) oxide is formed.[13]) The yield is quantitative except for manipulative losses.

Properties

Copper(I) iodide is a dense, pure white solid, crystallizing with a zinc-blende structure below 300°. It is less sensitive to light than either the chloride or bromide, although passage of air over the solid at room temperature in daylight for 3 hr results in the liberation of a small amount of iodine. It melts at 588°, boils at 1293°, and, unlike the other copper halides, is not associated in the vapor state. Being extremely insoluble (0.00042 g/L at 25°), it is not perceptibly decomposed by water. It is insoluble in dilute acids but dissolves in aqueous solutions of ammonia, potassium iodide, potassium cyanide, and sodium thiosulfate. It is decomposed by concentrated sulfuric and nitric acids.

References

1. C. M. Harris, *J. Proc. Roy. Soc. N. S. Wales*, **85**, 138 (1952).
2. C. Herbo and J. Sigalla, *J. Chim. Phys.*, **55**, 403 (1958).

3. A. M. Golub, S. M. Sazhienko, and L. I. Romanenko, *Ukr. Khim. Zh.*, **28**, 561 (1962).
4. C. Immerwahr, *Z. Anorg. Chem.*, **24**, 269 (1900).
5. G. Bodländer and O. Storbeck, *Z. Anorg. Chem.*, **31**, 458 (1902).
6. S. von Náray-Szabó and Z. Szabó, *Z. Physik. Chem.*, **A166**, 228 (1933).
7. Ya. D. Fridman and D. S. Sarbaev, *Zh. Neorg. Khim.*, **4**, 1849 (1959) (in Russian); *Russ. J. Inorg. Chem.*, **4**, 835 (1959) (in English).
8. M. L. Gavrish and I. S. Galinker, *Zh. Neorg. Khim.*, **9**, 1289 (1964).
9. G. B. Kauffman and R. P. Pinnell, *Inorg. Syn.*, **6**, 3 (1960).
10. G. B. Kauffman, *Inorg. Syn.*, **11**, 215 (1968).
11. I. Souberain, *J. Pharm. Chim.*, **13**, 427 (1832).
12. B. Lean and W. H. Whatmough, *J. Chem. Soc.*, **73**, 148 (1898).
13. M. Guichard, *Bull. Soc. Chim. France*, [4] **1**, 897 (1907).

21. COBALT(III) AMINE COMPLEXES WITH COORDINATED TRIFLUOROMETHANESULFONATE

Submitted by N. E. DIXON,* W. G. JACKSON,† G. A. LAWRANCE,* and A. M. SARGESON*
Checked by U. GOLI‡ and E. S. GOULD‡

Although the trifluoromethanesulfonate anion ($CF_3SO_3^-$) has been shown to be an excellent leaving group in nucleophilic substitution reactions in organic chemistry,[1] it has not been exploited as extensively in inorganic reactions. Recently, the coordination of $CF_3SO_3^-$ as a unidentate ligand in cobalt complexes has been demonstrated.[2] It has been shown to be labile to substitution in cobalt(III) ammine complexes and is only a little less labile than ClO_4^- ion in this situation.[3] For example, the $[Co(NH_3)_5(OClO_3)]^{2+}$ ion undergoes aquation ($t_{1/2} \approx 25$ s) only four times as fast as $[Co(NH_3)_5(OSO_2CF_3)]^{2+}$ at 25°. Although perchlorate complexes present an explosive hazard, no such problem has been encountered with the trifluoromethanesulfonate complexes. The advantages of the (trifluoromethanesulfonato)cobalt(III) trifluoromethanesulfonate complexes lie in the safety factor compared with perchlorato complexes, their solubilities in a wide range of common organic solvents, and the lability and ease of substitution of the coordinated $CF_3SO_3^-$. All these factors combine to make the trifluoromethanesulfonate complexes very useful precursors for syntheses requiring mild and non-aqueous conditions. Facile routes to the three cobalt(III) ammine complexes, $[Co(NH_3)_5(OSO_2CF_3)](CF_3SO_3)_2$, *cis*-$[Co(H_2NCH_2CH_2NH_2)_2(OSO_2CF_3)_2]$ ·

---

*Research School of Chemistry, The Australian National University, Canberra 2600, Australia.
†Chemistry Department, Faculty of Military Studies, University of New South Wales, Duntroon 2600, Australia.
‡Chemistry Department, Kent State University, Kent, OH 44242.

(CF_3SO_3) and fac-[Co($H_2NCH_2CH_2NHCH_2CH_2NH_2$)($OSO_2CF_3$)$_3$] are reported below.

■ **Caution.** *Copious fumes of gaseous HCl are evolved on addition of CF_3SO_3H. Reactions must be carried out in a well-ventilated hood. The compound CF_3SO_3H is a very strong acid and must be handled with care.*

Anhydrous trifluoromethanesulfonic acid (CF_3SO_3H, 3M Company), vacuum distilled (bp 52°, 10 torr) prior to use, is a stable, strong nonoxidizing acid.

A. PENTAMMINE(TRIFLUOROMETHANESULFONATO)COBALT(III) TRIFLUOROMETHANESULFONATE, [Co(NH$_3$)$_5$OSO$_2$CF$_3$](CF$_3$SO)$_2$

$$[Co(NH_3)_5Cl]Cl_2 + 3CF_3SO_3H \longrightarrow [Co(NH_3)_5OSO_2CF_3](CF_3SO_3)_2 + 3HCl \uparrow$$

Procedure

To [Co(NH$_3$)$_5$Cl]Cl$_2$ [4] (50 g, 0.20 mole) in a 2-L three-necked, round-bottomed flask fitted with a gas bubbler, anhydrous CF_3SO_3H (300 mL) is cautiously added. A steady stream of nitrogen is passed through the resulting solution and the flask is lowered into a silicone oil bath preheated to and maintained at 90-100°. After evolution of HCl (monitored by passing the effluent gas periodically through solutions of AgNO$_3$) has ceased (~1 hr), the flask is removed from the oil bath to an ice-water bath and allowed to cool to <5°. The gas flow is then discontinued, and diethyl ether (0.5 L) is added dropwise with vigorous mechanical stirring. The rate of addition of diethyl ether must be controlled since heat is evolved on formation of diethylhydroxonium trifluoromethanesulfonate. After further rapid dilution with diethyl ether to ~1.6 L, the violet suspension is transferred to screw-capped plastic bottles and centrifuged. The supernatant liquid is decanted and retained for recovery of the excess acid as the sodium salt.* The pellets are resuspended in diethyl ether (1 L) and recentrifuged (three times), then finally transferred to a large sintered-glass funnel (porosity grade 4) for filtration under vacuum. After further washings with diethyl ether, the violet solid is ground in a mortar and thoroughly dried in a lightly capped vessel

---

*Excess ethereal CF_3SO_3H solutions may be retained for recovery of $CF_3SO_3^-$ as the sodium salt. To the solution is slowly added an equal volume of cold water.
■ **Caution.** *Heating of the solution occurs and ether vapor is evolved. Operate in a fume hood.*
A stream of compressed air is slowly passed over the solution to cool it and remove diethyl ether as ~10 M NaOH is slowly added with vigorous mechanical stirring until pH ≃ 5. After evaporation to dryness, the salt may be isolated as NaCF$_3$SO$_3$ · H$_2$O from a minimum of hot water (~80°) upon cooling.

in a vacuum desiccator over phosphorous(V) oxide. The product is stable for at least 3 months if sealed from the atmosphere. Yield: 116 g, 98%.*

Anal. Calcd. for $C_3H_{15}N_5F_9O_9S_3Co$: C, 6.09; H, 2.56; N, 11.84; F, 28.92; S, 16.27; Co, 9.96. Found: C, 6.3; H, 3.0; N, 11.6; F, 28.5; S, 16.4; Co, 9.7.

There is an alterative procedure to centrifugation which is especially suited to smaller-scale preparations. The diluted diethyl ether suspension of the complex is poured onto a large sintered-glass funnel (porosity grade 3 or 4) and the solvent allowed to filter through under gravity over several hours. Further ether is added (twice) and allowed to filter through. Finally, a vacuum is applied and the filter cake is allowed to dry in air and then under vacuum over phosphorus(V) oxide as above.

B. cis-BIS(1,2-ETHANEDIAMINE)-DI(TRIFLUOROMETHANESULFONATO)COBALT(III) TRIFLUOROMETHANESULFONATE, cis-[Co(en)₂(OSO₂CF₃)₂](CF₃SO₃)

$$cis\text{-}[Co(en)_2Cl_2]Cl + 3CF_3SO_3H \longrightarrow$$

$$cis\text{-}[Co(en)_2(OSO_2CF_3)_2](CF_3SO_3) + 3HCl \uparrow$$

$$[Co(en)_2CO_3]Cl + 3CF_3SO_3H \longrightarrow$$

$$cis\text{-}[Co(en)_2(OSO_2CF_3)_2](CF_3SO_3) + CO_2 \uparrow + HCl \uparrow + H_2O$$

Method 1

Although cis-[Co(en)₂Cl₂]Cl [5] is the usual starting material, we have found that $trans$-[Co(en)₂Cl₂]Cl[5] may also be used successfully[2]; hence, isomeric purity is not required. A solution of cis-[Co(en)₂Cl₂]Cl (13.6 g) in anhydrous CF_3SO_3H (60 mL) under a stream of nitrogen is heated at 90-100° for 3 hr, essentially as described above for the pentaamminecobalt(III) complex. After cooling the mixture to <5°, diethyl ether (250 mL) is added slowly with vigorous mechanical stirring. The suspension is then filtered on a sintered-glass funnel (porosity grade 4), and the solid washed well with diethyl ether (250 mL), then triturated on the frit with diethyl ether (100 mL), and sucked dry. The solid is immediately transferred to a flask and boiled with dry chloroform (100 mL) for 10 min. After filtration and washing with diethyl ether, the purple product is ground in a

---

*The checkers obtained a yield of 92% using trifluoromethanesulfonic acid from a freshly opened bottle without redistillation.

mortar and dried over phosphorus(V) oxide in a vacuum desiccator. Yield: 27 g, 94%.

Anal. Calcd. for $C_7H_{16}N_4F_9O_9S_3Co$: C, 13.42; H, 2.57; N, 8.95; F, 27.30; S, 15.36. Found: C, 13.1; H, 2.6; N, 8.5; F, 26.8; S, 15.2.

Method 2

A convenient route requires $[Co(en)_2(CO_3)]Cl$,[6] or the analogous bromide, as starting material. Solid $[Co(en)_2(CO_3)]Cl$ (8 g) is slowly added to stirred anhydrous CF_3SO_3H (25 mL), initially at room temperature, resulting in some evolution of heat. Dry nitrogen is then bubbled through the solution for 15 min to ensure complete removal of CO_2 and HCl. After cooling the reaction mixture to <5°, diethyl ether (140 mL) is added dropwise and the product isolated and purified as described for Method 1. Yield: 15.6 g, 95%.

C. *fac*-TRIS(TRIFLUOROMETHANESULFONATO)[N-(2-AMINOETHYL)-1,2-ETHANEDIAMINE]COBALT(III), Co(dien)(OSO$_2$CF$_3$)$_3$

$$Co(dien)Cl_3 + 3CF_3SO_3H \longrightarrow \textit{fac-}Co(dien)(OSO_2CF_3)_3 + 3HCl \uparrow$$

Procedure

A mixture of $Co(dien)Cl_3$ [7] (12.8 g) and anhydrous CF_3SO_3H (50 mL) is heated under a stream of dry nitrogen at ~100° for 3 hr, as described in the above preparations. The bright purple product is then precipitated by slow addition of diethyl ether (200 mL) to the cooled and vigorously stirred solution. Isolation and purification then follow the procedure described in B, Method 1. Yield: 26 g, 90%.

Anal. Calcd. for $C_7H_{13}N_3F_9O_9S_3Co$: C, 13.80; H, 2.15; N, 6.90; F, 28.06; S, 15.78. Found: C, 13.9; H, 2.5; N, 6.5; F, 27.4; S, 15.8.

Properties

The trifluoromethanesulfonato complexes are isolated as free-flowing violet or purple powders. The yield in each case is essentially quantitative. As isolated, the products contain small amounts of $[(CH_3CH)_2OH]^+[CF_3SO_3]^-$ which may be simply removed by boiling suspensions of the solids in chloroform. Rapid recrystallization from cold aqueous $NaCF_3SO_3$ solutions is possible,[2] but losses through aquation are substantial. The crude products show acceptable analytical purity and are suitable for most preparative purposes.

Visible absorption spectra of solutions in anhydrous CF_3SO_3H are readily recorded.[2] Maxima (λ_{max}, nm(ϵ_{max}, M^{-1} cm^{-1})) are: [Co(NH$_3$)$_5$(OSO$_2$CF$_3$)] · (CF$_3$SO$_3$)$_2$, 524(45.8), 465 sh(33), 345(40.6); cis-[Co(en)$_2$(OSO$_2$CF$_3$)$_2$] · (CF$_3$SO$_3$) 590 sh(33), 508(80.1), 374(105.3); fac-Co(dien)(OSO$_2$CF$_3$)$_3$, 553(108.5), 385(57.3). Solvolysis of the complexes is generally rapid. As examples, the rate constant for aquation of [Co(NH$_3$)$_5$(OSO$_2$CF$_3$)]$^{2+}$ is $k = 2.7 \times 10^{-2}$ sec^{-1} (25°, 0.1 M CF$_3$SO$_3$H). Rate constants for the stereoretentive consecutive aquation reactions cis-[Co(en)$_2$(OSO$_2$CF$_3$)$_2$]$^+$ are 2.2×10^{-2} and 8.6×10^{-3} sec^{-1} (25°, 0.1 M CF$_3$SO$_3$H). Aminecobalt(III) complexes of coordinating solvents are simply prepared by dissolution of the trifluoromethanesulfonato complexes in the appropriate solvent. They may be isolated analytically pure by evaporation of the solvent or by crystallization. In the poorly coordinating solvents sulfolane (thiolane-1,1-dioxide) or acetone, neutral ligands such as urea readily substitute in the coordination sphere. Anionic ligands can also be readily inserted.

References

1. J. B. Hendrickson, D. D. Sternbach, and K. W. Bair, *Acc. Chem. Res.*, **10**, 306 (1977).
2. N. E. Dixon, W. G. Jackson, M. J. Lancaster, G. A. Lawrance, and A. M. Sargeson, *Inorg. Chem.*, **20**, 470 (1981).
3. J. MacB. Harrowfield, A. M. Sargeson, B. Singh, and J. C. Sullivan, *Inorg. Chem.*, **14**, 2864 (1975).
4. G. G. Schlessinger, *Inorg. Synth.*, **9**, 160 (1967).
5. J. C. Bailar, Jr., *Inorg. Synth.*, **2**, 224 (1946).
6. J. Springborg and C. E. Schäffer, *Inorg. Synth.*, **14**, 64 (1973).
7. J. P. Collman and W. L. Young, III, *Inorg. Synth.*, **7**, 211 (1963).

22. 2,9-DIMETHYL-3,10-DIPHENYL-1,4,8,11-TETRAAZACYCLOTETRADECA-1,3,8,10-TETRAENE, (Me$_2$Ph$_2$[14]-1,3,8,10-tetraeneN$_4$) COMPLEXES

Submitted by SUSAN C. JACKELS* and LOREN J. HARRIS*
Checked by RUDOLF THOMAS†

The template condensation of α-diketones with 1,3-propanediamine in the presence of divalent iron, cobalt, nickel, or copper leads to 14-membered-ring macrocyclic complexes containing two α-diimines. A number of complexes of this class, derived from 2,3-butanedione,[1-3] 1-phenyl-1,2-propanedione,[4,5] benzil,[6] and substituted benzils,[5] have been prepared. Preparation of the Ni(II)

*Department of Chemistry, Wake Forest University, Winston-Salem, NC, 27109.
†Department of Chemistry, University of Tasmania, Hobart 7001 Australia.

and Co(III) $Me_4[14]$-1,3,8,10-tetraeneN_4 complexes are included in *Inorganic Syntheses* Volume 18.[7] The present contribution discusses the preparation of the Fe(II), Cu(II) and Zn(II) complexes of the title ligand. The iron and copper complexes are prepared by a template reaction. However, under the template reaction conditions, zinc(II) catalyzes rapid polymerization of the reagents which leads to no macrocyclic product. Thus, at present the only route to the zinc complex is by a reductive transmetalation reaction in which copper(II) $Me_2Ph_2[14]$-1,3,8,10-tetraeneN_4 is reduced by metallic zinc to give metallic copper and a solution containing the macrocyclic zinc(II) complex. An interesting aspect of the template synthesis of complexes derived from 1-phenyl-1,2-propanedione is that in each case the reaction is regiospecific, yielding only the 2,9-dimethyl-3,10-diphenyl-substituted product.[4,8] The organic intermediate responsible for the specificity has been isolated and characterized.[4] Substituted [14]-1,3,8,10-tetraeneN_4 complexes have proven useful for many studies including the following: vitamin B_{12} model chemistry,[9] axial ligand exchange,[10-12] redox kinetics,[13] photolabilization of axial ligands,[14,15] photoassisted catalysis of the oxidation of methanol,[16] and correlation of metal redox potential with substituents.[5]

A. BIS(ACETONITRILE)(2,9-DIMETHYL-3,10-DIPHENYL-1,4,8,11-TETRAAZACYCLOTETRADECA-1,3,8,10-TETRAENE)IRON(II) HEXAFLUOROPHOSPHATE

$$2PhC-CCH_3 + 2NH_2(CH_2)_3NH_2 + Fe^{2+} \xrightarrow[MeOH]{acetate\ buffer}$$
$$\quad \| \quad \|$$
$$\quad O \quad O$$

$$\xrightarrow{CH_3CN, K[PF_6]} [Fe(Me_2Ph_2[14]\text{-}1,3,8,10\text{-tetraene}N_4)(CH_3CN)_2][PF_6]_2$$

$$+ 4H_2O + other\ products$$

Procedure

The solvents are prepared as follows. Anhydrous methanol is distilled under a blanket of nitrogen.* Acetonitrile is heated at reflux with CaH_2, then is distilled under nitrogen.

The following operations are carried out in an inert-atmosphere glove box (nitrogen) unless otherwise noted. A stirred mixture of $FeCl_3$ (1.00 g, 6.00 ×

---

*Methanol dried by refluxing over CaH_2 should be avoided since it may contain trace impurities which catalyze the rapid polycondensation of 1-phenyl-1,2-propanedione and 1,3-propanediamine.

10^{-3} mole), excess iron powder (about 0.4 g), and MeOH (30 mL) is heated at reflux for 5 min, cooled to room temperature, and gravity filtered. This results in a colorless solution containing approximately 0.009 mole Fe^{2+}. To this solution are added $SnCl_2 \cdot 2H_2O$ (0.13 g, 5.0×10^{-4} mole), glacial acetic acid (0.60 mL, 0.010 mole), and CH_3COOK (0.98 g, 0.010 mole), giving a milky white suspension. In a separate flask, 1-phenyl-1,2-propanedione (3.26 g, 2.20×10^{-2} mole) is added dropwise to a stirred solution of 1,3-propanediamine (1.48 g, 2.00×10^{-2} mole) in MeOH (30 mL). After being stirred for 15 min, the solution becomes golden yellow. The suspension containing ferrous ion is then added in bulk, giving a deep green solution. This solution is stoppered and stirred at room temperature for 48 hr, during which time the color changes to deep blue-green. Then excess CH_3CN (6 mL) and CH_3COOH (1.0 mL) are added, and the resulting purple solution is gravity filtered to remove a yellow solid. The solution is removed from the glove box and finely ground solid $K[PF_6]$ (2.76 g, 0.0150 mole) is added to the stirred solution. After about 15 min, the maroon microcrystalline product, $[Fe(Me_2Ph_2[14]-1,3,8,10$-tetraene$N_4)(CH_3CN)_2][PF_6]_2$, is collected by suction filtration. It is washed with absolute EtOH and dried under vacuum at room temperature over activated alumina. The yield is typically 2.6 g, 35%. The product can be recrystallized as follows: About 1.5 g crude product is dissolved in 15 mL CH_3CN and 1 mL CH_3COOH and the solution is filtered. About 125 mL hot MeOH is added to the stirred solution. Burgundy crystals of analytical purity form upon cooling. They are collected and dried as described above. The approximate yield is 1.2 g.

Anal. Calcd. for $C_{28}H_{34}N_6FeP_2F_{12}$: C, 42.02; H, 4.28; N, 10.50. Found: C, 42.12; H, 4.40; N, 10.42.

Properties

The complex $[Fe(Me_2Ph_2[14]-1,3,8,10$-tetraene$N_4)(CH_3CN)_2][PF_6]_2$ is air stable in acidic or neutral solution and has properties very similar to those of the tetramethyl-substituted analog.[1,4] The macrocyclic ligand is coordinated in a planar fashion, and the fifth and sixth axial coordination sites are occupied by acetonitrile.[17] The intense burgundy color of the compound arises from a charge transfer transition at 17.5×10^3 cm^{-1} (ϵ_m, 1.3×10^4 M^{-1} cm^{-1}) which has shoulders at 18.8×10^3 cm^{-1} and 20.0×10^3 cm^{-1} (ϵ_m, 6.8×10^3 and 2.8×10^3, respectively). The compound is diamagnetic and is a 1:2 electrolyte in acetonitrile. The proton NMR spectrum (CD_3CN, internal Me_4Si) features a methyl resonance at 2.44 ppm which is broadened into a poorly resolved triplet ($J \approx 1$ Hz) by homoallylic coupling with the α-methylene protons. The α-CH_2 and β-CH_2 patterns occur at 4.08 and 2.16 ppm, and the phenyl group gives rise to complex multiplets at 7.50 and 7.68 ppm. The CH_3CN axial ligands exhibit rapid exchange with solvent, hence only one resonance at 1.96 ppm is observed for CH_3CN.

B. (2,9-DIMETHYL-3,10-DIPHENYL-1,4,8,11-TETRAAZACYCLOTETRADECA-1,3,8,10-TETRAENE)COPPER(II) HEXAFLUOROPHOSPHATE

$$2PhC-CCH_3 + 2NH_2(CH_2)_3NH_2 + CuCl_2 \cdot 2H_2O \xrightarrow[MeOH]{ZnCl_2}$$
$$\underset{OO}{\|\|}$$

$[Cu(Me_2Ph_2[14]\text{-}1,3,8,10\text{-tetraeneN}_4)][ZnCl_4] + 6H_2O$

+ other products

$[Cu(Me_2Ph_2[14]\text{-}1,3,8,10\text{-tetraeneN}_4)][ZnCl_4] + 2K[PF_6] \xrightarrow{MeOH/H_2O}$

$[Cu(Me_2Ph_2[14]\text{-}1,3,8,10\text{-tetraeneN}_4)][PF_6]_2 + ZnCl_2 + 2KCl$

Procedure

1-Phenyl-1,2-propanedione (6.52 g, 4.40×10^{-2} mole) is added to a stirred solution of 1,3-propanediamine (2.96 g, 4.00×10^{-2} mole) in MeOH (65 mL).* The solution is stirred for 15 min and then a solution of $CuCl_2 \cdot 2H_2O$ (2.72 g, 1.60×10^{-2} mole) in MeOH (30 mL) is added in bulk. The resulting deep blue solution is stirred for 40 min. A $ZnCl_2$ solution, prepared by dissolving powdered anhydrous $ZnCl_2$ (8.00 g, 6.00×10^{-2} mole) in MeOH (10 mL) and diluting with absolute ethanol (30 mL), is added slowly to the stirred reaction mixture. As the $ZnCl_2$ solution is added, the color changes to purple as $[Cu(Me_2Ph_2[14]\text{-}1,3,8,10\text{-tetraeneN}_4)Cl]^+$ is converted to $[Cu(Me_2Ph_2[14]\text{-}1,3,8,10\text{-tetraeneN}_4)]^{2+}$. After the addition is complete, the product, $[Cu(Me_2Ph_2[14]\text{-}1,3,8,10\text{-tetraeneN}_4)][ZnCl_4]$, slowly begins to precipitate. The mixture is removed from stirring, stoppered, and allowed to stand at room temperature for 15-24 hr. It is then stored in a refrigerator for 24-48 hr until precipitation is complete. The purple microcrystalline product, which must be scraped off the sides of the flask, is collected by suction filtration. The product is washed with absolute EtOH and dried under vacuum at room temperature over activated alumina. The yield is about 3.0 g, 30%.

This tetrachlorozincate salt (1.0 g, 1.6×10^{-3} mole) is dissolved in a mixture of 15 mL MeOH and 15 mL H_2O by stirring for about 30 min and heating as

*The yield is comparable when the reagents are measured volumetrically: 5.93 mL 1-phenyl-1,2-propanedione and 3.34 mL 1,3-propanediamine.

necessary. The solution is filtered and heated to near boiling. Then a hot mixture of K[PF$_6$] (0.86 g, 4.7 × 10^{-3} mole) dissolved in 15 mL MeOH and 15 mL H$_2$O is added slowly. The solution is cooled to room temperature and is stored in a refrigerator for 12 hr. The pink crystalline product [Cu(Me$_2$Ph$_2$[14]-1,3,8,10-tetraeneN$_4$)] [PF$_6$]$_2$ is collected, washed with absolute EtOH, and dried under vacuum at room temperature over activated alumina. The yield is 0.75 g, 66%. Additional noncrystalline product can be recovered from the filtrate by rotoevaporation to about half volume, followed by refrigeration.

Anal. Calcd. for C$_{24}$H$_{28}$N$_4$CuP$_2$F$_{12}$: C, 39.71; H, 3.89; N, 7.72. Found: C, 39.83; H, 3.80; N, 7.90.

Properties

The compound [Cu(Me$_2$Ph$_2$[14]-1,3,8,10-tetraeneN$_4$)] [PF$_6$]$_2$ is mauve colored and has a solid-state visible absorption at 20.5 × 10^3 cm^{-1}. It is paramagnetic (μ_{eff} = 1.89 BM, solid state). The infrared spectrum shows a weak $\nu_{C=N(asym)}$ at 1609 cm^{-1}, a sharp $\nu_{C=N(sym)}$ at 1600 cm^{-1}, and the $\nu_{C=C}$ of phenyl at 1579 cm^{-1}. The band associated with coordinated α-diimines occurs at 1236 cm^{-1}. In CH$_3$NO$_2$ solution, it is a 1:2 electrolyte and has an absorption maximum at 19.6 × 10^3 cm^{-1} (ϵ_m = 136 M^{-1} cm^{-1}). This compound is a convenient starting material for the preparation of derivatives having one or two additional axial ligands.[8]

C. CHLORO(2,9-DIMETHYL-3,10-DIPHENYL-1,4,8,11-TETRAAZACYCLOTETRADECA-1,3,8,10-TETRAENE)ZINC(II) HEXAFLUOROPHOSPHATE

$$[Cu(Me_2Ph_2[14]\text{-}1,3,8,10\text{-tetraeneN}_4)][ZnCl_4] \xrightarrow[\text{MeOH}]{\text{Zn metal}}$$

$$[Zn(Me_2Ph_2[14]\text{-}1,3,8,10\text{-tetraeneN}_4)Cl]^+ + \text{other products} \xrightarrow{NH_4[PF_6]}$$

$$[Zn(Me_2Ph_2[14]\text{-}1,3,8,10\text{-tetraeneN}_4)Cl][PF_6]$$

Procedure

A 100-mL three-necked flask, equipped with N$_2$ inlet tube, N$_2$ outlet tube, and magnetic stirring bar, is charged with 50 mL MeOH and 1.00 g [Cu(Me$_2$Ph$_2$[14]-1,3,8,10-tetraeneN$_4$)] [ZnCl$_4$] (1.55 × 10^{-3} mole). The mixture is stirred and purged with N$_2$ gas for 15 min; then zinc metal (1.0 g, .015 mole, 20 mesh), activated by washing briefly with 2% aqueous HCl followed by rinsing with

MeOH, is added. The reaction mixture is stirred under N_2 for 1.5 hr. During this time, the purple starting material dissolves, metallic copper precipitates, and the solution becomes pale yellow-green. The mixture is removed from the N_2 blanket, filtered, and warmed to 40-45°. A warm solution of $NH_4[PF_6]$ in MeOH (0.67 g, 4.1 × 10^{-3} mole in 20 mL) is added to the stirred reaction mixture. After the addition is complete, a slight amount of flocculent precipitate is removed by rapid gravity filtration. The resulting solution is allowed to cool to room temperature. Then it is stored in a refrigerator for 24 hr or longer. The colorless crystalline product is collected, washed with absolute EtOH, and dried under vacuum over activated alumina. The yield is 0.38 g, 40%.

Anal. Calcd. for $C_{24}H_{28}N_4ZnClPF_6$: C, 46.62; H, 4.56; N, 9.06; Zn, 10.57. Found: C, 46.15; H, 4.40; N, 9.15; Zn, 10.78.

Properties

The compound [Zn(Me$_2$Ph$_2$[14]-1,3,8,10-tetraeneN$_4$)Cl][PF$_6$] is colorless. The only feature of its UV spectrum is a shoulder ($\epsilon_m \approx 4000\ M^{-1}\ cm^{-1}$) at 270 nm on the low-energy side of the intense intraligand bands which start at about 250 nm. It is diamagnetic and is a 1:1 electrolyte in CH_3CN and CH_3NO_2. The 90-mHz proton NMR spectrum in CD_3NO_2 (Me$_4$Si internal standard) has a sharp singlet methyl peak at 2.13 ppm and broad complex multiplets for the other types of protons (α-methylene, 3.7-4.1 ppm; β-methylene, 2.2-2.8 ppm; and phenyl, 7.4-7.8 ppm). The infrared spectrum has $\nu_{C=N(sym)}$ at 1615 cm^{-1}. This band is more intense and at higher energy than the Fe, Co, Ni or Cu Me$_2$Ph$_2$[14]-1,3,8,10-tetraeneN$_4$ complexes and is indicative of reduced π-backbonding. Chloride ion can be removed from the complex by precipitating with Ag[BF$_4$] from CH_3CN solution.

References

1. D. A. Baldwin, R. M. Pfeiffer, D. W. Reichgott, and N. J. Rose, *J. Am. Chem. Soc.*, **95**, 5152 (1973).
2. S. C. Jackels, K. Farmery, E. K. Barefield, N. J. Rose, and D. H. Busch, *Inorg. Chem.*, **11**, 2893 (1972).
3. R. R. Gagne, J. L. Allison, and D. M. Ingle, *Inorg. Chem.*, **18**, 2767 (1979).
4. D. S. Eggleston and S. C. Jackels, *Inorg. Chem.*, **19**, 1593 (1980).
5. R. G. Goel, P. M. Henry, and P. C. Polyzou, *Inorg. Chem.*, **18**, 2148 (1979).
6. W. A. Welsh, G. J. Reynolds, and P. M. Henry, *Inorg. Chem.*, **16**, 2558 (1977).
7. A. M. Tait and D. H. Busch, *Inorg. Synth.*, **18**, 22 (1978).
8. B. K. Coltrani and S. C. Jackels, *Inorg. Chem.*, **20**, 2032 (1981).
9. K. Farmery and D. H. Busch, *Inorg. Chem.*, **11**, 290 (1972).
10. D. P. Rillema, J. F. Endicott, and J. R. Barber, *J. Am. Chem. Soc.*, **95**, 6987 (1973).
11. C. E. Holloway, D. V. Stynes, and C. P. J. Vuik, *J. Chem. Soc., Dalton Trans.*, 124 (1979).

12. D. E. Hamilton, T. J. Lewis, and N. K. Kildahl, *Inorg. Chem.*, **18**, 3364 (1979).
13. B. Durham, J. F. Endicott, C. L. Wong, and D. P. Rillema, *J. Am. Chem. Soc.*, **101**, 847 (1979).
14. M. J. Incorvia and J. I. Zink, *Inorg. Chem.*, **16**, 3161 (1977).
15. G. Ferraudi, *Inorg. Chem.*, **18**, 1576 (1979).
16. D. W. Reichgott and N. J. Rose, *J. Am. Chem. Soc.*, **99**, 1813 (1977).
17. H. W. Smith, B. D. Santarsiero, and E. C. Lingafelter, *Cryst. Struct. Comm.*, **8**, 49 (1979).

23. DICHLOROBIS(3,3',3''-PHOSPHINIDYNETRIPROPIONITRILE)NICKEL(II) AND DIBROMOBIS(3,3',3''-PHOSPHINIDYNETRIPROPIONITRILE)NICKEL(II) MONOMERS AND POLYMERS

$$NiX_2 + 2P(CH_2CH_2CN)_3 \longrightarrow NiX_2[P(CH_2CH_2CN)_3]_2$$

$$NiX_2[P(CH_2CH_2CN)_3]_2 \longrightarrow \{NiX_2[P(CH_2CH_2CN)_3]_2\}_x$$

(square planar monomer)   (octahedral polymer)

Submitted by BRUCE M. FOXMAN,* JUDY JAUFMANN,* and BRIAN SULLIVAN*
Checked by BARRY A. SCHOENFELNER† and RICHARD L. LINTVEDT†

A unique class of thermal solid-state polymerization reactions of dihalobis(phosphine)nickel(II) complexes was discovered in 1968 by Walton and Whyman.[1] Studies of one such reaction demonstrated that it occurs with anisotropic front motion, high stereodirectionality, and chemical and crystallographic specificity.[2] The nature and concentration of the halide ligands have a striking effect on specificity, propagation, and nucleation in the reactions.[3] The materials are thus exemplary model systems for study and assessment of the relative importance of various chemical and physical factors in solid-state processes. The complexes may be prepared in high yield as *microcrystals* by the published procedure[1]; attempts to grow larger (at least 0.1 mm) crystals by recrystallization fail owing to simultaneous polymerization and crystallization[4] of solutions of the monomer.[3] The syntheses reported here, while producing the complexes in only moderate yield (30-60%), allow large crystals suitable for X-ray diffraction, optical, or other studies, to be obtained with ease.

*Department of Chemistry, Brandeis University, Waltham, WA 02254.
†Department of Chemistry, Wayne State University, Detroit, MI 48202.

A. DICHLOROBIS(3,3',3"-PHOSPHINIDYNETRIPROPIONITRILE)NICKEL(II)

Procedure

To a solution of 0.21 g (0.88 mmole) $NiCl_2 \cdot 6H_2O$, in 14 mL absolute ethanol, is added, at room temperature, 0.33 g (1.71 mmole) of $P(CH_2CH_2CN)_3$ (Strem Chemical Company, 7 Mulliken Way, Newburyport, MA 01950) dissolved in 15 mL acetone and 3 mL triethyl orthoformate. The solution is kept in a 50-mL beaker overnight at 5°. Large, well-defined, cherry-red crystals, 0.29 g (66%), are removed by filtration and stored in the freezer.

Properties

Dichlorobis(3,3',3"-phosphinidynetripropionitrile)nickel(II) forms cherry-red crystals, with 24 well-defined faces: {111}, {1̄11}, {110}, {210}, {012}, {010}, {100}. When the solid is heated carefully, three phenomena are observed: the color changes from (1) red to blue (polymerization) at 134°; (2) blue to red (depolymerization) at 183°; and (3) melting at 185°. These temperatures may vary somewhat from sample to sample owing to hysteresis. Occasionally, the solid does not undergo transformations (1) and (2); however, if the material is lightly ground in a mortar first, the transformations *will* occur. Heating the solid at temperatures near 130° rapidly produces the blue *polycrystalline* pseudomorph. However, slow polymerization over a period of several months at ~0°, shows that the reaction begins at many sites, both on the surface and in the bulk of the crystals. Crystals of the blue polymer produced in this fashion display the diffraction pattern of a single crystal of *solution-grown* polymer.[3]

Both the red monomeric complexes (chloride and bromide) have a *trans*-square planar configuration and are quite soluble in acetone. The monomers are produced through kinetic control and thus isomerization to the insoluble blue polymers will occur either in solution or in the solid state. The linear polymers consist of octahedral Ni(II) centers with Ni-P-CH_2-CH_2CN-Ni bridges; only one of the three 2-cyanoethyl chains on each phosphine ligand is involved in the bridging. Crystallographic and magnetic data for the complexes are given below.

Complex (L = $P(CH_2CH_2CN)_3$)	Space Group	a(Å)	b(Å)	c(Å)	β(deg)	V(Å)	μ_B(B.M.)
$NiCl_2L_2$	$P2_1/c$	11.982	22.203	9.033	96.0	2386	Diamagnetic
$(NiCl_2L_2)_x$	$B2_1/c$	13.684	21.101	8.351	95.0	2402	3.12
$NiBr_2L_2$	$Pbca$	13.169	22.004	8.487	(90)	2459	Diamagnetic
$(NiBr_2L_2)_x$	$B2_1/c$	13.907	21.055	8.496	97.5	2467	3.15

B. DIBROMOBIS(3,3',3"-PHOSPHINIDYNETRIPROPIONITRILE)NICKEL(II)

Procedure

To a solution of 0.20 g (0.73 mmole) $NiBr_2 \cdot 3H_2O$ in 14 mL absolute ethanol, is added, at room temperature, 0.28 g (1.45 mmole) of $P(CH_2CH_2CN)_3$ dissolved in 15 mL acetone and 3 mL triethyl orthoformate. The solution is kept at 5° overnight in a 50-mL beaker. Large, well-defined, nearly opaque crystals, 0.22 g (50%), are removed by filtration and stored in the freezer.

Properties

Dibromomobis(3,3',3"-phosphinidynetripropionitrile)nickel(II) forms wine-red crystals. Careful optical goniometry reveals the presence of 20 faces: {111}, {012}, {110}, {010}, {100}. When the solid is carefully heated, three phenomena are observed: the color changes from (1) red to blue (polymerization) at 132°; (2) blue to green (a partial decomposition reaction involving cross-linking) at 151°; and (3) melting at 164°. Again, some variation in temperature is expected owing to hysteresis. Heating the solid at temperatures near 130° rapidly produces the blue *polycrystalline* pseudomorph. However, slow polymerization over a period of several months at ~0° shows that polymerization begins uniquely on a [100] face or pair of {100} faces; anisotropic propagation of the polymerization then proceeds uniquely along [100]. Crystals of the blue polymer produced in this fashion are cracked and somewhat distorted, but nevertheless exhibit the X-ray diffraction pattern of a single crystal of *solution-grown* polymer.[3]

References

1. R. A. Walton and R. Whyman, *J. Chem. Soc.*, [A], 1394 (1968).
2. K. Cheng and B. M. Foxman, *J. Am. Chem. Soc.*, **99**, 8102 (1977).
3. K. Cheng, B. M. Foxman, and S. W. Gersten, *Mol. Cryst. Liq. Cryst.*, **52**, 77 (1979).
4. B. Wunderlich, *Angew. Chem.*, **80**, 1009 (1968); *Adv. Polym. Sci.*, **5**, 568 (1968).

24. POLYMER-STABILIZED DIVANADIUM

Submitted by MARK ANDREWS,* GEOFFREY A. OZIN,* and COLIN G. FRANCIS[†]
Checked by KENNETH KLABUNDE[‡] and THOMAS GRASHENS[‡]

There has been much interest in recent years in combining the high activity typically found for heterogeneous catalysts with the selectivity of homogeneous catalysts. This has been attempted by attaching the homogeneous catalyst to a polymer or oxide support to form heterogenized-homogeneous or "hybrid-phase" catalysts.[1] An alternative approach is to induce selectivity into an active heterogeneous catalyst by *homogenizing* it. Homogenization is illustrated in the title system, namely a vanadium dimer trapped within a liquid polymeric support.

The preparation of polysiloxane-V_2 via microscale metal vapor-matrix isolation[2] as well as gram, synthetic-scale rotary solution reactor[3] techniques is described. Consideration of the two different-scale processes reflects both the initial preparation of the molecule and demonstrates the utility of a combined, complementary matrix-/macrosynthetic approach.[4]

Procedure

A. MATRIX-SCALE SYNTHESIS

A metal vapor-matrix isolation spectroscopy apparatus of the Torromis type[§] is arranged as shown in Fig. 1a with the furnace positioned directly below the specimen and its support. The furnace can be rotated out of position to allow spectroscopy simply by twisting the shroud. The sample is supported on a Suprasil optical window sandwiched between indium gaskets within a copper housing. This assembly is attached to the second stage of a Displex‖ closed-cycle helium refrigeration system. The horizontal arrangement is required to maintain a liquid film of even thickness on the quartz window. Approxi-

---

*Lash Miller Chemistry Department, University of Toronto, Toronto, Ontario, Canada, M5S 1A1.
†Chemistry Department, University of Southern California, Los Angeles, CA 90007.
‡Department of Chemistry, Kansas State University, Manhattan, KS 66506.
§Torrovap Industries Inc., 3 Woodglen Way, Unionville, Ontario, Canada L3R 3A8.
‖Air Products, P.O. Box 2802, Allentown, PA 18105.

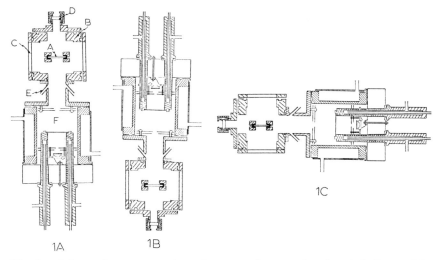

Fig. 1. Schematic representation of vacuum furnace closed-cycle helium refrigeration system used for metal vapor microsolution optical spectroscopy, as well as conventional metal vapor-matrix isolation experiments. (A) NaCl or Suprasil optical window, horizontal configuration; (B) stainless steel vacuum shroud; (C) NaCl or Suprasil optical viewing ports; (D) cajon-rubber septum, liquid or solution injection port; (E) gas deposition ports; (F) vacuum furnace quartz crystal microbalance assembly. With the optical window in a fixed horizontal configuration, liquid or solution sample injection onto the window at any desired temperature in the range 12–300 K is performed in position 1A, metal deposition is conducted in position 1B, and optical spectra are recorded in position 1C (see Procedure).

mately 0.3 μL of a liquid poly(dimethylsiloxane-co-methylphenylsiloxane)* (methyl:phenyl = 17:1) is applied directly to the upper surface of the optical window using a vacuum microsyringe. This in situ liquid injection procedure is accomplished through a side flange on the vacuum shroud, to which is attached a vacuum-tight, rubber septum-cajon assembly. This quantity of material is sufficient to create a film close to 1 μm in thickness, assuming an exposed quartz surface 1.9 cm in diameter. A resistively heated vanadium filament (28 mm × 5 mm × 0.254 mm)† is used as the source of vanadium atoms. The rate of metal deposition is monitored in situ by a line-of-sight quartz crystal microbalance protected by a radiation shield. A second radiation shield interposed

---

*Designated DC510/50 cs. Manufactured by Dow Corning. A comparable polymer is also available from Petrarch, 2731 Bartram Rd., Bristol, PA 19007.
†Purity 99.8%, available from A. D. Mackay, 10 Center St., Darien, CT 06820.

between the filament and the sample helps to collimate the vapor and effectively minimizes heating of the polymer film (Fig. 1). An absolute value of the number of moles of vanadium deposited into the sample can be obtained by calibration, in which the quartz sample window is replaced by a second mass monitor. This allows the ratio of the forward-backward metal flux to be obtained. The geometrical correction factor σ is then,

$$\sigma = \frac{\text{forward metal flux}}{\text{backward metal flux}}$$

After the injection of liquid polymer, the Displex shroud and its furnace are repositioned as in Fig. 1b and the system evacuated to about 5×10^{-6} torr (in the shroud). The vapor pressure of the polymer is sufficiently low that this step can be carried out with the film at room temperature. Evacuation is accompanied by bubbling of the liquid as the polymer rapidly outgasses for about 1 min, after which time the unbroken film reestablishes itself. Once a good vacuum has been achieved, the Displex is activated and the temperature of the optical window (and hence the polymer) lowered to 250 K.

With the furnace assembly aligned as in Fig. 1c, the UV-visible absorption spectrum of the polymer is scanned in the range 200-900 nm. The vanadium filament is then heated resistively ($\sim$35 V, $\sim$40 A) to first outgas the metal and vaporize surface impurities. (To prolong filament life, the metal should be heated/cooled gradually $\sim$5 min to a vaporization temperature corresponding to (20-40) σ Hz min^{-1}). This procedure requires roughly 20 min at a rate of (20-40) σ Hz min^{-1} (note: 1 Hz = 20 ng metal). A steady deposition rate of 800 σ ng $\approx$ sec^{-1} (40 σ Hz min^{-1}) is then obtained, the furnace rotated into position (Fig. 1b), and an increment of $\sim 10^4$ σ ng of vanadium deposited into the polymer film. The shroud is subsequently returned to the Fig. 1c position, heating of the filament is stopped, and an absorption spectrum of the products is recorded.

This procedure is repeated—essentially titrating known amounts of vanadium atoms into the polymer—until the spectrum shows a high concentration of the dimer absorbing at 453 nm[4,6] in addition to the (arene)$_2$V MLCT absorption at 324 nm.[4,6,7] By operating at 210-230 K, additional absorptions, at higher vanadium loadings, grow in at 550 and 635 nm and are associated with polymer-stabilized V_3 and V_4 cluster species.[4] (If the metal deposition rate is too high (>60 σ Hz min^{-1}) metal atom polymerization to produce colloid will compete effectively with product formation. In addition, the polymer can be titrated to saturation; i.e., there is a limit corresponding to a total metal loading of about 3.0×10^{-5} g under these conditions, beyond which the ratio of $V_1 : V_2$ no longer changes. In either case, the appearance of colloid is indicated by a broad underlying absorption in the range 200–300 nm. Similar considerations apply in the macrosynthesis of the products.)

B. MACROSCALE SYNTHESIS

Depending on whether vanadium is vaporized resistively or by means of an electron beam, the experimental procedures for metal preparation and deposition are sufficiently different that each merits a separate treatment. The type of apparatus used is a rotary reactor of the Torrovap design (Fig. 2).

1. Vaporization by Resistive Heating

Since the vapor pressure of vanadium is 10^{-2} torr at 2100 K (mp = 2175 K), it is possible to generate atomic vanadium by resistive heating of a tungsten spiral* containing vanadium pieces (0.5 g). With the tungsten basket fixed in position between two water-cooled copper electrodes, a 2-L reactor vessel (5-L flasks have also been used with an appropriate increase in sample volume) is bolted in place and the system evacuated to 5×10^{-7} torr. A 100-mL quantity of the polymer is degassed in a 500-mL flask fitted with a Rotaflo stopcock and B19 male joint. This flask is inverted over a 10-cm column, equipped with a B19 female joint at the top, a side-arm with a J. D. Young Teflon stopcock, and tapered at the other end to fit the swagelock connection of a Nupro brass/stainless steel metering needle valve. The latter is connected to a 3-mm stainless-steel transfer tube terminating near the bottom of the reaction vessel. Once the flask containing the polymer is in place, the short column is separately pumped free of air. The metering valve is opened to complete the evacuation, and the polymer is then gravity fed into the reaction flask by opening the Rotaflo stopcock. This technique satisfactorily prevents any contamination of the evaporation source with polymer. Rotation of the reactor is begun and a refrigerant bath (250 K: dry ice in a 1 : 1 mixture of ethylene glycol and water) raised around the reactor. The fluid is spun into a 8–10-cm band when the flask is rotating at ~60 rpm. It is important that the flask be angled no more than 15° from horizontal to ensure that the band of fluid finds itself directly over the zone of vaporization. It is important to rotate the flask at a rate sufficient to prevent the polymer from falling and cracking on the source. Rotation also facilitates cooling of the fluid, mixing, and renewal of fresh fluid.

Power is applied to the tungsten spiral (~40 V, ~50 A, ac) gradually over 10–15 min. Reaction is accompanied by color changes from colorless to yellow

*Available from Sylvania Emissive Products, Box 220, Exeter, NH 03833.

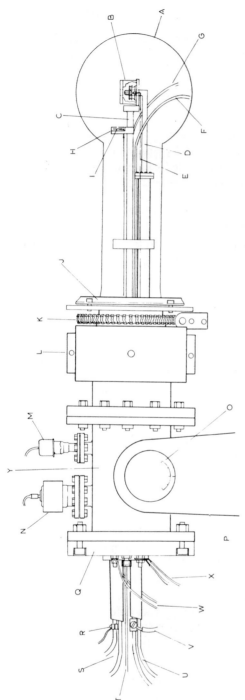

Fig. 2. Schematic of reverse-polarity electron gun-metal vapor rotary reactor equipment (Torrovap) used for preparing liquid poly(dimethyl-co-methylphenylsiloxane)-supported divanadium: (A) 2-5-L Glass reaction vessel; (B) electrongun; (C) liquid-N_2-cooled electron-gun support; (D) water-cooled hearth (anode); (E) thoriated tungsten filament power leads; (F) rotatable liquid-product outlet tube; (G) rotatable liquid-reagent inlet tube; (H) quartz crystal support; (I) 5 MHz AT cut shear-mode quartz crystal; (J) rotary vacuum seal-reaction vessel mating flange; (K) motor-driven worm and gear assembly; (L) rotary vacuum seal; (M) thermocouple vacuum gauge; (N) cold cathode vacuum gauge; (O) pivot point for setting angle of inclination of reaction vessel; (P) reactor support; (Q) Plexiglass back flange; (R) external dc power supply for in situ electron beam focusing; (S) liquid nitrogen feed through; (T) liquid-reagent inlet tube; (U) electron-gun water supply; (V) high-voltage connection to electron-gun hearth; (W) quartz crystal microbalance oscillator-power supply; (X) filament power supply; (Y) main vacuum manifold connected to angle-adjustable vacuum seal and diffusion pump-rotary backing pump assembly.

to dark red-brown. Approximately 0.3 g of vanadium is deposited over 40 min ($\sim$0.15 mmole min^{-1}) after which time the source is slowly cooled and rotation halted. Nitrogen or argon is introduced into the flask through the stainless steel transfer tube. A second transfer tube dipping into the fluid is used to siphon the product back into a Schlenk tube. The fluid is then filtered through a fine-porosity frit using standard Schlenk techniques and stored under an inert atmosphere.

2. Vaporization by Electron Beam

It is easier to vaporize larger quantities of vanadium over a wider range of rates by means of electron beam heating using a positive hearth electrostatically focussed electron gun. Vanadium rod,* 6.2 mm in diameter, is pretreated under vacuum by electron beam to degas and shape the metal into a domed bead. Approximately 1.0-1.5 g of metal is centered on the water-cooled hearth, and filament and can heights are adjusted to maximize focusing of the electron beam (filament height = 9 mm from the top of the hearth; can height = 4 mm from the top of the bead). As the bead size diminishes during the vaporization an adjustable (0-200 V) potential from an external dc power supply is applied to the can to maintain a focused beam throughout the experiment. Following the vanadium pretreatment, the polymer is introduced as described for resistive heating. Note that the e gun is operated in the near horizontal mode for reasons detailed previously.

While the cooled flask is rotating (pressure $<10^{-6}$ torr) the 1% thorium-doped tungsten filament is brought carefully (outgassing) to 16-18 A. A 2-kV hearth voltage giving rise to 30-mA emission current is sufficient to vaporize 0.3 g of vanadium from a 0.7-g bead over a 30-min period ($\sim$0.2 mmole min^{-1}). After the vaporization has been terminated, the product is isolated as described above.

Properties

The product fluid is clear red-brown in color and is thermally stable at 300 K but is extremely air sensitive. The UV-visible spectrum shows bands at 222 and 262 nm due to uncoordinated

$$\begin{array}{c} CH_3 \\ | \\ -Si-O- \\ | \\ C_6H_5 \end{array}$$

*Purity 99.8%, available from A. D. Mackay, 10 Center St., Darien, CT 06820.

and bands at 324 and 453 nm. The absorption at 324 nm is due to a MLCT transition of $(arene)_2V$[4,6,7] ($\epsilon \approx 10^4$) associated with a weaker absorption close to 450 nm ($\epsilon \approx 10^3$). The latter transition is masked by the more intense band at 453 nm due to polymer-supported divanadium. This absorption is assigned on the basis of quantitative metal atom concentration studies and kinetic analysis of the growth of mononuclear and binuclear metal-arene complexes in both arene/rare gas matrices at 12 K, and in the thin film poly(dimethylsiloxane-co-methylphenylsiloxane) at 200-270 K.[4,9]

The infrared spectrum of the products of the reaction of vanadium atoms with DC 510 is congested with polymer absorptions which obscure metal-ring stretching and ring deformation modes expected at approximately 430 and 478 cm^{-1} for a bis(arene) vanadium complex. Microsolution infrared experiments of vanadium arene complexes produced in Santovac-5[4] reveal a strong absorption at 432 cm^{-1}. The higher-energy arene-ring deformation mode (470-480 cm^{-1}) is concealed by a substrate absorption centered at 470 cm^{-1}. Whereas it is difficult to distinguish between the polymer-supported bis(arene)vanadium and the divanadium species in the infrared region, such distinctions can be made by means of Raman spectroscopy. The major Raman lines for the neat polymer (100-200 K) below 500 cm^{-1} are rather broad (bwhh ~30 cm^{-1}) but are few in number, occurring at 490 (pol.) and 180 (depol.) cm^{-1}. For the polymer-supported bis(arene)vanadium and divanadium compounds, new Raman bands (Ar ion laser excitation: 4880 Å or 5145 Å, 50-150 mW) appear at 520, 430, 270, and 155 cm^{-1}. All of the lines are polarized by comparison with the depolarized 180 cm^{-1} polymer reference line. The 155 cm^{-1} band is the most intense, followed in order of decreasing intensity by the 430, 270, and 520 cm^{-1} absorptions. These latter bands are, by analogy with the vibrational spectra of discrete bis(arene)metal sandwich complexes,[10] assigned to arene ring-deformation and metal-arene ring-stretching modes. The very intense band at 155 cm^{-1}, which definitely has no counterpart in, for example, bis(toluene)vanadium (generated under mononuclear vanadium conditions and observed in situ in V/toluene/77 K matrices; strongest Raman line at 288/274 (pol.) cm^{-1}— site symmetry or ring conformation splitting effort[10]) nor in, for example, polymer-supported bis(arene)chromium (the latter generated under mononuclear chromium conditions in Cr/poly(dimethyl siloxane-co-methylphenylsiloxane)/ 250 K films; strong Raman line at 288 (pol.) cm^{-1} with weaker polarized lines at 450 and 423 cm^{-1}), appears to be characteristic of the polymer-supported divanadium species. Consistent with the low-frequency Raman spectra of other metal-metal bonded cluster complexes,[11] the intense, strongly polarized Raman line at 155 cm^{-1} of the polymer-supported divanadium is assigned to the symmetric V-V stretching mode.

A fairly concentrated sample of polymer-supported bis(arene) vanadium at room temperature shows an anisotropic epr spectrum due to a reduced tumbling rate in the viscous liquid polymer matrix. The spectrum displays in essence 8 vanadium hyperfine lines (average $\langle A \rangle = \frac{1}{3} \langle A_{\parallel} + 2A_{\perp} \rangle = 78$ G) due to coupling of the single unpaired electron in a mainly metal d_{z^2} orbital with the spin $I = \frac{7}{2}$ vanadium nucleus. No proton superhyperfine can be resolved for the polymer anchored bis(arene) vanadium complex. The shape of the spectrum and the measured hyperfine coupling are markedly temperature dependent (through the polymer viscosity) and can be easily understood by comparison with the epr spectrum of the corresponding unsupported bis(benzene)vanadium complex in pentane over the temperature range 120-300 K. Values of the hyperfine coupling constants and the g-value ($\langle g \rangle = \frac{1}{3}(g_{\parallel} + 2g_{\perp}) = 1.9850$) viewed in this way are in agreement with the literature values for unsupported bis(arene)vanadium complexes.[12] Epr signals from polymer samples containing both the bis(arene)vanadium and dimer are essentially the same as those described above, implying diamagnetism and epr silence for the V-V bonded species.

References

1. E. Bayer and V. Schurig, *Chemtech*, 212 (1976); R. H. Grubbs, *Chemtech*, 512 (1977); and references therein.
2. S. Cradock and A. J. Hinchcliffe, *Matrix Isolation*, Cambridge University Press, Cambridge, 1975; M. Moskovits and G. A. Ozin (eds.), *Cryochemistry*, Wiley, New York, 1976, and references cited therein.
3. R. MacKenzie and P. L. Timms, *J. Chem. Soc. Chem. Commun.*, 605 (1974).
4. G. A. Ozin and C. G. Francis, *J. Mol. Struct.*, 59, 55 (1980). The same experiments can be carried out at 290 K using a meta-substituted pentaphenyl ether (Santovac-5) manufactured by Monsanto. The spectroscopic properties of the products up to a vanadium cluster nuclearity, $n = 3$ (microsolution) or $n = 2$ (macroscale), are identical to those of the DC510 supported vanadium.
5. M. Moskovits and G. A. Ozin, *Appl. Spectroscop.*, 26, 481 (1972).
6. C. G. Francis, H. Huber, and G. A. Ozin, *Angew. Chem. Int. Ed. Engl.*, 19, 402 (1980).
7. C. G. Francis and P. L. Timms, *J. Chem. Soc. Chem. Commun.*, 466 (1977); idem, *J. Chem. Soc. Dalton Trans.*, 1401 (1980).
8. For additional details regarding this type of evaporation see P. L. Timms, *Cryochemistry*, M. Moskovits and G. A. Ozin (eds.)., Wiley, New York, 1976. For use of the technique see F.G.N. Cloke, M.L.H. Green, and D. A. Price, *J. Chem. Soc. Chem. Commun.*, 431 (1978); F.G.N. Cloke and M.L.H. Green, *J. Chem. Soc. Chem. Commun.*, 127 (1979).
9. E. P. Kundig, M. Moskovits, and G. A. Ozin, *Angew. Chem. Int. Ed.*, 14, 292 (1975); M. Andrews and G. A. Ozin, *Angew. Chem.*, 94, 219 (1982); and references cited therein.
10. E. Maslowsky, Jr., *Vibrational Spectra of Organometallic Compounds*, Wiley, New York, 1977, and references cited therein.
11. D. F. Shriver and C. W. Cooper, III, *Adv. Infrared Raman Spectrosc.*, vol. 6, Heisden, London, 1981, and references cited therein.
12. G. H.-Olive and S. Olive, *Z. Phys. Chem.*, 56, 223 (1967).

25. TRIAMMINECHLOROPLATINUM(II) CHLORIDE

$trans\text{-}[Pt(NH_3)_2Cl_2] + KI \longrightarrow trans\text{-}[Pt(NH_3)_2ClI]$

$trans\text{-}[Pt(NH_3)_2ClI] + AgNO_3 \longrightarrow trans\text{-}[Pt(NH_3)_2Cl(H_2O)]NO_3 + AgI$

$trans\text{-}[Pt(NH_3)_2Cl(H_2O)]NO_3 + NH_3 \longrightarrow [Pt(NH_3)_3Cl]NO_3 + H_2O$

$[Pt(NH_3)_3Cl]NO_3 + HCl \longrightarrow [Pt(NH_3)_3Cl]Cl + HNO_3$

Submitted by HIDEYOSHI MORITA* and JOHN C. BAILAR, JR.*
Checked by J. JEFFREY MACDOUGALL†

Triamminechloroplatinum(II) choride was first prepared by Cleve by the reaction of the *cis*-dichlorodiammine complex with the calculated amount of ammonia.[1] Chugaev later prepared the compound by an improved method using the *cis* isomer and potassium cyanate as starting materials.[2] However, this method was still tedious, and the yield was low.

More recently the same compound was prepared in about 40% yield by Gel'fman using the *trans* isomer as a starting material.[3]

A modification of the method of Gel'fman is described in this synthesis.

Procedure

There are three steps in the preparation of triamminechloroplatinum(II) chloride: preparation of *trans*-diamminechloroiodoplatinum(II), conversion to *trans*-diammineaquachloroplatinum(II) nitrate, and replacement of the coordinated water molecule by ammonia to give triamminechloroplatinum(II) chloride.

A. *trans*-DIAMMINECHLOROIODOPLATINUM(II) CHLORIDE

To a cold suspension obtained by mixing 2.29 g (7.6 mmole) *trans*-[Pt(NH$_3$)$_2$Cl$_2$] with 20 mL of water is added, with stirring, a solution of 1.26 g (7.6 mmole) of KI in 10 mL of water. An immediate color change of the suspended solid from yellow to gray occurs. After the mixture has been stirred for about 24 hr, the gray precipitate is separated by filtration and washed successively with three 3-mL portions of water, ethanol, and diethyl ether. The yield is 2.81 g (95%).

Anal. Calcd. for *trans*-[Pt(NH$_3$)$_2$ClI]: N, 7.16; H, 1.54. Found: N, 7.14; H, 1.46.

*Department of Chemistry, University of Illinois, Urbana, IL 61801.
†Chemistry Department, University of Nevada–Reno, Reno, NV 89557.

B. trans-DIAMMINEAQUACHLOROPLATINUM(II) NITRATE

In a 250-mL Erlenemeyer flask, 2.8 g (7.2 mmole) of trans-[Pt(NH$_3$)$_2$ClI] is suspended in 200 mL of water. To the mixture is added 10 mL of aqueous solution containing 1.20 g (7.1 mmole) of AgNO$_3$. The mixture is boiled on a hot plate with stirring for about 2 hr. A color change of the suspended solid from gray to yellow occurs during the heating. The yellow precipitate of AgI is separated by filtration of the hot mixture and washed with three 10-mL portions of hot water. The wash water is added to the filtrate. The amount of silver iodide is 1.70 g, which corresponds to the theoretical.

C. TRIAMMINECHLOROPLATINUM(II) CHLORIDE

To the filtrate containing trans-[Pt(NH$_3$)$_2$Cl(H$_2$O)]NO$_3$ is added 21 mL of 0.5 M ammonium hydroxide (1.5 mole per mole of complex) with stirring. The solution is heated on a water bath with stirring for about 2 hr. It is then cooled, neutralized with 5 mL of 0.5 M HCl, and evaporated under vacuum to a volume of about 20 mL. This results in the precipitation of a small amount (~0.2 g) of trans-[Pt(NH$_3$)$_2$Cl$_2$]. After separation of the precipitate by filtration, 10 mL of concentrated HCl, 30 mL of ethanol, and 50 mL of diethyl ether are added to the mixture.

The resulting precipitate is separated by filtration and washed with three 3-mL portions of ethanol and then three of diethyl ether. This crude [Pt(NH$_3$)$_3$Cl]Cl is dissolved in a minimum amount of cold water (about 25 mL), the solution is filtered to remove the remaining diammine complex, and a second precipitation with a mixture of alcohol and diethyl ether (volume ratio 3:5) is carried out. The material separates as a very pale yellow powder. The yield of [Pt(NH$_3$)$_3$Cl]Cl after two recrystallizations is 0.85 ≈ 0.95 g (about 37%).

Anal. Calcd. for [Pt(NH$_3$)$_3$Cl]Cl: N, 13.25; H, 2.86; Cl, 22.36. Found: N, 13.27; H, 2.91; Cl, 22.38.

References

1. P. T. Cleve, *On Ammoniacal Platinum Bases*, Stockholm (1872).
2. L. A. Chugaev, *J. Chem. Soc.*, **107**, 1247 (1915).
3. M. I. Gel'fman, *Dokl. Akad. Nauk SSSR*, **167**, 819 (1966).

26. CHLOROTRIS(DIMETHYL SULFIDE) PLATINUM(II) COMPOUNDS

$$cis\text{-}[Pt(Me_2S)_2Cl_2] + AgNO_3 \cdot SMe_2 \longrightarrow [Pt(Me_2S)_3Cl]NO_3 + AgCl$$

$$[Pt(Me_2S)_3Cl]NO_3 + HBF_4 \longrightarrow [Pt(Me_2S)_3Cl]BF_4 + HNO_3$$

$$[Pt(Me_2S)_3Cl]NO_3 + NaB(C_6H_5)_4 \longrightarrow [Pt(Me_2S)_3Cl]B(C_6H_5)_4 + NaNO_3$$

Submitted by HIDEYOSHI MORITA* and JOHN C. BAILAR, JR.*
Checked by NORMAN C. SCHROEDER† and ROBERT J. ANGELICI†

Chlorotris(dimethyl sulfide)platinum(II) tetrafluoroborate was prepared by Goggin et al. using cis-[Pt(Me$_2$S)$_2$Cl$_2$] as a starting material.[1] The corresponding tetraphenylborate was prepared like the tetrafluoroborate.

A modification of the method of Goggin et al. for the preparation of chlorotris(dimethyl sulfide)platinum(II) complexes is described in this synthesis.

Procedure

cis-Dichlorobis(dimethyl sulfide)platinum(II) (398 mg, 1.02 mmole) and AgNO$_3$ · Me$_2$S (232 mg, 1.00 mmole; prepared by addition of dimethyl sulfide to concentrated aqueous solution of silver nitrate, and recrystallized from ethanol[2]) are suspended in about 40 mL of acetone, and the mixture is stirred for about 24 hr at room temperature. After removal of the precipitated AgCl by filtration, the solvent is evaporated under vacuum at room temperature by use of a rotary evaporator. The yellow residue is then dissolved in a small amount of warm water (70°, 5 mL), and after removal of undissolved materials by filtration, about 0.5 mL of 48% aqueous hydrogen tetrafluoborate is added. The mixture is cooled in an ice bath for about 3 hr. The pale yellow needles thus obtained are separated by filtration and washed with three 1-mL portions of cold ethanol and three 2-mL portions of ether, then dried under vacuum. These needles are recrystallized from a minimum amount of warm water (~70°). The yield is 275 mg (55%).

Anal. Calcd. for [Pt(Me$_2$S)$_3$Cl]BF$_4$: C, 14.31; H, 3.60; Cl, 7.04; S, 19.10; F, 15.09. Found: C, 14.33; H, 3.69; Cl, 7.09; S, 19.03; F, 14.86.

The corresponding tetraphenylborate is prepared by dropwise addition of the filtrate containing the nitrate complex to a cold ethanol solution containing the calculated amount of sodium tetraphenylborate instead of hydrogen tetrafluo-

---
*Department of Chemistry, University of Illinois, Urbana, IL 61801.
†Department of Chemistry, Iowa State University, Ames, IA 50011.

borate. The white precipitate which immediately appears is separated by filtration and washed with three 5-mL portions of cold water, three 2-mL portions of cold ethanol, three 3-mL portions of diethyl ether, and dried under vacuum. The yield at this point is 567 mg (77%). The crude precipitate (~100 mg) is put on a G-4 glass filter and leached with 10 mL of hot (not boiling) methanol by stirring for several seconds; then the mixture is filtered with suction. Two more 10-mL portions of hot methanol are used for similar extractions. The filtrates are put together and cooled in an ice bath for 1 hr. The white precipitate thus obtained is separated by filtration and dried under vacuum. The yield is about 20 mg (~20%).

Anal. Calcd. for $[Pt(Me_2S)_3Cl]B(C_6H_5)_4$: C, 48.95; H, 5.20; Cl, 4.82; S, 13.07. Found: C, 48.96; H, 5.15; Cl, 4.74; S, 12.59.

Properties

The tetrafluoroborate is readily soluble in chloroform, dichloromethane, acetone, and water, but only slightly soluble in ethanol and benzene. In CD_2Cl_2, the ^{1}H-NMR spectrum of the compound shows two methyl signals due to the Me_2S (*trans* to Me_2S) and Me_2S (*trans* to Cl) groups which integrate as 2:1 at 2.62 ($^3J_{PtH}$ = 43.8 Hz) and 2.63 ($^3J_{PtH}$ = 51.0 Hz) ppm from Me_4Si, respectively.[1] On the other hand, the tetraphenylborate is readily soluble in dichloromethane and acetone but only slightly soluble in ethanol and insoluble in water and benzene. This compound is unstable in solution and gradually decomposes with time. Immediately after the dissolution of the compound in CD_2Cl_2, the ^{1}H-NMR spectrum shows two methyl signals due to the Me_2S (*trans* to Me_2S) and Me_2S (*trans* to Cl) groups in the ratio 2:1 at 2.42 ($^3J_{PtH}$ = 44.2 Hz) and 2.13 ($^3J_{PtH}$ = 50.5 Hz) ppm from Me_4Si, respectively.

References

1. P. L. Goggin, R. J. Goodfellow, S. R. Haddock, F.J.S. Reed, J. G. Smith, and K. M. Thomas, *J. Chem. Soc. Dalton Trans.*, 1904 (1972).
2. P. C. Ray, N. Adhikari, and H. Ray, *J. Indian Chem. Soc.*, 8, 689 (1931).

27. TETRABUTYLAMMONIUM TRICHLORO(DIMETHYL SULFIDE)PLATINUM(II)

$$[Pt(C_2H_4)Cl_2]_2 + 2Me_2S \longrightarrow [Pt(Me_2S)Cl_2]_2 + 2C_2H_4$$

$$[Pt(Me_2S)Cl_2]_2 + 2[(C_4H_9)_4N]Cl \longrightarrow 2[(C_4H_9)_4N][Pt(Me_2S)Cl_3]$$

Submitted by HIDEYOSHI MORITA* and JOHN C. BAILAR JR.*
Checked by NORMAN C. SCHROEDER[†] and ROBERT J. ANGELICI[†]

Tetraalkylammonium trichloro(dimethyl sulfide)platinate(II) has been prepared as $Pr_4N[Pt(Me_2S)Cl_3]$[1] and $Et_4N[Pt(Me_2S)Cl_3]$[2] by the reaction of di-μ-chloro-dichlorobis(dimethyl sulfide)diplatinum(II) with tetra-(propyl- and ethyl) ammonium chloride. However, the yields of these complexes are not good, and the method of Kennedy et al. for obtaining $(Et_4N)[Pt(Me_2S)Cl_3]$[2] is very tedious.

In this synthesis tetrabutylammonium trichloro(dimethyl sulfide) platinate(II) is prepared by the reaction of di-μ-chloro-dichlorobis(dimethyl sulfide)diplatinum(II) with tetrabutylammonium chloride by a modification of the method of Goggin et al.[1]

Procedure

Di-μ-chloro-dichlorobis(dimethyl sulfide)diplatinum(II) is prepared by a modification of the method of Chatt and Venanzi[3] using $[Pt(C_2H_4)Cl_2]_2$[4] instead of $[Pt(C_3H_6)Cl_2]_2$ as a starting material. Thus, 2.84 mL of 0.1 M dimethyl sulfide in dichloromethane, cooled to about 0°, is added to 50 mL of a dichloromethane solution, cooled to -70°, and containing 834 mg (1.42 mmole) of $[Pt(C_2H_4)Cl_2]_2$. The mixture is concentrated to about 15 mL using a rotary evaporator at room temperature. Then 35 mL of toluene is added to the solution, and the mixture is boiled under reflux for about 3 hr. After being concentrated to about 25 mL in a rotary evaporator, the mixture is cooled and allowed to stand overnight in a refrigerator. The mustard-yellow precipitate thus obtained is collected by filtration, washed with three 5-mL portions of boiling chloroform, and dried under vacuum. The yield is 813 mg (87%).

Anal. Calcd. for $[Pt(Me_2S)Cl_2]_2$: C, 7.32; H, 1.84; Cl, 21.61. Found: C, 7.54; H, 1.88; Cl, 21.40.

Tetrabutylammonium chloride (818 mg, 2.50 mmole, 85% tetrabutylammonium chloride, remainder butanol) is dissolved in about 40 mL of dichloro-

---
*Department of Chemistry, University of Illinois, Urbana, IL 61801.
[†]Department of Chemistry, Iowa State University, Ames, IA 50011.

methane. In this solution di-μ-chlorodichlorobis(dimethyl sulfide)diplatinum(II) (813 mg, 1.24 mmole) is dissolved by stirring at room temperature. After it is stirred overnight, a large amount of diethyl ether (about 400 mL) is added to the orange-yellow solution to form a precipitate. Sometimes oil is formed by this treatment, but the oil gradually changes to a solid with time in this case. The orange precipitate thus obtained is separated by filtration, washed with three 5-mL portions of diethyl ether, and dried under vacuum. The specimen is recrystallized as orange needles from a small amount of acetone by addition of diethyl ether. The yield is 1.20 g (79%).

Anal. Calcd. for $[(C_4H_9)_4N][PtCl_3(Me_2S)]$; N, 2.31; C, 35.67; H, 6.99; Cl, 17.55; S, 5.29. Found: N, 2.25; C, 35.53; H, 7.11; Cl, 17.25; S, 5.11.

Properties

This compound is only slightly soluble in water, but is readily soluble in benzene, chloroform, dichloromethane, acetone, and ethanol. In CD_2Cl_2, the ^{1}H-NMR spectrum of the compound shows a sharp methyl signal due to dimethyl sulfide at 2.34 ($^3J_{PtH}$ = 47.3 Hz) ppm from tetramethylsilane.

References

1. P. L. Goggin, R. J. Goodfellow, S. R. Haddock, F.J.S. Reed, J. G. Smith, and K. M. Thomas, *J. Chem. Soc. Dalton Trans.*, 1904 (1972).
2. R. P. Kennedy, R. Gosling, and M. L. Tobe, *Inorg. Chem.*, **16**, 1744 (1977).
3. J. Chatt and L. M. Venanzi, *J. Chem. Soc.*, 2787 (1955).
4. G. W. Littlecott, F. J. McQuilline, and K. G. Powell, *Inorg. Synth.*, **16**, 114 (1976).

Chapter Three
MAIN GROUP, LANTHANIDE, ACTINIDE, AND ALKALI METAL COMPOUNDS

28. TRIBUTYL PHOSPHOROTRITHIOITE $(C_4H_3S)_3P$ FROM ELEMENTAL PHOSPHORUS

$$P_4 + 6NaSBu + 6BuSH + 6CCl_4 \longrightarrow 4(BuS)_3P + 6CHCl_3 + 6NaCl$$

Submitted by C. BROWN,* R. F. HUDSON,* and G. A. WARTEW*
Checked by M. S. HOLT† and J. H. NELSON†

The conventional route to phosphorothioite esters involves the conversion of elemental phosphorus to the trichloride, followed by reaction with a thiol in presence of base.[1] This can be reduced to a one-stage process by treating white phosphorus with the thiol in the presence of an electrophile, in particular a polychlorinated hydrocarbon.

The reactions of nucleophiles with elemental phosphorus usually lead to complex mixtures derived from the intermediate phosphide ion, which is captured by a proton or Lewis acid to give mixtures of products derived from phosphorous acid and phosphine.[2]

---

*The Chemical Laboratory, University of Kent at Canterbury, Canterbury, Kent, CT2 7NH, England.
†Chemistry Department, University of Nevada–Reno, NV 89557.

The combined action of nucleophile and electrophile, in the present example thiolate ions and carbon tetrachloride, essentially eliminates side reactions and leads to high yields of trithioite esters, as in the analogous reaction of alcohols to give trialkyl phosphites.[3]

Procedure

■ **Caution.** *White phosphorus must be handled with care, for it ignites spontaneously in air, causes burns on the skin, is toxic, and can cause cumulative poisoning. Minor skin burns can be treated with a 5% aqueous solution of copper(II) sulfate. All operations should be carried out in a well ventilated hood, with due regard for the toxic and malodorous nature of 1-butanethiol and the dangers associated with the handling of metallic sodium.*

A 1-L round-bottomed flask is equipped with a magnetic stirrer and a reflux condenser fitted with a drying tube. Sodium 1-butanethiolate is prepared by adding clean sodium metal (11.20 g, 0.49 mole) in 1-g pieces down the condenser to an excess of 1-butanethiol (56.0 g, 0.54 mole) in diethyl ether (400 mL) at such a rate as to maintain a gentle reflux. When addition is complete, stirring is continued for 48 hr and the product is collected on a sintered-glass funnel, washed with diethyl ether, and dried in a desiccator under reduced pressure.

Phosphorus sand is prepared by adding a clean dry piece of white phosphorus (ca. 1.3 g, 0.01 mole) to carbon tetrachloride (100 mL) contained in a three-necked 1-L round-bottomed flask equipped with a reflux condenser and drying tube, a nitrogen inlet, and a stopper and heating the mixture gently under nitrogen until the phosphorus melts. The mixture is then stirred rapidly and cooled in ice to give the finely divided form.

To the above solution is added additional carbon tetrachloride (300 mL) containing 1-butanethiol (25 mL, 0.23 mole). The stopper is then replaced with an L-shaped tube terminating in a ground-glass joint. Sodium 1-butanethiolate (13.9 g, 0.12 mole) is then added slowly under nitrogen from this tube to the stirred solution of phosphorus sand and 1-butanethiol. The addition takes 5 min, after which the apparatus is sealed under nitrogen, and the mixture is stirred magnetically at room temperature for 48 hr. The mixture is filtered through kieselguhr under nitrogen and the residue, obtained from rotary evaporation of solvent for the solution, is distilled under reduced pressure to give tributyl phosphorotrithioite (8.68 g. 71%), bp 125-7°/0.5 torr (Reference 1: bp 144°/0.3 torr); m/z 298(M$^{+\cdot}$).

Anal. Calcd. for $C_{12}H_{27}PS_3$: C, 48.3; H, 9.1. Found: C, 48.1; H. 9.2.

Properties

The product is a colorless oil with a characteristic powerful odor, insoluble in water, but soluble in most organic solvents: bp 174-180°/15 torr, n_D^{25} = 1.5305.

A noise-decoupled ^{31}P-NMR spectrum shows one line at δ_p(CDCl$_3$) +116.7 ppm (lit.^{4} 116.1 ppm): δ(CCl$_4$) 2.73 (6H, dt, $^3J_{PH}$ 9.0, $^3J_{HH}$ 7.1 Hz, S-CH$_2$).

The trithioite ester is stable in air but decomposes above 100°. It is stable towards water and acids but is rapidly decomposed by alkali. Oxidation with 3% hydrogen peroxide in acetic acid gives S,S,S,-tributyl phosphorotrithioate (yield 58%, bp 162-164°/1 torr. (lit.^{5} bp 166.5°/1 torr)).

References

1. A. Lippert and E. E. Ried, *J. Am. Chem. Soc.*, **60**, 2370 (1938).
2. M. M. Rahut, *Topics in Phosphorus Chemistry*, Vol. 1, M. Grayson and J. E. Griffith (eds.), Interscience, New York, 1964, p. 1.
3. C. Brown, R. F. Hudson, G. A. Wartew, and H. Coates, *J. Chem. Soc. Commun.*, **7**, (1978).
4. K. Moedritzer, L. Maier, and L.C.D. Groenweghe, *J. Chem. Eng. Data*, **7**, 307 (1962).
5. D. Voight, M. C. Labarre, and M. Therasse, *C. R. Hebd. Services Acad. Sci.*, **260**, 2210 (1965).

29. PREPARATION OF DIMETHYLPHENYLPHOSPHINE

$$(C_6H_5)Cl_2P + 2CH_3MgBr \longrightarrow (C_6H_5)(CH_3)_2P + 2MgBrCl$$

Submitted by C. FRAJERMAN* and B. MEUNIER†
Checked by M. E. STREM ‡

Dimethylphenylphosphine is a ligand commonly used to prepare a large range of transition metal-organophosphorus complexes. This basic phosphine is easy to handle and presents a small cone angle.[1,2] It is usually prepared in 40-50% yield by reaction of a solution of methylmagnesium halide with phenylphosphonous dichloride followed by hydrolysis with an excess of saturated aqueous ammonium chloride solution.[3,4] In the present preparation the yield has been greatly increased (to 85%).

Procedure

A 2-L three-necked flask is equipped with a condenser, a mechanical stirrer, and a pressure-equalized dropping funnel. The vessel is flushed with nitrogen and

---

*Institute de Chimie des Substances Naturelles, CNRS, 91190 Gif-sur-Yvette, France.

†Present address: Laboratoire de Chimie de Coordination, CNRS, 205, route de Narbonne, 31400 Toulouse, France.

‡Strem Chemicals, Inc., 7 Mulliken Way, Dexter Industrial Park, P.O. Box 108, Newburyport, MA 01950.

kept under an inert atmosphere throughout the preparation. The flask is charged with 790 mL of a 1.59M diethyl ether solution of bromomethylmagnesium (1.25 mole). (This Grignard reagent is prepared in 91% yield from 206 g of bromomethane and 48 g of magnesium in 1 L of diethyl ether. After iodometric titration, 790 mL of this solution is used. The commercial availability of this reagent makes it a desirable reactant for the synthesis presented here.) A solution of 67.8 mL of phenylphosphonous dichloride (89.5 g, 0.5 mole) in 300 mL of anhydrous diethyl ether is added dropwise to the cooled solution (ethanol/dry ice, bath temperature $-78°$) of the Grignard reagent (an efficient stirrer is required owing to the thickness of the mixture). After an additional period of 3.5 hr, an extra 50 mL of diethyl ether, used to rinse the dropping funnel, is added and the mixture is stirred at room temperature for 4.5 hr. Instead of the usual hydrolysis with a saturated aqueous ammonium chloride solution, the reaction mixture is hydrolyzed with 400 mL of $4M$ deoxygenated hydrochloric acid solution which is added through the dropping funnel. The flask is cooled with an ice bath during this vigorous reaction. After the addition (1 hr), the aqueous layer is clear. The ethereal layer is separated from the aqueous phase by decantation and discarded by the usual rubber septum and cannula technique. After addition of 300 mL of fresh diethyl ether,* the reaction mixture, stirred and cooled with an ice bath, is treated with 204 mL of a 16% ammonia solution over a period of 0.6 hr. The ethereal layer is separated and kept under nitrogen. The aqueous phase is extracted with 5 X 150 mL of diethyl ether. All the ethereal solutions are collected in the same flask and dried over sodium sulfate. After filtration most of the diethyl ether is evaporated at normal pressure. The crude phosphine is distilled under reduced pressure. The fraction 108-110°/55-60 torr is collected. The yield is 58.7 g (85% based on phenylphosphonous dichloride).

Properties

Dimethylphenylphosphine is kept under nitrogen to avoid atmospheric oxidation. The purity of the phosphine can be checked by ^{1}H-NMR. A concentrated solution (60% v:v) of phosphine in $CDCl_3$, with Me_4Si as internal standard, shows two sets of peaks, a complex multiplet at $\delta = 7.0$-7.5 ($i = 5H$) and a doublet centered at $\delta = 1.2$ ($i = 6H$ and $J_{P-H} = 1.9$ Hz). The ^{31}P-NMR (^{1}H decoupled) shows a singlet at -45.89 ppm in $CDCl_3$ (25% v:v) (85% H_3PO_4 as external reference). Additional properties are $d_4^{26}: 0.9670^5$, $n_D^{19.5}: 1.5673.^6$ The IR and UV spectra can be found in references 7 and 8.

---

*Throughout the neutralization and extraction procedures, all the solutions and solvents used are carefully deoxygenated with nitrogen.

References

1. C. A. Tolman, *Chem. Rev.*, **77**, 313 (1977).
2. A. Immirz and A. Musco, *Inorg. Chim. Acta*, **25**, L41 (1977).
3. K. Bowden and E. A. Braude, *J. Chem. Soc.*, 1952, 1068.
4. M. A. Mathur, W. H. Myers, H. H. Sisler, and G. E. Ryschkewitsch, *Inorg. Synth.*, **15**, 132 (1974).
5. E. A. Yakovleva, E. N. Tsvetkov, D. I. Lobanov, M. I. Kabachnik, and A. I. Shatenshtein, *Dokl. Akad. Nauk. SSSR.*, **170**, 1103-6 (1966); *Chemical Abstracts*, **66**, 37038g (1967).
6. J. M. Angeleilli, R.T.C. Brownlee, A. R. Katritzky, R. D. Topsomand, and L. Yaknontov, *J. Am. Chem. Soc.*, **91**, 4500 (1969).
7. H. Schindlbauer, *Monatsh. Chem.*, **94**, 99 (1963).
8. G. Aksnes and L. J. Brudvik, *Acta Chem. Scand.*, **17**, 1616 (1963).

30. ELECTROCHEMICAL SYNTHESIS OF SALTS OF HEXAHALODIGALLATE(II) AND TETRAHALOGALLATE(III) ANIONS

Submitted by MICHAEL J. TAYLOR* and DENNIS G. TUCK†
Checked by BRUNO J. JASELSKIS‡

The 2+ oxidation state of gallium has been identified in a number of systems, including aqueous and non-aqueous solutions,[1] and halide melts,[2] in which it is a moderately powerful reducing agent. Some low-valent species of gallium are known to be mixed-valence compounds, such as "gallium dichloride," which has the Ga(I)/Ga(III) structure $Ga^+[GaCl_4]^-$, but a number of gallium halide solid phases and other systems, including a few incompletely characterized organometallic derivatives, provide evidence of authentic Ga(II) compounds.[3] In particular, crystallographic and other evidence relating to salts of the $[Ga_2X_6]^{2-}$ ions[1,4-6] and to the uncharged complexes $Ga_2X_4(dioxane)_2$ [7] (X = Cl, Br or I) prove these to be metal-to-metal bonded gallium(II) species containing a dinuclear $[Ga_2]^{4+}$ core analogous to the mercury(I) state in $[Hg_2]^{2+}$ compounds.

Development of the chemistry of the Ga(II) state is hampered by a lack of convenient synthetic routes to pure gallium(II) compounds. In the following preparation of salts of the ions $[Ga_2X_6]^{2-}$ (X = Cl, Br or I), a simple low-voltage electrochemical technique is used to obtain crystalline salts of a suitable counterion from the metal in a single stage. Elaborate precautions to exclude oxygen or moisture from the electrolytic solutions are unnecessary, and flushing the cell

---
*Department of Chemistry, University of Auckland, Auckland, New Zealand.
†Department of Chemistry, University of Windsor, Windsor, Ontario N9B 3P4, Canada.
‡Department of Chemistry, Loyola University, Chicago, IL 60626.

with nitrogen, and cooling to avoid disintegration of the low-melting gallium anode, are the only requirements.

The electrochemical (or current) efficiency, E_F, defined as moles of metal dissolved per Faraday, can be readily determined by recording the current, total time of electrolysis, and weight of gallium dissolved. For a more accurate determination, a silver voltammeter can be connected in series with the cell. Current efficiencies measured in this way are close to the theoretical value of 0.5 for the fundamental process $Ga \rightarrow Ga^{2+} + 2e^-$. The method given here for preparing gallium(II) compounds uses an electrolyte based on acetonitrile and is a marked improvement on earlier chemical or electrolytic routes to low-valent gallium compounds,[1] from which mixed products, partially oxidized to the Ga(III) state, were often obtained.

Electrolytic procedures employing gallium metal in the form of a sacrificial anode can also be conveniently used to synthesize salts of the gallium(III) complex halide ions, $[GaX_4]^-$. A straightforward procedure described here involves electrolysis of gallium in aqueous acid, HX, followed by heating to oxidize any Ga(II) species to Ga(III):

$$[Ga_2X_6]^{2-} + 2HX \longrightarrow 2[GaX_4]^- + H_2$$

Alternatively, a single-stage electrochemical route can be used, in which a nonaqueous electrolyte containing an alkylammonium salt and dissolved halogen, for example bromine or iodine, is employed. This method has the advantage of yielding a crystalline product which requires no further purification or recrystallization.

General Electrochemical Procedures

Commercial gallium (99.9%) is formed into a suitably sized electrode by casting approximately 3 g of the metal around a platinum wire. A simple method is to melt the gallium (mp 30°) under warm water in a glass container, pour the gallium into a length of soft plastic tubing which is pinched off at one end, and then introduce the platinum wire. After the gallium has solidified, the tubing can be cut away to provide the desired length of exposed metal. This electrode, which forms the anode of the cell, is rinsed with water, then acetone, and accurately weighed before and after each electrolysis. No surface coating builds up, and provided the cell current is regulated to avoid surface disintegration or overheating and accidental melting of the gallium, the electrode can be used repeatedly. The cell, a 100-mL tall-form beaker closed with a rubber bung (Fig. 1), uses a stout platinum wire as cathode and is continuously flushed with nitrogen, the flow of which is regulated with the aid of a bubbler in the exit line.

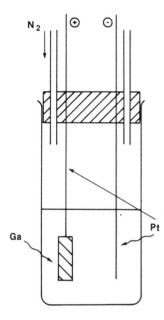

Fig. 1. Electrolysis cell.

A. TRIPHENYLPHOSPHONIUM HEXAHALODIGALLATE(II), [(C$_6$H$_5$)$_3$PH]$_2$[Ga$_2$X$_6$] (X = Cl, Br, or I)

$$2Ga(anode) + 6HX(sol) + 2(C_6H_5)_3P(sol) \longrightarrow$$
$$[(C_6H_5)_3PH]_2[Ga_2X_6](s) + 2H_2(g)$$

Procedure

The electrolyte phase consists of acetonitrile (30 mL) containing 0.5 g triphenylphosphine and concentrated aqueous halogen acid, HX (2 mL). Reagent grade solutions of HBr or HI are frequently yellow or orange due to the formation of small amounts of halogen, which should be reduced to halide before use by shaking the contaminated acid with ~1 mL of mercury under nitrogen in a separating funnel until the acid is colorless. Solvent extraction[8] or ion exchange[9] procedures are also available for this purpose.

Slight chemical attack of the gallium anode by this electrolyte, evidenced by evolution of hydrogen, is suppressed when the dc voltage is applied. An electromotive force of 3-5 V gives a current of 50 mA and results in the loss of ~0.2 g of metal from the anode during electrolysis lasting 2.5 hr. Crystals of the product begin to appear after about 30 min, both at the anode and from the bulk

TABLE I Gallium(II) Compounds Prepared

	$[(C_6H_5)_3PH]_2[Ga_2Cl_6]$	$[(C_6H_5)_3PH]_2[Ga_2Br_6]$	$[(C_6H_5)_3PH]_2[Ga_2I_6]$
Weight of Ga dissolved (g)	0.25^a	0.172	0.190
Current efficiency (mol F^{-1})	0.77^a	0.53	0.59
Yield (g) (%, based on metal dissolved)	0.61 (39)	0.87 (62)	0.70 (36)
Melting point (°C)	159–161	160–162	ca. 225
Analytical results (%) (Calculated values in parentheses)	Cl = 24.1 (24.2)	Br = 41.2 (41.8)	I = 54.7 (53.3)
	C = 49.14 (49.16)	C = 37.77 (37.72)	C = 30.11 (30.26)
	H = 3.65 (3.67)	H = 2.81 (2.82)	H = 2.31 (2.26)

^{a}Some Ga dissolves by chemical, as well as electrochemical, reaction.

solution phase. The products, $[(C_6H_5)_3PH]_2[Ga_2X_6]$, colorless in the case of chloride and bromide and pale yellow for iodide, are in good crystalline form and of high purity. Table I compares the current efficiencies during the electrolysis, yields, melting points, and analytical figures for a series of typical preparations.

B. TETRAPHENYLPHOSPHONIUM HEXABROMODIGALLATE(II)

$$2Ga(anode) + 2[(C_6H_5)_4P]Br(sol) + 4HBr(sol) \longrightarrow$$
$$[(C_6H_5)_4P]_2[Ga_2Br_6](s) + 2H_2(g)$$

Procedure

The electrolyte consists of 30 mL of acetonitrile containing 2 mL of concentrated aqueous hydrobromic acid and 1.0 g of $[(C_6H_5)_4P]Br$. Electrolysis in the cell shown (Fig. 1), using a potential of 4V and current of 50 mA for 120 min, yields 0.80 g of colorless cube-shaped crystals of $[C_6H_5)_4P]_2[Ga_2Br_6]$. A further 0.4 g of crystals can be recovered by allowing part of the solvent to evaporate under a stream of nitrogen. Yield 52%; mp ~ 215°. The crystals of $[(C_6H_5)_4P]_2[Ga_2Br_6]$ dissolve in nitromethane on warming and recrystalline on cooling.

Anal. Calcd. for $C_{48}H_{40}P_2Ga_2Br_6$; Br, 37.0. Found: Br, 36.6%.

C. TETRABUTYLAMMONIUM TETRACHLORO- AND TETRABROMOGALLATE(III)

$$2Ga(anode) + 6HX(sol) \longrightarrow [Ga_2X_6]^{2-}(sol) + 2H_2(g)$$
$$[Ga_2X_6]^{2-}(sol) + 2HX(sol) \longrightarrow 2[GaX_4]^-(sol) + H_2(g)$$
$$[GaX_4]^-(sol) + [Bu_4N]^+(sol) \longrightarrow [Bu_4N][GaX_4](s)$$

Procedure

These complexes can be prepared from gallium metal by anodic oxidation in aqueous halogen acid solution, followed by heating during which any Ga(II) species are oxidized to Ga(III): the salts are then precipitated by adding $[Bu_4N]X$ to the solution.

As in the previous experiments, the cell consists of an open beaker, cooled in ice, with the supported gallium anode and platinum wire cathode dipping into the electrolyte, 20 mL of 6 M aqueous acid, HCl or HBr. Electrolysis for 2 hr at a potential of 1 V (current ≈ 100 mA) results in a weight loss of 0.30–0.33 g from the anode. (*Note:* Gallium is slowly attacked by acid of this concentration, forming a mixture of $[Ga_2X_6]^{2-}$ and $[GaX_4]^-$ complexes, but making it the anode of a cell serves to increase substantially the rate of dissolution of the metal.) Following electrolysis, the solution is heated with stirring until no further evolution of hydrogen occurs.

The salt $[Bu_4N][GaCl_4]$ is precipitated from this electrochemically prepared solution by the addition of 2.8 g (0.5 mmole) of 50% aqueous solution of $[Bu_4N]Cl$. The precipitated solid (1.5 g) is subsequently dissolved in 10 mL of chloroform, from which a colorless crystalline product is obtained by the addition of 30 mL of diethyl ether (yield 1.40 g, ca. 72% based on gallium dissolved).

Anal. Calcd. for $C_{16}H_{36}NGaCl_4$: Cl, 31.2. Found: Cl, 30.3%.

In the case of the bromide, 1.61 g (0.5 mmole) of $[Bu_4N]Br$ in 5 mL of water is added. The copious precipitate is rinsed with water, collected, and air-dried, giving a crude yield of 2.50 g. Purification can be achieved by dissolving the solid in warm chloroform (20 mL), cooling, and adding diethyl ether (30 mL) to obtain 2.35 g of cream solid as small crystals (yield ca. 86%, based on gallium dissolved).

Anal. Calcd. for $C_{16}H_{36}NGaBr_4$: Br, 50.6. Found: Br, 49.7%.

D. TETRABUTYLAMMONIUM TETRAIODOGALLATE(III)

$$Ga(s) + 4HI(sol) \longrightarrow [GaI_4]^-(sol) + 2H_2(g)$$
$$[GaI_4]^-(sol) + [Bu_4N]^+(sol) \longrightarrow [Bu_4N][GaI_4](s)$$

Procedure

The anodic oxidation of gallium in aqueous hydriodic acid is not a successful route to the anionic iodo-complex because of the competing reaction $2I^- \rightarrow I_2 + 2e^-$, but a simple and direct chemical route is available.

Finely divided gallium (0.20 g) is dissolved in 5 mL of concentrated hydriodic acid. Dissolution is complete after some hours, and a solution of 1.1 g of $[Bu_4N]I$ in 5 mL of concentrated hydriodic acid is then added. The precipitated product is collected, dried, dissolved in 10 mL chloroform/diethyl ether (50:50 v:v), reprecipitated with 80 mL diethyl ether, and dried under vacuum. Yield: 0.85 g.

Anal. Calcd. for $C_{16}H_{36}NGaI_4$: I, 61.9. Found: I, 61.4%.

E. TETRAETHYLAMMONIUM TETRABROMOGALLATE(III)

$$2Ga(anode) + 3Br_2(sol) + 2[Et_4N]Br(sol) \longrightarrow 2[Et_4N][GaBr_4](s)$$

Procedure

■ **Caution.** *This procedure involves liquid bromine, and all operations should be carried out in a well-vented hood. Liquid bromine can cause serious chemical burns and should be handled with extreme care.*

This compound can be obtained in a one-stage direct electrochemical synthesis using techniques successful in the preparation of other anionic halo complexes of main group and transition elements.[10,11] The electrolyte is a solution of 0.5 g of bromine and 0.6 g of [Et$_4$N]Br in a mixture of 6 mL of dry methanol and 24 mL of dry benzene. A gallium anode and platinum wire cathode are used, as in the electrolyses described previously, and the cell is cooled in ice to insure that the gallium anode does not melt. Electrolysis proceeds at a voltage of 50 V over a period of 2 hr, during which time the cell current falls gradually from 50 mA to 20 mA. The yellow crystals which form on the anode are scraped off at intervals. The product is washed, first with benzene/methanol and then with diethyl ether.

Anal. Calcd. for $C_8H_{20}NGaBr_4$: Br, 61.5. Found: Br, 61.2%.

Properties: Gallium(II) Compounds

The gallium(II) compounds, $[(C_6H_5)_3PH]_2[Ga_2X_6]$ (X = Cl, Br, or I) and $[(C_6H_5)_4P]_2[Ga_2Br_6]$, are air-stable crystalline solids, which melt with decomposition at the temperatures given in Table I. They are insoluble in cold water but react with hot water, liberating hydrogen and forming a gelatinous precipitate, probably gallium(III) hydroxide. Aqueous silver nitrate solution is initially reduced to silver, but the expected AgX precipitate forms on heating, and satisfactory halide analyses can be performed by the Volhard method. The compound $[(C_6H_5)_4P]_2[Ga_2Br_6]$ is moderately soluble in nitromethane. All are slightly soluble in tetrahydrofuran, a convenient solvent in which to conduct reactions aimed at forming other Ga(II) derivatives by replacement of halides by pseudohalides, organic substitutents, or neutral ligands. Reaction with halogen leads to scission of the metal-metal bond:

$$[Ga_2X_6]^{2-} + X_2 \longrightarrow 2[GaX_4]^-$$

The structures of the $[Ga_2X_6]^{2-}$ anions have been determined by X-ray crystallography.[4-6] The vibrational spectra have also been reported;[1] a characteristic Raman feature is the intense band due to in-phase stretching of Ga–X and

Ga-Ga bonds, at 233 cm^{-1} in the spectrum of $[Ga_2Cl_6]^{2-}$, at 164 cm^{-1} in $[Ga_2Br_6]^{2-}$, and at 118 cm^{-1} in $[Ga_2I_6]^{2-}$.

Properties: Gallium(III) Compounds

The known compounds $[R_4N][GaX_4]$, for which convenient preparative procedures are described here, were characterized by analysis for halogen, and by their infrared spectra, in which a strong Ga-X stretching band occurs at 378 cm^{-1} (X = Cl), 278 cm^{-1} (X = Br) or 222 cm^{-1} (X = I).

References

1. C. A. Evans, K. H. Tan, S. P. Tapper, and M. J. Taylor, *J. Chem. Soc., Dalton Trans.*, 988 (1973).
2. M. J. Taylor, *J. Chem. Soc.* (A), 2812 (1970).
3. M. J. Taylor, *Metal-to-metal Bonded States of the Main Group Metals*, Academic Press, London, 1975.
4. K. L. Brown and D. Hall, *J. Chem. Soc., Dalton Trans.*, 1843 (1973).
5. J. C. Beamish, R.W.H. Small, and I. J. Worrall, *Inorg. Chem.*, **18**, 220 (1979).
6. M. Khan, C. Oldham, M. J. Taylor, and D. G. Tuck, *Inorg. Nucl. Chem. Lett.*, **16**, 469 (1980).
7. H. J. Cumming, D. Hall, and C. E. Wright, *Cryst. Struct. Comm.*, **3**, 107 (1974).
8. D. G. Tuck, R. M. Walters, and E. J. Woodhouse, *Chem. Ind.*, 1352 (1963).
9. C. M. Davidson and R. F. Jameson, *Chem. Ind.*, 1686 (1963).
10. J. J. Habeeb, L. Neilson, and D. G. Tuck, *Syn. React. Inorg. Metal-Org. Chem.*, **6**, 105 (1976).
11. J. J. Habeeb and D. G. Tuck, *Inorg. Synth.*, **19**, 257 (1979).

31. POLYMERIC SULFUR NITRIDE (POLYTHIAZYL), $(SN)_x$

$$S_4N_4(s) \xrightarrow{75-80°} S_4N_4(v) \xrightarrow[Ag_2S]{250°} 2S_2N_2(v) \longrightarrow 2S_2N_2(s)$$

$$\frac{x}{2} S_2N_2(s) \xrightarrow{25°} (SN)_x(s)$$

Submitted by ALAN G. MACDIARMID,* CHESTER M. MIKULSKI,† A. J. HEEGER,‡
and A. F. GARITO ‡
Checked by DAVID C. WEBER §

Polymeric sulfur nitride (polythiazyl), $(SN)_x$, is the first example of a covalent polymer that displays metallic properties even though it contains no metal atoms; it may be the forerunner of a new class of material–polymeric metals.[1] It is prepared in the form of lustrous, golden monoclinic crystals by the solid-state polymerization of molecular crystals of disulfur dinitride, S_2N_2, at room temperature.[2,3] It may also be prepared in the form of golden epitaxial films, in which the $(SN)_x$ chains are aligned parallel to each other.[4] The polymer has electrical and optical properties characteristic of a three-dimensional anisotropic metal, the metallic properties being more pronounced along the direction of the $(SN)_x$ polymer chain than in directions perpendicular to the chain.[5] At room temperature, the conductivity is the same order of magnitude as that of a metal such as mercury. The material is reported to become superconducting near 0.3 K.[6] Polymeric sulfur nitride can be used as an electrode in the electrolysis of aqueous solutions of metal ions.[7] It may be used as a precursor for the syntheses of metallic, covalent, polymeric derivatives of $(SN)_x$, such as the poly(thiazyl bromides), $(SNBr_y)_x$, some of which, for example $(SNBr_{0.4})_x$, are significantly better conductors than the parent compound.[8]

Polymeric sulfur nitride was first synthesized in 1910 by F. P. Burt,[9] who passed the vapor of S_4N_4 over silver gauze or quartz wool at 100–300°. In 1956, M. Goehring prepared S_2N_2 from S_4N_4 vapor and silver wool at 300°[10,11] and found that the S_2N_2 polymerized spontaneously in the solid state to $(SN)_x$ at room temperature. The real catalyst is apparently Ag_2S, which is formed from S_4N_4 and the silver wool at the beginning of the reaction.[12]

Polycrystalline films of $(SN)_x$ may be conveniently synthesized directly from

---

*Department of Chemistry, University of Pennsylvania, Philadelphia, PA 19104.
† Department of Chemistry and Physics, Beaver College, Glenside, PA 19038.
‡ Department of Physics, University of Pennsylvania, Philadelphia, PA 19104.
§ Naval Research Laboratory, Washington, DC 20375.

S_4N_4 vapor at ~260°[13] or from S_2N_2 vapor at ~100–130° or by the action of UV light on S_2N_2 vapor at room temperature.[14]

Well-formed crystals of $(SN)_x$ are best synthesized by first growing crystals of S_2N_2 of the desired size and quality and then permitting them to undergo spontaneous solid-state polymerization at room temperature during several days. The colorless crystals of S_2N_2 first turn dark blue-black and become paramagnetic, but after 2–3 days at room temperature they again become diamagnetic and golden.[2] Even at this stage, the crystals still contain small but significant amounts of unpolymerized S_2N_2 which must be removed before the crystals are exposed to air.[15] Up to 10 weeks at room temperature may be required for all the S_2N_2 to undergo spontaneous polymerization to $(SN)_x$.

■ **Caution.** *Tetrasulfur tetranitride, S_4N_4, has been reported to explode under certain conditions if subjected to percussion, friction, sudden heating, or temperatures close to its melting point.[16] It has also been reported[2] that S_2N_2 may explode spontaneously when subjected to slight mechanical shock or rapid heating. Polymeric sulfur nitride, $(SN)_x$, decomposes violently when heated much above 200°.[2] On occasions it has exploded when subjected to high pressures.[2] These substances should therefore be handled with appropriate caution.*

Procedure

Tetrasulfur tetranitride,[11] S_4N_4, is purified by recrystallization from chloroform, is pumped for several hours on the high-vacuum system and is finally sublimed under vacuum at 80–90° before use.[16] The purity of the compound is checked by its melting point (found, 185.5–186°; literature values; 178–187°).[9,10]

The apparatus shown in Fig. 1 is constructed from Pyrex glass and attached to a standard vacuum system equipped with a mechanical oil pump and mercury vapor diffusion pump. A detachable trap cooled in liquid nitrogen is inserted between the apparatus and the vacuum system so that any volatile condensible material escaping from the reactors during the experiment can be collected and weighed.

In a typical experiment, 0.60 g (3.26 mmole) of S_4N_4 is placed in each of the two reactors shown in Fig. 1. One gram of silver wool (obtained from Fischer Scientific Co.) is then pulled apart gently so as to occupy a larger volume and is placed in each reactor to fill a length of approximately 10 cm.* Care must be taken to ensure even packing of the wool. If open channels are present in the wool, some S_4N_4 will pass through without reacting. Three chromel-alumel thermocouples are attached by means of electrical tape to the outside glass surface of the tube at the bottom, middle, and upper portions of the silver wool zone. These are then wrapped with a layer of aluminum foil and two layers of

---

*The checker finds that better yields are obtained if 1.0 g S_4N_4 and 1.2 g of silver wool are employed.

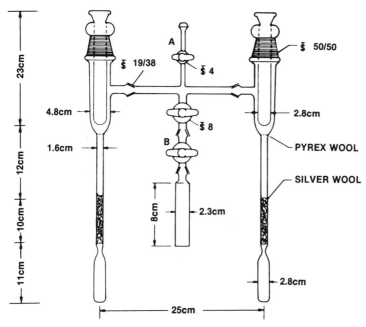

Fig. 1. Apparatus for preparation of SN_x.

heating tape. Small tufts of Pyrex wool are placed below the cold fingers, as shown in Fig. 1, to prevent S_2N_2 or $(SN)_x$ flakes from falling off the cold finger and back onto the hot silver wool.

The apparatus is then attached to the vacuum system through the detachable trap and evacuated to $<10^{-4}$ torr with all stopcocks open to the pump. The heating tapes are turned on so that a temperature of $\sim 250°$ is recorded by the middle thermocouple. The upper and lower thermocouples should not differ by more than $\pm 10°$ from this temperature. The temperature of the oil bath is then raised to 75-80° with the surface of the oil almost touching the lower portion of the heating tape. The S_4N_4 vapor is pumped through the heated silver wool for approximately 1 hr. The cold fingers are then filled with liquid nitrogen. The material which first collects on the cold fingers is white, but after several hours it usually appears tan. The S_2N_2 is collected on the cold fingers for ~ 12 hr, after which time all the S_4N_4 should have sublimed. If the experimental conditions are correct, no unreacted S_4N_4 should have deposited on the glass tubing immediately above the heating tape. An ice bath is placed around the rectangular trap constructed from square Pyrex glass tubing* obtained from Ace Glass Co. (A rectangular trap is used for growing the S_2N_2 crystals since it is be-

*The checker reports that larger crystals are obtained if only the very bottom of the trap is immersed in the ice bath.

lieved that fewer distortions are introduced in the S_2N_2 crystals if their (flat) faces grow on a flat rather than on a rounded glass surface.) The liquid nitrogen is evaporated from the cold fingers with a stream of air. Acetone is then added to each cold finger. By the time the temperature of the acetone has risen to approximately $-10°$ as measured by a low-temperature thermometer, the tan color of the deposit on the cold fingers should have faded. It has been suggested[12] that this color may be due to a form of NS radical or short-chain species which polymerizes on warming.

During the warming process a small amount of an unidentified volatile material is liberated. If collected in the detachable trap immersed in liquid nitrogen between the apparatus and the pump, it condenses to give a solid which is orange-tan in color at room temperature. Before continuing with the synthesis, it is desirable to check that all this unidentified material has been removed from the cold fingers. This may be done by pumping a small amount of material from the cold fingers on to a fresh glass surface in the detachable trap. This solid, after warming to room temperature, should now consist of colorless S_2N_2, which leaves only a characteristic thin blue film of $(SN)_x$ upon evaporation. At this stage in the synthesis, stopcock A is closed, the acetone from the cold fingers is removed using a syringe, and the cold fingers are permitted to warm to room temperature. By the time room temperature is reached, the cold fingers are covered with a pale grayish blue deposit composed of white S_2N_2 mixed with a small quantity of blue $(SN)_x$, which begins to form as soon as the temperature rises from liquid nitrogen to room temperature. If a red color persists at room temperature, it may be due to S_4N_2[2,13] impurity, and the experiment should be discontinued, since this material appears to interfere with the polymerization of S_2N_2 and will contaminate the $(SN)_x$. If the synthesis is carried out according to the method described, none of this impurity should, however, be produced. It has been suggested[2,12] that S_4N_4 vapor may partly decompose in the silver wool under certain conditions to give sulfur and nitrogen and that the sulfur then combines with S_2N_2 to give S_4N_2.

After 48 hr, during which time ice is replenished around the rectangular trap as required, dark blue-black crystals of partly polymerized S_2N_2 will have formed on the walls of the trap. Stopcock B is then closed, the ice bath is removed, and the exterior of the rectangular trap is immediately dried to prevent temperature variations on the surface of the trap due to evaporation of water. Such variations may cause unpolymerized S_2N_2 to sublime from one portion of the trap to another and hence cause imperfections in the crystals. After an additional 60 hr at room temperature golden crystals of $(SN)_x$ will have formed, most of which will have fallen to the bottom of the trap. At this stage the $(SN)_x$ crystals will normally contain some unpolymerized S_2N_2 which is removed by pumping the rectangular trap at room temperature for 2 hr, then for 1 hr as the temperature is raised from room temperature to $75°$, and then finally for 4 hr at

75°. The rectangular trap is then detached and weighed. At this stage no white deposit of S_2N_2 should form on the cold glass surface of the rectangular trap after cotton soaked in liquid nitrogen is held to a wall for approximately 1 min while the bottom of the trap is at 75°. During the initial stages of the heating process, a very thin film of orange-yellow solid (possibly S_4N_4) usually sublimes from the heated portion of the rectangular trap onto the upper cooler walls. This is pumped away during the remainder of the heating process.

In a typical experiment all the S_4N_4 was consumed and provided a 57.8% yield of crystalline $(SN)_x$ and an 8% yield of blue $(SN)_x$ flakes on the cold finger.

The resulting lustrous, golden $(SN)_x$ crystals are analytically pure.

Anal. Calcd. for $(SN)_x$: S, 69.59; N, 30.41. Found: S, 69.63; N, 30.43. When a 50 mg sample of $(SN)_x$ is ground with 0.2 mL of chloroform, no S_4N_4 should be extracted.

A typical experiment normally yields several well-faceted single crystals of $(SN)_x$ ranging in size from 0.5 to 1.5 mm; these represent approximately 10% of the total mass. The remainder consists of less perfect crystals and also polycrystalline agglomerates. Crystals up to 4-5 mm in length have been obtained by this method.

Properties

Polymeric sulfur nitride forms well-defined diamagnetic monoclinic $(P2_{1/c})$ crystals,[2,17] which are a lustrous golden color on the faces and dark blue-black on the ends. The fibrous crystals consist of nearly planar, parallel $(SN)_x$ chains having the conformation[2,17] in which the bond lengths are intermediate between those expected for single and double S-N bonds (Fig. 2).[2,17] The material is relatively inert chemically at room temperature to air and water,[14] and it may be heated in air for short periods without tarnishing, although it decomposes when heated in a sealed container for one to two days at ~150°. It is insoluble in all solvents with which it does not react chemically. It reacts only slowly with chlorine at room temperature to give NSCl,[12] but it reacts rapidly with bromine vapor to give the highly metallic polythiazyl bromides.[8] It decomposes violently when heated in excess of 200° in the absence of air.[2] It also explodes on occasions when compressed strongly in, for example, a potassium bromide pellet press.[2]

The purity of the compound is best confirmed by an elemental analysis for sulfur and nitrogen. If the sum of the sulfur and nitrogen percent compositions is less than 100, this strongly suggests that oxygen impurity is present.[2,15] The infrared spectrum of $(SN)_x$ can be measured on a thin film sublimed[4] onto a sodium chloride window. Principal absorption maxima in the 600-1200 cm^{-1} region are 620-625(w), 670(vs), 720-840(, broad), and 995(s) cm^{-1}.

Fig. 2. Structure SN_x.

Polymeric sulfur nitride also exists in the form of as yet ill-defined dark blue-black powders that are either amorphous or only poorly crystalline. Both this material and the golden crystalline $(SN)_x$ may be sublimed[4] at about 145° to yield golden films of crystalline $(SN)_x$ which deposit on cooler surfaces.

References

1. V. V. Walatka, Jr., M. M. Labes, and J. H. Perlstein, *Phys. Rev. Lett.*, **31**, 1139 (1973); C. Hsu and M. M. Labes, *J. Chem. Phys.*, **61**, 4640 (1974); A. A. Bright, M. J. Cohen, A. F. Garito, A. J. Heeger, C. M. Mikulski, P. J. Russo, and A. G. MacDiarmid, *Phys. Rev. Lett.*, **34**, 206 (1975); R. L. Greene, P. M. Grant, and G. B. Street, *Phys. Rev. Lett.*, **34**, 89 (1975).
2. C. M. Mikulski, P. J. Russo, M. S. Saran, A. G. MacDiarmid, A. F. Garito, and A. J. Heeger, *J. Am. Chem. Soc.*, **97**, 6358 (1975).
3. G. B. Street, H. Arnal, W. D. Gill, P. M. Grant, and R. L. Greene, *Mater. Res. Bull.*, **10**, 877 (1975).
4. A. A. Bright, M. J. Cohen, A. F. Garito, A. J. Heeger, C. M. Mikulski, and A. G. MacDiarmid, *Appl. Phys. Lett.*, **26**, 612 (1975).
5. C. K. Chiang, M. J. Cohen, A. F. Garito, A. J. Heeger, C. M. Mikulski and A. G. MacDiarmid, *Solid State Commun.*, **18**, 1451 (1976); H. P. Geserich and L. Pintschovius, in *Festkörperprobleme* (Advances in Solid State Physics), Vol. XVI, J. Treusch (ed.), Vieweg, Braunschweig, 1976, p. 65.
6. R. L. Greene, G. B. Street, and L. J. Suter, *Phys. Rev. Lett.*, **34**, 577 (1975).
7. R. J. Nowak, H. B. Mark, Jr., A. G. MacDiarmid, and D. Weber, *J. Chem. Soc. Chem. Commun.*, 9 (1977).
8. M. Akhtar, C. K. Chiang, A. J. Heeger, J. Milliken, and A. G. MacDiarmid, *Inorg. Chem.*, **17**, 1539 (1978); G. B. Street, W. D. Gill, R. H. Geiss, R. L. Greene, and J. J. Mayerle, *J. Chem. Soc. Chem. Commun.*, 407 (1977); C. K. Chiang, M. J. Cohen, D. L. Peebles, A. J. Heeger, M. Akhtar, J. Kleppinger, A. G. MacDiarmid, J. Milliken, and M. J. Moran, *Solid State Commun.*, 607 (1977).
9. F. P. Burt, *J. Chem. Soc.*, **97**, 1171 (1910).
10. M. Goehring and D. Voigt, *Z. Anorg. Allgem. Chem.*, **285**, 181 (1956).
11. M. Goehring, *Q. Rev. Chem. Soc.*, **10**, 437 (1956); M. Becke-Goehring, *Inorg. Synth.*, **6**, 123 (1960); M. Villena-Blanco and W. L. Jolly, *Inorg. Synth.*, **9**, 98 (1967).
12. R. L. Patton, "Studies of Disulfur Dinitride, S_2N_2, and Trithiazyl Trichloride, $S_3N_3Cl_3$," Ph.D. Thesis, University of California, Berkeley, 1969.
13. E. J. Louis, A. G. MacDiarmid, A. F. Garito, and A. J. Heeger, *J. Chem. Soc. Chem. Commun.*, 426 (1976).
14. J. Kleppinger, A. G. MacDiarmid, A. J. Heeger and A. F. Garito, Abstracts, 11th Mid-

dle Atlantic Regional Meeting, American Chemical Society, Newark, DE, April 20-23, 1977, page IN6.
15. C. M. Mikulski, A. G. MacDiarmid, A. F. Garito, and A. J. Heeger, *Inorg. Chem.*, 15, 2943 (1976).
16. A. J. Banister, *Inorg. Synth.*, 17, 197 (1977).
17. M. J. Cohen, A. F. Garito, A. J. Heeger, A. G. MacDiarmid, C. M. Mikulski, M. S. Saran, and J. Kleppinger, *J. Am. Chem. Soc.*, 98, 3844 (1976); G. Heeger, S. Klein, L. Pintschovius, and H. Kahlert, *J. Solid State Chem.*, 23, 341 (1978).

32. PYRIDINIUM HEXACHLOROPLUMBATE(IV)

$$PbCl_2 + 2HCl + Cl_2 \longrightarrow H_2[PbCl_6]$$

$$H_2[PbCl_6] + 2C_5H_5N \longrightarrow (C_5H_5NH)_2[PbCl_6]$$

**Submitted by GEORGE B. KAUFFMAN,* LEO KIM,† and DEAN F. MARINO ‡
Checked by WILLIAM B. WITMER§ and JOHN H. TURNEY§**

All the elements of the main periodic group IV(A) except carbon form hexahalide complex anions of type $[MX_6]^{2-}$. A solution of dihydrogen hexachloroplumbate(IV) can be prepared at low temperatures by saturating a hydrochloric acid suspension of lead(II) chloride with chlorine,[1] by electrolyzing a hydrochloric acid solution of lead(II) chloride,[2,3] by adding lead(IV) oxide to hydrochloric acid,[4-7] or by dissolving lead(IV) acetate in hydrochloric acid.[8] The first method is used in this synthesis. Although the free acid $H_2[PbCl_6]$ is unstable above $-5°$, the $[PbCl_6]^{2-}$ anion attains a maximum stability in combination with large, poorly polarizing cations, and the ammonium,[1,3,9,10a] potassium,[3,6,7,8,10b] rubidium,[6,8,11] cesium,[6,8] diazonium,[12,13] and especially pyridinium[3,6,8,10c,14] and quinolinium[3] salts are fairly stable. These salts are precipitated by addition of the alkali metal halide, diazonium salt, or free organic base to the $H_2[PbCl_6]$ solution.

Only a few simple compounds of lead(IV) are known, for example, the acetate, fluoride, and chloride. Lead(IV) chloride, an unstable substance, may be prepared by treatment of one of the hexachloroplumbates mentioned above with cold concentrated sulfuric acid.[1,11,14]

---

*California State University, Fresno, Fresno, CA 93740.
†Shell Oil Company, 901 Grayson, Berkeley, CA 94710.
‡Experimental Station, E. I. duPont de Nemours & Co., Wilmington, DE 19899.
§Chemstrand Research Center, Research Triangle Park, Box 731, Durham, NC 27702.

Procedure

Twenty grams (0.0719 mol) of very finely powdered lead(II) chloride is suspended in 400 mL of concentrated hydrochloric acid contained in an all-glass 500-mL gas washing bottle whose inlet tube terminates in a high-porosity fritted-glass cylinder. The bottle and contents are maintained at 10-15° by means of an ice bath, the suspension is magnetically stirred at a high rate, and a stream (2-3 bubbles/sec) of chlorine (*Hood!*) is introduced until all the lead(II) chloride has dissolved (2-3 hr).

After excess chlorine has been removed by bubbling air through the pale lemon-yellow solution, it is cooled to 0°, and 10 mL (0.124 mole) of pyridine is added *dropwise* with rapid stirring. The resulting bright lemon-yellow precipitate is collected on a fine sintered-glass funnel, washed with five 10-mL portions of ice-cold ethanol, and air-dried. The powdered product is then dried at 50° for 1 hr. The yield is 18.4 g (44.1%).

Anal. Calcd. for $C_{10}H_{12}N_2PbCl_6$: Cl, 36.68; Pb, 35.73. Found: Cl, 37.30; Pb, 35.77.

Properties

Pyridinium hexachloroplumbate(IV) is a lemon-yellow salt which is stable on exposure to air, being neither hydrolyzed nor losing chlorine, but it is hydrolyzed by water to give lead(IV) oxide and hydrochloric acid. Since it reacts with concentrated sulfuric acid to yield lead(IV) chloride, it is a convenient source of this unstable explosive compound.

■ **Caution.** *Lead(IV) chloride is explosive and is potentially very hazardous.*

References

1. H. Friedrich, *Monats. Chem.*, **14**, 505 (1893); *Ber.*, **26**, 1434 (1893).
2. F. Försted, *Z. Elektrochem.*, **3**, 507 (1897).
3. K. Elbs and R. Nübling, *Z. Elektrochem.*, **9**, 776 (1903).
4. N. A. E. Millon, *J. Pharm. Chem.*, [3] **18**, 299 (1850).
5. J. Strachan, *Chem. News*, **98**, 102 (1908).
6. H. L. Wells, *Am. J. Sci.*, [3] **46**, 180, 186 (1893); *Z. Anorg. Chem.*, **4**, 335 (1893).
7. J. Nikoliukin, *J. Russ. Phys. Chem. Soc.*, **17**, 200 (1885).
8. A. Gutbier and M. Wissmüller, *J. Prakt. Chem.*, [2] **90**, 491 (1914).
9. V. F. Postnikov and A. I. Speranskii, *J. Gen. Chem. (U.S.S.R.)*, **10**, 1328 (1940).
10. G. Brauer, *Handbuch der präparativen anorganischen Chemie*, 2d ed., Band I, Ferdinand Enke Verlag, Stuttgart, 1960, (a) p. 667, (b) p. 669, (c) p. 666.
11. H. Erdmann and P. Köthner, *Liebigs Ann. Chem.*, **294**, 71 (1896).
12. E. Sakellarios, *Ber.*, **56B**, 2536 (1923).
13. F. D. Chattaway, F. L. Garton, and G. D. Parkes, *J. Chem. Soc.*, **125**, 1980 (1924).
14. W. Biltz and E. Meinecke, *Z. Anorg. Allgem. Chem.*, **131**, 1 (1923).

33. BIS[(4,7,13,16,21,24-HEXAOXA-1,10-DIAZABICYCLO[8.8.8]HEXACOSANE) POTASSIUM] TETRABISMUTHIDE(2−)

$$KBi_2(s) + 2,2,2\text{-crypt} \xrightarrow{en} [K(2,2,2\text{-crypt})]_2Bi_4$$

Submitted by A. CISAR* and J. D. CORBETT*
Checked by B. VAN ECK† and J. L. DYE†

A number of the intermetallic phases formed between the alkali metals and the heavier posttransition elements (Sn, Pb, As, Sb, Bi, Se, Te) have the very remarkable property of dissolving in liquid ammonia to form very intensely colored solutions. The most extensive characterization of these solutions has been by Zintl and co-workers, who by potentiometric titrations established the apparent formation of polyatomic ions such as Sn_9^{4-}, Bi_5^{3-}, and Te_3^{2-}.[1,2] Unfortunately, on evaporation of solvent these salts always revert to the more stable intermetallic phases, for example,

$$6(Na(NH_3)_n^+)_3Bi_5^{3-} \longrightarrow Na_5Bi_4 + 13NaBi_2 + 6nNH_3(g).$$

This reversion to the simple metallic phases can be prevented by blocking the (partial) electron transfer or delocalization from the anion back onto the alkali metal ion. The key to this[3] is the very strong complexing of the alkali metal ion provided by 2,2,2-crypt, (4,7,13,16,21,24-hexaoxa-1,10-diazabicyclo-[8.8.8]hexacosane), $(N(C_2H_4OC_2H_4OC_2H_4)_3N)$. This is one of a group of cage-like polycyclic amine ethers[4] which complex alkali and alkaline earth metal ions extremely well. The compound 2,2,2-crypt is the ligand of choice for it is commercially available‡ at a price significantly below that of any other analog, and potassium is selected because it forms the most stable complex among the alkali metals.[5] The crypt has a sufficient effect on stability that the more convenient 1,2-ethanediamine can be used as the solvent.

Four compounds are known in the potassium-bismuth system: K_3Bi, K_3Bi_2, K_5Bi_4, and KBi_2, melting at 671, 442, 381 (incongruently) and 565°, respec-

---

*Ames Laboratory and Department of Chemistry, Iowa State University, Ames, IA 50011. The Ames Laboratory is operated for the U.S. Department of Energy by Iowa State University under contract No. W-7405-Eng-82.

†Department of Chemistry, Michigan State University, East Lansing, MI 48824.

‡Parish Chemical Company, Provo, Utah; PCR, Gainesville, FL; E. Merck, Darmstadt, West Germany.

tively.[6] All may be conveniently prepared by fusing together stoichiometric amounts of the elements.[7] Only K_5Bi_4 requires any special care to obtain a pure phase, namely fusion at 440°, quenching to near room temperature and then annealing for several days at 370-375°. Although any of these phases as well as their mixtures may be employed for the crypt synthesis of Bi_4^{2-}, the two lower-melting and less stable compounds, K_3Bi_2 and K_5Bi_4, give a faster reaction. Fortunately these two may also be prepared in a fused silica container.

Procedure

The air-sensitive nature of the compounds requires that all manipulations be carried out either in a dry box or on a high vacuum line.

An alloy composition near K_3Bi_2 or K_5Bi_4 is desired, but a single phase material is not necessary. A bulk form of bismuth* is usually lower in nonmetals and may be easily broken up to small pieces with a hammer. Nonmetal impurities in the less pure, granular variety can be easily removed for the main part by one or two fusions (mp 271°) in vacuum or under inert gas followed by scraping or filing to remove the dross. The best material will result if care is taken to ensure that unreacted potassium does not come in contact with the silica container before it reacts with the bismuth. (An alumina crucible [Coors, cylindrical, 3-5 mL] inside the SiO_2 tube may be used.) The silica tube (8-12 cm long, 12-18 mm diameter) is closed on one end and connected at the other end through a length of 6-7 mm tubing to a standard taper or ball joint.

Bismuth (6.41 g, 0.0307 mole) is placed in the silica tube (or alumina crucible) and taken into the dry box along with the potassium.

■ **Caution.** *Potassium ignites on contact with moisture and forms a highly caustic KOH layer on air exposure. Scraps should be disposed of by reaction with 1-butanol or a higher alcohol.*

Potassium is cut, scraped clean, and weighed (1.50 g, 0.0384 mole for K_5Bi_4) in the dry box, and placed on top of the bismuth. The reaction container is capped by a stopcock between two suitable ground joints, removed from the dry box, evacuated, and sealed off below the joint.

■ **Caution.** *Eye protection is required.*

The container is heated in a furnace to 80°, where the potassium melts, and after some minutes is heated to ~280° to melt the bismuth. The potassium has sufficiently reacted at this point that the tube may be heated to above the melting point of the particular mixture (500-550°), its contents briefly mixed, and then cooled. Though the reaction is quite exothermic, it is perfectly safe in a sealed, *evacuated* tube since no significant pressures develop.

The potassium-bismuth compounds are all friable and range in color from metallic green for K_3Bi through bluish-silver for K_3Bi_2 and K_5Bi_4, both of

*United Mineral and Chemical Corp., New York, NY; Cerac, Inc. Inc., Butler, WI.

which develop a golden surface coating after handling in the dry box, to silver for KBi_2.

1,2-Ethanediamine (en) is dried by stirring over CaH_2 for about 2 days, followed by heating at reflux at reduced pressure over fresh CaH_2 for 24 hr and distilling onto dried 4-A molecular sieves in a flask equipped with a Teflon needle valve (Fisher-Porter) for storage.* For use in a reaction, the solvent is distilled directly from the storage flask into the reaction vessel, using an ice-water bath. Joints between Teflon valves should be greased with a silicone. The 2,2,2-crypt is used as received and is handled only in the dry box. Protection from strong light is advised.

The tetrabismuthide salt $[K(2,2,2-crypt)]_2Bi_4$ is the unique crystalline product when any of the compounds in the potassium-bismuth system or mixtures of them are allowed to react with 2,2,2-crypt in en at room temperature. In a typical reaction, stoichiometric amounts (with respect to K) of the ground intermetallic compound and crypt are loaded into section A of the reaction vessel in Fig. 1, using a funnel which extends beyond the valve seat. A scale based on 0.1-0.2 g of 2,2,2-crypt is convenient, which requires 0.05-0.11 g K_5Bi_4. The vessel is evacuated and 40-50 mL solvent distilled into it.

The best and fastest reaction is obtained at room temperature with either of the intermediate compositions K_3Bi_2 or K_5Bi_4. After the solvent has been condensed, a clear, deep-green solution rapidly forms at room temperature, and overnight the solution becomes intensely colored with a dichroic character, green in thin layers and red in thick layers. This solution does not appear to change further for many weeks, although a slight amount of gas evolution, presumably H_2 from solvent reduction, is detectable (≤ 100 torr total pressure). After 7-10 days tiny black crystals with a hexagonal shape appear on the walls just above the alloy. After an additional week, these are about 0.2 mm in diameter, and the solution may be decanted into B. The solution then can be frozen to $-80°$ and the apparatus evacuated and sealed off between A and B prior to removal to the dry box. If the reaction is allowed to continue, the crystals either grow larger or more nucleate until all the starting intermetallic species are consumed, at which time the gas evolution ends and the solution fades to a deep clear emerald green. This final color comes from a more reduced solution in equilibrium with the solid salt and is necessary if the salt is to be stable in the presence of en. If fresh en is condensed onto crystals of the tetrabismuthide salt, they disproportionate to form this colored solution and bismuth.

The yield is >75% based on bismuth. The product is best characterized by X-ray diffraction either as single crystal lattice constants or by a powder pattern.

---

*Large amounts of 1,2-ethanediamine should be moved around and condensed with a $-80°$ bath even though en retains a small vapor pressure at that temperature. Glass traps containing a plug of en condensed at $-196°$ often crack on warmup, perhaps owing to expansion accompanying a phase transition.

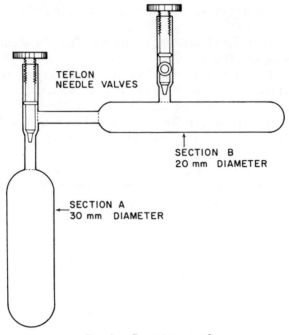

Fig. 1. Reaction vessel.

Distances (Å) for the strongest third of the lines at low angles are (intensities in parentheses): 10.95(6), 10.32(8), 10.15(5), 9.92(9), 7.85(5), 7.77(5), 7.21(6), 6.82(10), 5.83(5), 4.86(6). The inferred reaction equation with K_5Bi_4 as an example is, ignoring solution intermediates,

$$K_5Bi_4 + 5(2,2,2\text{-crypt}) + 3en \longrightarrow$$
$$[K(2,2,2\text{-crypt})]_2Bi_4 + 3[K(2,2,2\text{-crypt})]en^- + 3/2H_2(g)$$

where "en$^-$" represents the anion formed by removing a proton from an en molecule. Similar equations apply with other starting compositions for the intermetallic reactant.

If K_3Bi is used, the reaction follows the same course but much more slowly, the entire reaction requiring several months and with more gas evolution, as expected. Finely divided bismuth (or intermediate K-Bi phases) formed by oxidation by solvent are believed to speed equilibration and formation of product in the foregoing. With KBi_2, which already has the correct stoichiometry, the initial solution is brown but becomes dichroic after about a week, and crystals appear on the surface of the intermetallic compound after about a month. Some of the starting material remains even after a year. Evaporation of any of the intensely colored solutions yields only a dark brown paste and no crystalline material. Condensation of en on the intermetallic compounds or K_5Bi_4 in the absence of crypt gives a purple solution, the color reported by Zintl for Bi_3^{3-}.[1] When this solution is poured off the intermetallic compound and onto crypt, it immediately reacts to form the green solution with sufficient heat evolution to boil the solvent in the vacuum.

Properties

Only a single phase containing Bi_4^{2-} has been found to separate spontaneously from these solutions, in spite of suggestions that several more exist in the solutions. The crystals appear black under the microscope but give a very dark green streak when ground. The phase has been analyzed by X-ray crystallography, which shows the presence of an essentially square planar Bi_4^{2-} therein, d_{Bi-Bi} = 2.936(2) and 2.941(2) Å and angles within 0.15° (2.5 σ) of 90°.[3] The ion is thus isoelectronic with Te_4^{2+},[8] where stabilization depends on very different factors.[9] The bonding in these square planar ions can be described in terms of four sigma and one bonding pi orbitals, the last four electrons existing in a slightly antibonding (e_g) pi orbital. An apparent relationship of this ion to the square planar bismuth units which occur in some complex intermetallic phases (e.g., $Ca_{11}Bi_{10}$) has also been discussed.[3]

Techniques similar to those described can be utilized to obtain alkali metal-crypt salts of particularly the trigonal bipyramidal Pb_5^{2-} [10] and, by solvent evaporation, Sb_7^{3-} (C_{3v})[11] and Te_3^{2-} (C_{2v}).[12]

References

1. E. Zintl, J. Goubeau, and W. Dullenkopf, Z. Phys. Chem., Abt. A, **154**, 1 (1931).
2. E. Zintl and W. Dullenkopf, Z. Phys. Chem., Abt. B, **16**, 183 (1932).
3. A. Cisar and J. D. Corbett, Inorg. Chem., **16**, 2482 (1977).
4. J. M. Lehn, Struct. Bond., **16**, 1 (1973).
5. J. M. Lehn and J. P. Sauvage, J. Am. Chem. Soc., **97**, 6700 (1975).
6. R. P. Elliott, Constitution of Binary Alloys, First Supplement, McGraw-Hill Book Co., New York, New York, 1965, p. 185.
7. M. Okada, R. A. Guidotti, and J. D. Corbett, Inorg. Chem., **7**, 2118 (1968).
8. T. W. Couch, D. A. Lokken, and J. D. Corbett, Inorg. Chem., **11**, 357 (1972).
9. J. D. Corbett, Prog. Inorg. Chem., **21**, 129 (1976).
10. P. A. Edwards and J. D. Corbett, Inorg. Chem., **16**, 903 (1977).
11. D. G. Adolphson, J. D. Corbett, and D. J. Merryman, J. Am. Chem. Soc., **98**, 7234 (1976).
12. A. Cisar and J. D. Corbett, Inorg. Chem., **16**, 632 (1977).

34. [5,10,15,20-TETRAPHENYLPORPHYRINATO(2−)] LANTHANIDES AND SOME [5,10,15,20-TETRAPHENYLPORPHYRINATO(2−)] ACTINIDES

$$H_2tpp + Ln(acac)_3 \cdot xH_2O \xrightarrow[\Delta]{\text{trichlorobenzene}} Lntpp(acac) + 2Hacac$$

(H_2tpp = free base 5,10,15,20-tetraphenylporphyrin (*meso*-tetraphenylporphyrin); Hacac = 2,4-pentanedione; Ln = lanthanide metals)

Submitted by CHING-PING WONG*
Checked by GRAHAM BISSET†

Despite considerable research interest in metalloporphyrin chemistry, general preparative methods for the formation of lanthanide and actinide complexes of this important class of macrocycle have not been available until recently demonstrated by Wong and co-workers.[1-5] A general technique for the preparation of [tetraarylporphyrinato(2−)] lanthanide and actinide complexes is described here.

Procedure

■ **Caution.** *Trichlorobenzene is toxic and a good hood should be used.* To a 150-mL round-bottomed flask equipped with a nitrogen inlet and reflux condenser, Fig. 1, are added H_2ttp (1.0 g, 1.63 × 10^{-3} mole), hydrated Ln(acac)$_3$ (2.0 g, 4.0 × 10^{-3} mole), and 75 mL of 1,2,4-trichlorobenzene. (The tetraarylporphyrin free base is obtained commercially or prepared by a literature method.[6]) Lanthanide complexes of other β-diketonates such as β-acetylcamphor and 2,2,6,6-tetramethylheptanedione (dpm) may be equally well employed as reactants. The solution is stirred magnetically and heated to reflux (~215°) while a slow flow of nitrogen is passed through the reaction vessel. The course of the reaction is followed by examining the visible absorption spectra of periodically withdrawn aliquots. Completion of the reaction, which generally occurs within 3–4 hr,‡ is indicated by the complete disappearance of the four visible absorptions of the free base at (relative intensities indicated) 510 > 550 > 590 ~

---

*Western Electric Company, Engineering Research Center, P.O. Box 900, Princeton, NJ 08540.
† Chemistry Department, University of British Columbia, Vancouver, BC, Canada V6T1Y6.
‡ Lanthanum and thorium complexes may require 11–12 hr of reflux to reach complete reaction.

[5,10,15,20-Tetraphenylporphyrinato(2−)]Lanthanides and Actinides 157

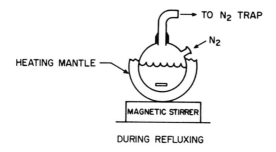

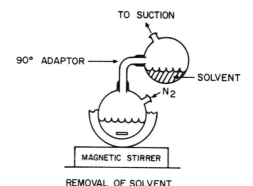

Fig. 1. Apparatus set-up for synthesis of metalloporphyrin.

645 nm. (Fig. 2 shows the behavior of the visible absorption spectrum during the course of a typical reaction.) A shift in the Soret band from 415 nm (H_2tpp) to 420 nm (Lntpp) may also be noted.

Upon completion of the reaction, the solvent is distilled away under reduced pressure (<2 mm/Hg) with the aid of a slow stream of dry nitrogen. [During this part of the procedure, care must be exercised to avoid overheating the porphyrin complex because demetalation will occur. It is important, however, to make certain that all traces of solvent are eliminated as any trace of 1,2,4-trichlorobenzene creates difficulty with drying and considerably hampers purification. As a consequence, the preferred procedure is to commence removal of the solvent when the reaction is ~75% complete (Fig. 2c).] During the final stage of the distillation, the porphyrin complex solution becomes a very viscous fluid. The flask must be cooled immediately after all solvent is removed. To minimize decomposition, magnetic stirring and nitrogen gas bleed are continued until the reaction mixture has cooled to ambient temperature. Prolonged heating of the product at this stage causes conversion of the metalloporphyrin

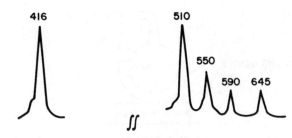
(a) BEFORE REFLUX (FREE BASE PORPHYRIN ONLY)

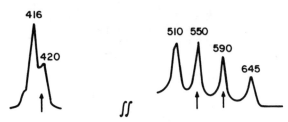
(b) AFTER 1 HOUR REFLUX (SMALL PORTION PRODUCT FORMED)

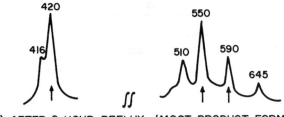
(c) AFTER 2 HOUR REFLUX (MOST PRODUCT FORMED)

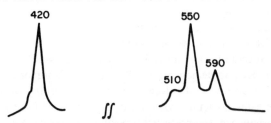
(d) AFTER 4 HOUR REFLUX (ALL PRODUCT FORMED)

Fig. 2. Typical UV-vis spectrum (nm) of the formation of metalloporphyrin from free-base prophyrin and metal acetylacetonate.

back to free base porphyrin. Finally, the crude product, which may still contain excess Ln(acac)$_3$, is vacuum dried overnight.

Porphyrin complexes of yttrium, lanthanum, cerium,* praesodymium, samarium, europium, gadolinium, terbium, dysprosium, holmium, erbium, thulium, ytterbium, lutetium, and thorium can be synthesized using this method.

Unreacted H$_2$tpp and Ln(acac)$_3$ are removed by column chromatography, using a neutral aluminum oxide (obtained from Merck Co., Germany) column (2.5 × 15 cm) saturated with toluene. The crude product† is dissolved in a minimum volume of toluene (100–150 mL) or dichloromethane/hexane, the solution filtered, applied to the column, and the H$_2$tpp eluted with toluene. The remaining H$_2$tpp, a trace of metalloporphyrin, and the toluene solvent residues are then eluted with acetone. Finally, dimethyl sulfoxide is used to elute the majority of the reddish-pink metalloporphyrin. The unreacted Ln(acac)$_3$ apparently remains at the top of the column. The dimethyl sulfoxide fractions are treated with equal volumes of freshly distilled chloroform (previously treated with a 0.1 M NaOH solution) and washed three times with equal volumes of water. The metalloporphyrin is recovered from the chloroform solution by rotary evaporation under reduced pressure, approximately 2 torr. The product is further purified by dissolving in base-treated chloroform, followed by precipitation with distilled methanol or cooling.‡ The purple crystals so obtained are dried under vacuum. Yields as high as 90% are indicated from the visible spectra; however, final yields of only 30–60% are to be expected because of losses during the purification process.

Anal. Calcd. for (2, 4-pentanedionato)[5,10,15, 20-tetrakis(4-methylphenyl) porphyrinato(2−)] ytterbium, Yb(p-Me$_3$) tppacac, YbC$_{53}$H$_{43}$N$_4$O$_2$: C, 67.66; H, 4.57; N, 5.96. Found: C, 67.56; H, 4.70; N, 5.80.

Anal. Calcd. for Yb(p-Me$_3$)tppdpm, YbC$_{59}$H$_{55}$N$_4$O$_2$: C, 69.14; H. 5.37; N, 5.48. Found: C, 69.70; H, 5.74; N, 4.89.

Products can also be analyzed spectrophotometrically by quantitative conversion to the dication (H$_4$tpp^{2+}) of the free base H$_2$tpp by bubbling hydrogen chloride gas through solutions of the metalloporphyrin complexes in organic solvents.[7] Extinction coefficients for the Soret band(s) at 420 nm and the three visible bands at 510 (shouderband), 550 (β-band), and 590 nm (α-band) are very similar to those of Sn(IV) and Cd(II) tpp.[8] Thin layer chromatography (10% dimethyl sulfoxide in MeOH on polyamide plates) gives R_f values of 0.5.

---

*The checker could not detect the formation of Ce porphines even after 26 hr.

†La, Pr, and Ce porphyrins tend to demetalate on purification. Thorium porphyrins are difficult to isolate in a pure state.

‡The checker used an alternative crystallization method which calls for dissolution of the porphyrin in a minimum amount of base-treated dichloromethane and methanol (3 volumes). The dichloromethane is then distilled off using a rotary evaporator and the precipitated porphyrin collected by filtration.

Properties

Visible spectra of various metalloporphyrin complexes are tabulated in Table I. (S stands for Soret band. In general the β band is of higher intensity than the α. All bands are in nanometers.) Some molar extinction coefficients $(M^{-1}\,cm^{-1}) \times 10^{-3}$ given for the four bands in order of increasing wavelengths are: Sm(381, 2.8, 15.2, 4.6); Eu(365, 2.8, 13.2, 4.8); Tb(491, 3.0, 18.4, 5.3); Dy(401, 2.7, 16.8, 4.3; Ho(348, 2.4, 16.8, 3.9); Er(397, 3.3, 17.3, 5.5); Tm(365, 2.3, 18.0, 3.7); Yb(346, 3.3, 15.8, 6.6); Lu(380, 3.7, 18.9, 4.1).

TABLE I UV-VIS Spectra of Newly Synthesized Metalloporphyrin Complexes[a]

	Metalloporphyrin	UV-VIS Absorption (nm)		
		(s)	(α)	(β)
1	Latpp(acac)	422(S),	562 > 608	
2	Latpp(dpm)	424(S),	544 > 590	
3	Cetpp(acac)$_2$	409(S),	554 > 600	
4	Pr(p-CH$_3$)tpp(acac)	426(S),	556 > 597	
5	Prtpp(acac)	421(S),	554 > 593	
6	Pr(p-CH$_3$)tpp(0 − <)	424.5(S),	556 > 597	
7	Ndtpp(acac)	423(S),	553 > 590	
8	Ndtpp(dpm)	427(S),	558 > 594	
9	Smtpp(dpm)	425(S),	547 > 589	
10	Smtpp(acac)	423(S),	552.5 > 590	
11	Eu(p-CH$_3$)tpp(acac)	426(S),	556 > 597	
12	Eutpp(acac)	426(S),	556 > 597	
13	Eu(3,5-Cl$_2$)tpp(acac)	426(S),	556 > 597	
14	Gdtpp(acac)	422(S),	556 > 596	
15	Tbtpp(acac)	424(S),	549 > 587	
16	Tbtpp(dpm)	424(S),	555 > 593	
17	Dytpp(acac)	420(S),	550 > 587	
18	Dytpp(dpm)	424(S),	544 > 590	
19	Hotpp(acac)	422(S),	544 > 592	
20	Hotpp(dpm)	424(S),	552 > 590	
21	Er(m-F)tpp(acac)	423(S),	553 > 592	
22	Ertpp(acac)	423(S),	553 > 592	
23	Tmtpp(dpm)	424(S),	555 > 592	
24	Ybtpp(acac)	422(S),	553 > 590	
25	Yb(m-F)tpp(dpm)	422(S),	533 > 590	
26	Lutpp(acac)	418(S),	553 > 592	
27	Thtpp(acac)$_2$	422(S),	550 > 599	
28	Ytpp(acac)	422(S),	552 > 591	

[a]Vis spectra were taken from Cary 14, 15, and 17 at room temperature in acetone solution. Different organic solvents give slightly variable absorptions (±15 nm).

(Mass spectra data were obtained using a MS 902 instrument operating at 70 eV with a chamber temperature of 350-400° and perfluorokerosene or tris(perfluoroheptyl)-s-triazine as a fragment reference.) In all cases, molecular parent ions ($M^{+\cdot}$) with satellite peaks consistent with the known isotopic abundance for each metal were observed as well as peaks corresponding to the loss of the β-diketonate anion. Often a minor peak corresponding to the parent ion of an acetate complex was observed, perhaps formed by pyrolysis in the high-temperature probe with loss of a C_3H_4 fragment. The fragmentation patterns are quite similar to those reported in the literature for other metalloporphyrins.[9,10] The mass spectra are listed as follows.

Latpp(acac); 850 ($M^{+\cdot}$). Cetpp(acac)$_2$; 950 ($M^{+\cdot}$). Pr(p-CH$_3$)tpp(acac); 908 ($M^{+\cdot}$), 809 (M-acac), 868 (M-ac*). Ndtpp(acac); 855 ($M^{+\cdot}$), 756 (M-acac), 815 (M-ac*). Smtpp(acac); 861 ($M^{+\cdot}$), 762 (M-acac). Eutpp(acac); 862 ($M^{+\cdot}$), 763 (M-acac). Gd(p-Me)tpp(acac); 924 ($M^{+\cdot}$), 825 (M-acac), 884 (M-ac*). Tbtpp(acac); 870 ($M^{+\cdot}$), 771 (M-acac), 830 (M-ac*). Dytpp(acac); 873 ($M^{+\cdot}$), 777 (M-acac), 836 (M-ac*). Ertpp(acac); 878 ($M^{+\cdot}$), 779 (M-acac). Tmtpp (acac); 880 ($M^{+\cdot}$), 781 (M-acac). Tmtpp(dpm); 964 ($M^{+\cdot}$), 781 (M-dpm). Yb(p-Me)tpp(acac); 942 ($M^{+\cdot}$), 843 (M-acac), 902 (M-ac*). Yb(p-Me)tpp(dpm); 1024 ($M^{+\cdot}$), 841 (M-dpm). Lutpp(acac); 886 ($M^{+\cdot}$), 787 (M-acac), 826 (M-ac*). Thtpp(acac)$_2$; 1024 ($M^{+\cdot}$), 943 (M-acac), 1002 (M-acac).

Infrared spectra (in KBr) show bands at ~1510 and ~1595 cm^{-1} characteristic of metalloporphyrin 2,4-pentanedione complexes:[10]

Prtpp(acac); 1512, 1552. Pr(p-Me)tpp(acac); 1512, 1555. Ndtpp(acac); 1511, 1591. Smtpp(acac); 1512, 1590. Eutpp(acac); 1575, 1600. Tbtpp(acac); 1510, 1590. Dytpp(acac); 1510, 1590. Hotpp(acac); 1520, 1600. Ertpp(acac); 1510, 1590. Er(m-F)tpp(acac); 1500, 1600. Yb(p-Me)tpp(acac); 1512, 1595. Yb(p-Me)tpp(dpm); 1500, 1510. Tbtpp(acac)$_2$; 1520, 1595.

Most of the lanthanide porphyrin complexes are paramagnetic, except those of lanthanum and lutetium which are diamagnetic. They all show paramagnetic behaviors in the NMR spectra similar to those of the lanthanide β-diketonate complexes, except that the shift capabilities are less.[4] Some of these paramagnetic lanthanide complexes (i.e., that of ytterbium) show Curie behavior at low temperature. Diamagnetic lutetium shows some "ring current" phenomena.[4]

The proton NMR spectrum of the tetra-p-tolyl derivative, Eu(p-Me)tpp-(acac), at -21° exhibits peaks (ppm from TMS) at -13.31 d (ortho-endo), -9.33 d (meta-endo), -8.30 (pyrrole), -8.13 (ortho- and meta-exo), -3.44 (p-Me), and 0.88 (acac-Me).

This synthesis is a general method for preparing metalloporphyrins. It works on the whole series of lanthanide and some actinide metals. With this new procedure, virtually all lanthanide metals can form metalloporphyrin complexes.[3]

*acac = acetylacetonate anionic ligand; ac = acetate anionic ligand; dpm = tris(2,2,6,6-tetramethylheptane-3,5-dionato) anionic ligand.

References

1. C. P. Wong, R. F. Ventichler, and W. D. Horrocks, Jr., *J. Am. Chem. Soc.*, **96**, 7149 (1974).
2. C. P. Wong and W. D. Horrocks, Jr., *Tetrahedron Lett.*, 2637 (1975).
3. C. P. Wong, "Syntheses, Characterization and Applications of Some Lanthanide and Actinide Porphyrin Complexes," Ph.D. thesis, The Pennsylvania State University, University Park, PA 16802, Part I, 1975.
4. W. D. Horrocks, Jr. and C. P. Wong, *J. Am. Chem. Soc.*, **98**, 7157 (1976).
5. L. A. Martarano, C. P. Wong, and W. D. Horrocks, Jr., A. M. P. Goncalves, *J. Phys. Chem.*, **80**, 2389 (1976).
6. A. D. Alder, F. R. Longo, J. D. Finarelli, J. Goldmatcher, J. Assour, and T. Korsakoff, *J. Org. Chem.*, **32**, 476 (1967).
7. A. Stone and E. B. Fleischer, *J. Am. Chem. Soc.*, **90**, 2735 (1968).
8. G. D. Dorough, J. R. Miller, and F. M. Huennekens, *J. Am. Chem. Soc.*, **73**, 4315 (1951).
9. A. Treibs, *Justus Liebigs Ann. Chem.*, **728**, 115 (1969).
10. J. W. Buchler, G. Eikelmann, L. Puppe, K. Rohback, H. H. Schneehage, and W. Weck, *Justus Liebigs Ann. Chem.*, **745**, 135 (1971).

Chapter Four

ORGANOMETALLIC COMPOUNDS

35. [μ-NITRIDO-BIS(TRIPHENYLPHOSPHORUS)](1+) TRICARBONYLNITROSYLFERRATE(1−) AND [μ-NITRIDO-BIS(TRIPHENYLPHOSPHORUS)](1+) DECACARBONYL-μ-NITROSYL-TRIRUTHENATE(1−), [(Ph$_3$P)$_2$N][Fe(CO)$_3$(NO)] and [(Ph$_3$P)$_2$N][Ru$_3$(CO)$_{10}$(NO)]

$$[(Ph_3P)_2N](NO_2) + Fe(CO)_5 \xrightarrow{THF}$$
$$[(Ph_3P)_2N][Fe(CO)_3(NO)] + CO \uparrow + CO_2 \uparrow$$

$$[(Ph_3P)_2N](NO_2) + Ru_3(CO)_{12} \xrightarrow{THF}$$
$$[(Ph_3P)_2N][Ru_3(CO)_{10}(NO)] + CO \uparrow + CO_2 \uparrow$$

Submitted by R. E. STEVENS,[*] T. J. YANTA,[*] and W. L. GLADFELTER[*]
Checked by W. F. ENRIGHT,[†] C. M. JENSEN,[†] and H. D. KAESZ[†]

The substitution of an appropriate number of nitric oxide ligands for carbon monoxide is one possible method of activating metal carbonyl clusters for homo-

---

[*]Department of Chemistry, University of Minnesota, Minneapolis, MN 55455.
[†]Department of Chemistry and Biochemistry, Univ. of California, Los Angeles, CA 90024.

geneous catalysis. We have found that μ-nitrido-bis(triphenylphosphorus)(1+) nitrite ((PNP)NO$_2$), prepared by the method of Martinsen and Songstad,[1] is a useful reagent in the preparation of nitrosyl carbonyl compounds.[2] In particular, the previous preparation by Hieber and Beutner[3] of [Fe(CO)$_3$(NO)]$^-$ involved the use of NaNO$_2$ in methanolic methoxide at 100°. The (PNP)$^+$ salt was then prepared by metathesis with (PNP)Cl.[4] This procedure took approximately 2 days, giving a yield of 50%. In contrast, the method reported here gives basically quantitative yields in approximately 45 min. Further, because of nitrite's increased reactivity in dipolar aprotic solvents, several new nitrosyl carbonyl compounds can be prepared. These include (PNP)[Ru$_3$(CO)$_{10}$(NO)] (presented here), (PNP)[Os$_3$(CO)$_{10}$(NO)], and (PNP)[Mn(CO)$_2$(NO)$_2$]. Other known compounds such as Mn(CO)$_4$(NO) and Fe(CO)(PPh$_3$)(NO)$_2$ are also formed in high yields using this reagent.[2]

General Procedure

Tetrahydrofuran (THF) and diethyl ether are freshly distilled from sodium benzophenone ketyl under nitrogen. Hexane is freshly distilled from potassium metal under nitrogen. Pentacarbonyl iron (Strem Chemical Company), sodium nitrite, and μ-nitrido-bis(triphenylphosphorus)(1+) chloride, (PNP)Cl (Alfa Chemicals) are used as obtained. Dodecacarbonyl triruthenium is synthesized from ruthenium trichloride trihydrate (Alfa Chemicals) in the manner described by Mantovani and Cenini.[5]

A. μ-NITRIDO-BIS(TRIPHENYLPHOSPHORUS(1+) NITRITE)

Procedure

μ-Nitrido-bis(triphenylphosphorus)(1+) chloride, (5.02 g, 8.74 mmole) is placed in a 500-mL flask and dissolved in 150 mL of hot distilled water. Sodium nitrite (30 g) is dissolved in 60 mL of hot distilled water. The nitrite solution is added dropwise to the (PNP)Cl solution and white crystals form. When all of the nitrite has been added, the solution is cooled in an ice bath (0°) for 30 min. The product is collected by filtration, washed with water and diethyl ether, and dried in air overnight to give 4.78 g (7.93 mmole) of μ-nitrido-bis(triphenylphosphorus)(1+) nitrite monohydrate, (PNP)NO$_2 \cdot$ H$_2$O. The water of crystallization is removed via an azeotropic distillation with benzene to give anhydrous (PNP)NO$_2$.

Anal. Calcd. for C$_{36}$H$_{30}$N$_2$P$_2$O$_2$: C, 73.96; H, 5.18; N, 4.79. Found: C, 73.69; H, 5.27; N, 4.81. Yield of purified product 4.62 g (0.790 mmole), 90%.

Properties

Anhydrous (PNP)NO$_2$ is a white, air-stable solid. It is very soluble in CH$_2$Cl$_2$, acetonitrile, methanol and ethanol; slightly soluble in acetone THF and water; and insoluble in diethyl ether and hexane.

B. μ-NITRIDO-BIS(TRIPHENYLPHOSPHORUS)(1+) TRICARBONYLNITROSYLFERRATE(1-)

■ **Caution.** *Because of the high toxicity of carbon monoxide and iron carbonyls, this reaction should be carried out in an efficient fume hood.*

A sample of 503.4 mg (0.861 mmole) anhydrous (PNP)NO$_2$ is placed in a 150-mL Schlenk tube with a magnetic stirrer. The reaction vessel is evacuated and filled with nitrogen three times to remove oxygen. Then 60 mL THF is added to the flask. Very little of the (PNP)NO$_2$ dissolves until a sample of 0.116 mL (0.863 mmole) of iron pentacarbonyl is added to the reaction vessel via syringe. Immediate bubbling occurs and the solution turns yellow as most of the (PNP)NO$_2$ dissolves. After 10 minutes, all of the solid (PNP)NO$_2$ is gone and a clear yellow solution remains. The solvent is removed on a rotary evaporator to give bright yellow crystals of μ-nitrido-bis(triphenylphosphorus)(1+) tricarbonylnitrosylferrate(1-). The crude product is recrystallized from dichloromethane and diethyl ether to give an 87% yield of purified product, 527.3 mg (0.774 mmole).

Properties

μ-Nitrido-bis(triphenylphosphorus)(1+) tricarbonylnitrosylferrate(1-), (PNP)-[Fe(CO)$_3$(NO)], is obtained as bright-yellow, plate-like crystals. The solid is air stable, but solutions decompose slowly when exposed to air. It is soluble in tetrahydrofuran, dichloromethane, and acetonitrile and insoluble in diethyl ether and hexane. The IR spectrum of the compound contains two CO stretching vibrations (THF): 1982(m) and 1876(s) cm^{-1} and one NO stretching vibration (THF): 1650(m) cm^{-1}. (PNP)[Fe(CO)$_3$(NO)] itself is a useful reagent in condensation reactions to give carbonyl nitrosyl clusters.[6]

C. μ-NITRIDO-BIS(TRIPHENYLPHOSPHORUS)(1+) DECACARBONYL-μ-NITROSYL-TRIRUTHENATE(1-)

☐ **Caution.** *Owing to the high toxicity of carbon monoxide and ruthenium carbonyls, this reaction should be carried out in an efficient fume hood.*

A 40-mL Schlenk tube with magnetic stirring bar is charged with 74.8 mg (0.117 mmole) of $Ru_3(CO)_{12}$ and 94.0 mg (0.161 mmole) of $(PNP)NO_2$. If the $(PNP)NO_2$ is not already a free-flowing powder, it should be gently ground on weighing paper with a spatula. The reaction vessel is flamed out under evacuation and filled with nitrogen three times to remove oxygen. The solvent (17 mL THF) is added via syringe. Effervescence immediately follows, and the solution turns yellow-red. The mixture is stirred for 10 min and then the solvent is removed under vacuum to give a red-brown oil. The oil is triturated with 2 × 10 mL hexane to extract any unreacted $Ru_3(CO)_{12}$. The resulting yellow powder is dissolved in 6 × 10 mL diethyl ether and the solution filtered. The ether is removed under vacuum to give deep yellow crystals. The product is washed with hexane and dried in air.

Anal. Calcd. for $C_{46}H_{30}N_2P_2Ru_3O_{11}$: C, 47.96; H, 2.62; N, 2.43. Found: C, 47.73; H, 2.80; N, 2.38. A typical yield of the product is 121 mg (90%).

Properties

μ-Nitrido-bis(triphenylphosphorus)(1+) decacarbonyl-μ-nitrosyltriruthenate(1-), $(PNP)[Ru_3(CO)_{10}(NO)]$, is obtained as deep yellow, needle-like crystals. The solid is air stable, but solutions decompose in minutes when exposed to air. It is soluble in tetrahydrofuran, diethyl ether, acetone, and dichloromethane and insoluble in hexane and water. The IR spectrum of the compound contains the following CO stretching vibrations (THF): 2066(w), 2009(vs), 2000(s), 1982(s), 1945(s) cm^{-1}. The NO stretch is not normally observed because of cation absorptions but has been found to appear at 1479 cm^{-1}.

References

1. A. Martinsen and J. Songstad, *Acta Chem. Scand. A.*, **31**, 645 (1977).
2. R. E. Stevens, T. J. Yanta, and W. L. Gladfelter, *J. Am. Chem. Soc.*, **103**, 4981 (1981).
3. W. Hieber and H. Beutner, *A. Anorg. Allgem. Chem.*, **320**, 101 (1963).
4. N. G. Connelly and C. Gardner, *J. Chem. Soc., Dalton Trans.*, 1525 (1976).
5. A. Mantovani and S. Cenini, *Inorg. Synth.*, **16**, 47 (1976).
6. D. E. Fjare and W. L. Gladfelter, *J. Am. Chem. Soc.*, **103**, 1572 (1981).

36. METALLACYCLOPENTANE DERIVATIVES OF PALLADIUM

Submitted by P. DIVERSI,* G. INGROSSO,* and A. LUCHERINI*
Checked by K. A. ERNER† and T. H. TULIP†

So far, palladacyclopentane complexes, unlike the nickel and platinum analogs, have been relatively little explored, one of the reasons probably being the lack of general synthetic routes.

Described here is the synthesis of some palladacyclopentanes starting from easily accessible salts of palladium(II) with nitrogen and phosphorus ligands.

A. (1,4-BUTANEDIYL)[1,2-ETHANEDIYLBIS(DIPHENYLPHOSPHINE)]-PALLADIUM(II)

$$PdCl_2(diphos) + LiCH_2(CH_2)_2CH_2Li \longrightarrow \overline{PdCH_2(CH_2)_2CH_2}(diphos) + 2LiCl$$

Dichloro[1,2-ethanediylbis(diphenylphosphine)]palladium(II) is prepared according to the literature[1] by adding to K_2PdCl_4 (2 g, 6.127 mmole) in anhydrous N,N-dimethylformamide (20 mL) a solution of 1,2-ethanediylbis(diphenylphosphine) (diphos) (2.44 g, 6.127 mmole) in dichloromethane (12 mL). The resulting yellow suspension is refluxed for 2 hr, under stirring. Upon cooling and dilution with water, a white solid forms which is separated from the solution, washed with diethyl ether, and dried under vacuum. The yield is 3.34 g (96%), mp 326°.

All the following reactions and operations should be carried out under argon, and all solvents must be saturated with argon before use.

A Schlenk tube (100-mL), equipped with a magnetic stirrer, is charged with $PdCl_2(diphos)$ (0.5 g, 0.86 mmole) and freshly distilled diethyl ether (40 mL) previously dried over $LiAlH_4$. The Schlenk tube is then equipped with a pressure-equalizing dropping funnel containing 2.62 mmole (7.5 mL of 0.35 M) of 1,4-butane-diyldilithium in diethyl ether, prepared from 1,4-dichlorobutane and lithium by established methods.[2] After cooling the yellow suspension in a solid CO_2-acetone bath (-78°), the (1,4-butanediyl)dilithium solution is added dropwise with rapid stirring for about 1 hr. The suspension is stirred for 2 hr, then

---

*Istituto di Chimica Organica Industriale, Università di Pisa, Via Risorgimento 35, 56100. Pisa, Italy.

†Central Research and Development, E. I. du Pont de Nemours and Co., Wilmington, DE 19898.

the bath is removed, and the mixture is stirred for a further 3 hr, by which time it has warmed to room temperature and changed to a white (occasionally grayish) suspension. The solid is separated from the supernatant solution by filtration, then washed with diethyl ether (10 mL) and dried under vacuum. The solid is extracted with toluene (2 × 20 mL). Pentane (30 mL) is carefully added to the combined extracts in such a way to form a double layer. Very pale yellow crystals are slowly formed at the interface. The crystallization is completed overnight. The crystals are then separated from the mother liquor, washed with pentane, and dried under vacuum. The yield is 0.21 g (40%). Approximately 50 mg of additional material can be obtained by keeping the mother liquor at −30° for about 12 hr.

Anal. Calcd. for $C_{30}H_{32}P_2Pd$: C, 64.20; H, 5.75; P, 11.04. Found: C, 64.74; H, 5.65; P, 10.08.

Properties

(1,4-Butanediyl)[1,2-ethanediylbis(diphenylphosphine)]palladium(II) is an off-white crystalline compound moderately air stable as dry solid, but rather sensitive in solution. It can be stored for months in a refrigerator, under dinitrogen, the color slowly changing to light pink. The compound is slightly soluble in benzene, and readily soluble in pyridine. The ^{1}H-NMR spectrum (60 MHz, pyridine-d_5, δ in ppm downfield from TMS) shows two broad multiplets at 2.13 and at 2.66 (methylene protons of the palladacyclopentane moiety). The diphos ligand gives rise to signals at 2.31 (methylene protons, J_{PH} = 18 Hz), and 7.2–8.1 (phenyl protons). Other chemical properties are given in the literature.[3]

B. (1,4-BUTANEDIYL)(*N*,*N*,*N′*,*N′*,-TETRAMETHYL-1,2-ETHANEDIAMINE)-PALLADIUM(II)

$$PdCl_2(tmeda) + LiCH_2(CH_2)_2CH_2Li \longrightarrow \overline{PdCH_2(CH_2)CH_2}(tmeda) + 2LiCl$$

Procedure

Dichloro(*N*,*N*,*N′*,*N′*-tetramethyl-1,2-ethanediamine)palladium(II) is prepared by adding dropwise a solution of 0.355 g of *N*,*N*,*N′*,*N′*-tetramethyl-1,2-ethanediamine (tmeda) (3.06 mmole) in water (5 mL) to a stirred aqueous solution of $K_2[PdCl_4]$ (1 g, 3.06 mmole, in 20 mL of water.)[4] Upon addition, a deep-yellow solid forms; this suspension is stirred for 60 min and then the solid is separated by filtration, washed with diethyl ether, and dried under vacuum. The yield is 0.7 g (80%), mp 246° (decomp., darkening at 220°).

All the following reactions and operations are carried out under argon, and all solvents must be saturated with argon before use.

A 100-mL Schlenk tube, equipped with a magnetic stirrer and charged with $PdCl_2$(tmeda) (0.43 g, 1.46 mmole) and diethylether (40 mL) previously dried over $LiAlH_4$, is cooled in a dry ice/acetone bath ($-78°$). The mixture is treated dropwise, under stirring, with 2.92 mmole (9.7 mL of $0.3\,M$) of (1,4-butanediyl)-dilithium in diethyl ether, prepared from 1,4-dichlorobutane and lithium by established methods.[2] The reaction mixture is stirred for a further 30 min, then the temperature is slowly raised to room temperature. After being stirred for 1 hr at this temperature, the mixture is evaporated to dryness; this leaves an off-white residue which is extracted with pentane (50 mL). After filtration, the solution is cooled overnight to $-70°$ to give 0.2 g of the title compound (50%).

Anal. Calcd. for $C_{10}H_{24}N_2Pd$: C, 43.05; H, 8.60; N, 10.05; MW, 278.71. Found: C, 42.45; H, 8.49; N, 9.07; MW, 278 (mass spectrometry, for ^{106}Pd).

Properties

(1,4-Butanediyl)(N, N, N', N'-tetramethyl-1,2-ethanediamine)palladium(II) is a white crystalline solid, highly hygroscopic and air-sensitive, readily losing tmeda. It must be handled and stored under argon. It is moderately soluble in pentane and very soluble in benzene. The ^{1}H-NMR spectrum (60 MHz, benzene-d_6, δ in ppm downfield from TMS) must be recorded immediately after the preparation of the sample, at temperatures lower than 10°. It shows a broad multiplet ranging from 0.9-1.6 (methylene protons of the palladacyclopentane moiety), and resonances at 2.2-2.3 (methyl and methylene protons of tmeda). Other chemical properties are given in the literature.[3]

C. (1,4-BUTANEDIYL)BIS(TRIPHENYLPHOSPHINE)PALLADIUM(II)

$$\overline{PdCH_2(CH_2)_2CH_2}(tmeda) + 2P(C_6H_5)_3 \longrightarrow$$
$$\overline{PdCH_2(CH_2)_2CH_2}[P(C_6H_5)_3]_2 + tmeda$$

Procedure

All reactions and operations should be carried out under argon and all solvents must be saturated with argon before use.

To a pentane (10 mL) solution of $\overline{PdCH_2(CH_2)_2CH_2}$(tmeda) (0.1 g, 0.36 mmole) triphenylphosphine (0.19 g, 0.72 mmole) in pentane (10 mL) is added with rapid stirring, over a period of 3 min, at room temperature. Almost immediately, a pale-yellow crystalline solid forms. The mixture is stirred for 2 hr, then filtered, and the crystals are washed several times with pentane and dried under vacuum. The yield is 0.204 g (90%).

Anal. Calcd. for $C_{40}H_{38}P_2Pd$: C, 69.88; H, 5.57; P, 9.01. Found: C, 69.39; H, 5.48; P, 8.85.

Properties

(1,4-Butanediyl)bis(triphenylphosphine)palladium(II) is a cream-colored crystalline compound, moderately air-stable as a solid and quite unstable in solution. It is insoluble in pentane but soluble in benzene. Its ^{1}H-nmr spectrum (60 MHz, benzene-d_6, δ in ppm downfield from TMS) shows a complex multiplet centered at 2.33 (methylene protons of the palladacyclopentane moiety) and multiplets at 6.80-7.80 (aromatic protons).

D. (2,2'-BIPYRIDINE)(1,4-BUTANEDIYL)PALLADIUM(II)

$$\overline{PdCH_2(CH_2)_2CH_2}(\text{tmeda}) + \text{bpy} \longrightarrow \overline{PdCH_2(CH_2)_2CH_2}(\text{bpy}) + \text{tmeda}$$

Procedure

This preparation is carried out similarly to that of (1-4-butanediyl)bis(triphenylphosphine)palladium(II), using 0.1 g of $\overline{PdCH_2(CH_2)_2CH_2}$(tmeda) (0.36 mmole) in pentane (10 mL) and 56 mg of 2,2'-bipyridine (0.36 mmole) in pentane (10 mL). The reaction product, after washing with pentane, is dissolved in acetone (3 mL), and the filtered solution is kept overnight at −25° to give 0.105 g of the title compound (91%).

Anal. Calcd. for $C_{14}H_{16}N_2Pd$: C, 52.7; H, 5.05; N, 8.78. Found: C, 52.50; H, 4.95; N, 8.70.

Properties

(2,2'-Bipyridine)(1,4-butanediyl)palladium(II) is an orange crystalline air-stable solid. It can be stored for months under dinitrogen in a refrigerator without appreciable decomposition. It is insoluble in pentane, and soluble in acetone and chloroform. Its ^{1}H-NMR spectrum (60 MHz, CDCl$_3$, δ in ppm downfield from TMS) shows two multiplets at 1.72 and 2.25 (methylenes of the palladacyclopentane moiety) and multiplets at 7.42, 8.06, and 8.76 (bipy).

References

1. A. D. Westland, *J. Chem. Soc.*, 3060 (1965).
2. J. X. McDermott, J. F. White, and G. M. Whitesides, *J. Am. Chem. Soc.*, **98**, 6521 (1976).
3. P. Diversi, G. Ingrosso, A. Lucherini, and S. Murtas, *J. Chem. Soc., Dalton Trans.*, 1633 (1980).
4. F. G. Mann and H. R. Watson, *J. Chem. Soc.*, 2772 (1958).

37. METALLACYCLOPENTANE DERIVATIVES OF COBALT, RHODIUM, AND IRIDIUM

Submitted by P. DIVERSI,* G. INGROSSO,* and A. LUCHERINI*
Checked by K. A. ERNER† and T. H. TULIP†

A. $CoI_2(\eta^5\text{-}C_5H_5)[P(C_6H_5)_3] + BrMgCH_2(CH_2)_2CH_2MgBr \longrightarrow$

$\overline{CoCH_2(CH_2)_2CH_2}(\eta^5\text{-}C_5H_5)[P(C_6H_5)_3] + 2MgBrI$

B. $RhCl_2[\eta^5\text{-}C_5(CH_3)_5][P(C_6H_5)_3] + LiCH_2(CH_2)_2CH_2Li \longrightarrow$

$\overline{RhCH_2(CH_2)_2CH_2}[\eta^5\text{-}C_5(CH_3)_5][P(C_6H_5)_3] + 2LiCl$

C. $IrCl_2[\eta^5\text{-}C_5(CH_3)_5][P(C_6H_5)_3] + BrMgCH_2(CH_2)_2CH_2MgBr \longrightarrow$

$\overline{IrCH_2(CH_2)_2CH_2}[\eta^5\text{-}C_5(CH_3)_5][P(C_6H_5)_3] + 2MgBrCl$

The interest[1] in transition-metal metallacycles has led to an intensive study of a number of such complexes from the structural and chemical point of view. In this context, the availability of metallacyclopentane derivatives of the cobalt-triad metals, which are implicated in so many catalytic and stoichiometric reactions, enables one to extend the range of studies of the chemistry of this class of organometallics.

Described here is the preparation of (cyclopentadienyl)(triphenylphosphine)-metallacyclopentanes of cobalt(III), rhodium(III), and iridium(III).

A. (1,4-BUTANEDIYL)(η^5-CYCLOPENTADIENYL)-(TRIPHENYLPHOSPHINE)COBALT(III)

Procedure

All reactions and operations should be carried out under dinitrogen, and all solvents must be saturated with dinitrogen before use.

(η^5-Cyclopentadienyl)diiodo(triphenylphosphine)cobalt(III) is prepared by a two-step synthesis.[2-5]

■ **Caution.** *Carbon monoxide and metal carbonyls are highly toxic and this preparation should be performed in an efficient fume hood.*

*Istituto di Chimica Organica Industriale, Università di Pisa, Via Risorgimento 35, 56100 Pisa, Italy.

†Central Research and Development, E.I. duPont de Nemours and Co., Wilmington, DE 19898.

172 *Organometallic Compounds*

Dicarbonyl(η^5-cyclopentadienyl)cobalt (7.00 g, 38.9 mmole) is dissolved in diethyl ether (50 mL) which has been dried over Li[AlH$_4$]. To this solution, iodine (10.00 g, 39.4 mmole) in diethyl ether (200 mL) is added dropwise over a period of 2 hr. After the addition, the mixture is stirred for further 72 hr, at room temperature. The CoI$_2$(CO)(η^5-C$_5$H$_5$) so obtained (13.17 g, 83%) is dissolved in CH$_2$Cl$_2$ (300 mL) and treated dropwise over 1 hr with a solution of triphenylphosphine (9.28 g, 35.4 mmole) in CH$_2$Cl$_2$ (100 mL). The mixture is stirred overnight, at room temperature, then filtered, and worked up exactly as it is described in the literature:[2] 18.67 g of CoI$_2$(η^5-C$_5$H$_5$)[P(C$_6$H$_5$)$_3$] is obtained (90%).

μ-1,4-Butanediyl-bis(bromomagnesium)[6] is freshly prepared by adding dropwise over 1 hr a solution of 1,4-dibromobutane (4.10 g, 19 mmole) in tetrahydrofuran* (35 mL) to a suspension of magnesium turnings (2.00 g, 82.3 mmole) in tetrahydrofuran (10 mL) containing a small iodine crystal.

A 100-mL Schlenk tube, equipped with a magnetic stirrer, is charged with CoI$_2$(η^5-C$_5$H$_5$)[P(C$_6$H$_5$)$_3$] (0.4 g, 0.625 mmole) and freshly distilled diethyl ether (50 mL) which has been dried over LiAlH$_4$. The Schlenk tube is then equipped with a pressure-equalizing dropping funnel containing 1.875 mmole (7.5 mL of 0.25 M) of μ-1,4-butanediyl-bis(bromomagnesium) in tetrahydrofuran, as prepared above. The di-Grignard reagent is added dropwise to the stirred suspension at room temperature, over a period of 20 min. The mixture is stirred for further 3 hr, then filtered through a sintered-glass filter, and the resulting red solution is evaporated under vacuum leaving a dark red oil. This is treated with pentane (80 mL) with stirring, for 10 hr. The pentane extract is decanted and reduced to 8 mL, by evaporation under vacuum, and applied to a 12 × 1.3 cm chromatographic column filled with Merck alumina 90 (activity grade II-III) as absorbent (Fig. 1), and developed with pentane. The first pink band is collected and discarded. The second, orange, band is eluted by a solvent mixture of 6:1 pentane-diethyl ether. After evaporation to dryness, the residue is dissolved in pentane (2 mL) and cooled to -30° to give, overnight (occasionally a longer period is necessary), 41.1 mg of the title compound (15%).

Anal. Calcd. for C$_{27}$H$_{28}$CoP: C, 69.95; H, 7.27; P, 7.00. Found: C, 69.70; H, 7.29; P, 6.80.

Properties

(1,4-Butanediyl)(η^5-cyclopentadienyl)(triphenylphosphine)cobalt(III) is a red crystalline compound which remains unchanged when stored under an inert atmosphere, at -25°. It is soluble in common organic solvents. The ^{1}H-NMR spectrum (60 MHz, acetone-d_6, δ in ppm downfield from TMS) shows broad multi-

*All tetrahydrofuran used in this synthesis is purified by distillation from sodium potassium alloy and benzophenone.

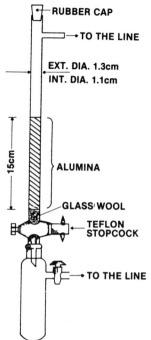

Fig. 1. Chromatographic column.

plets in the range 0.6–1.5 (methylenes), a singlet at 4.41 (cyclopentadienyl), and a broad multiplet at 7.38 (phenyls).

B. (1,4-BUTANEDIYL)(η^5-PENTAMETHYLCYCLOPENTADIENYL)- (TRIPHENYLPHOSPHINE)RHODIUM(III)

Procedure

Dichloro(η^5-pentamethylcyclopentadienyl)(triphenylphosphine)rhodium(III) is prepared by a two-step synthesis according to the literature.[7] Rhodium trichloride trihydrate (2.06 g, 7.81 mmole) and hexamethylbicyclo[2.2.0] hexa-2,5-diene (4.6 mL, 10 mmole) are refluxed in methanol (60 mL), under dinitrogen, to give 1.91 g (3.09 mmole) of di-μ-chlorobis[chloro(η^5-pentamethylcyclopentadienyl)rhodium(III)] (79%), which, by further reaction with triphenylphosphine (1.82 g, 6.94 mmole) in refluxing ethanol (80 mL), under dinitrogen, gives 3.36 g of $RhCl_2[\eta^5\text{-}C_5(CH_3)_5][P(C_6H_5)_3]$.

All the following reactions and operations must be carried out under argon, and all solvents must be saturated with argon before use. A Schlenk tube (100 mL), equipped with a magnetic stirrer and degassed through several pump and flush cycles, is charged with di-μ-chloro(η^5-pentamethylcyclopentadienyl)(triphenylphosphine)rhodium (0.3 g, 0.525 mmole) and Li[AlH$_4$]-dried freshly distilled diethyl ether (80 mL). The Schlenk-tube is then equipped with a pressure-equalizing dropping funnel containing 1.577 mmole (8.3 mL of 0.19 M) of 1,4-butanediyldilithium in diethyl ether, prepared from 1,4-dichlorobutane and lithium by established methods.[8] After cooling the suspension to 0°, 1,4-butanediyldilithium is added dropwise with a rapid stirring, over 30 min. Stirring is continued at 0° for 1.5 hr. The mixture is filtered through a sintered-glass filter, and the resulting orange-red solution is evaporated to dryness under vacuum. The residue is dissolved in 3 mL of a solvent mixture of 3:2 pentane-benzene (benzene is distilled from sodium-potassium alloy and benzophenone), applied to a 15 × 1.3-cm chromatographic column constructed with Merck alumina 90 (activity grade II–III) as the absorbent, and developed with pentane. (A solvent mixture of 1:2.5 benzene-pentane can be alternatively used as eluant to speed up the elution of the product.) A yellow band is collected as it elutes. After evaporation to dryness, an oil is obtained which is dissolved in 60 mL of pentane. The solution is reduced to 30 mL and cooled to –30° overnight to give 62 mg of the title compound (21%).

Anal. Calcd. for $C_{32}H_{38}PRh$: C, 69.06; H, 6.88; P, 5.56; MW, 556.53. Found: C, 69.13; H, 6.90; P, 5.40; MW, 556 (mass spectroscopy).

Properties

(1,4-Butanediyl)(η^5-pentamethylcyclopentadienyl)(triphenylphosphine)rhodium-(III) is a yellow crystalline solid that remains unchanged when stored under an inert atmosphere at –25°. It is soluble in common organic solvents. The ^{1}H-NMR spectrum (100 MHz, benzene-d_6, δ in ppm downfield from TMS) shows two broad multiplets centered at 1.38 and at 1.98 (methylenes), a doublet at 1.46 (methyl protons, J_{PH} = 2 Hz), and two multiplets in the range 7.14–7.60 (phenyl protons). The ^{13}C-NMR spectrum (25 MHz, benzene-d_6, δ in ppm downfield from TMS) shows signals at: 9.62 (singlet, methyls), 24.28 (double doublet) J_{PC} = 75 Hz, J_{RhC} = 13.24 Hz (α-methylenes of the rhodacyclopentane moiety), 34.21 (singlet, β-methylenes carbon atoms of the cyclopentadienyl ring), 127.12–135.86 (multiplet, phenyl carbon atoms). Other chemical properties are given in the literature.[9,10]

C. (1,4-BUTANEDIYL)(η^5-PENTAMETHYLCYCLOPENTADIENYL)-(TRIPHENYLPHOSPHINE)IRIDIUM(III)

Procedure

Dichloro(η^5-pentamethylcyclopentadienyl)(triphenylphosphine)iridium(III) is prepared by a two-step synthesis.[7] 1-(1-Chloroethyl)pentamethylcyclopentadiene (4.1 g, 20.6 mmole), obtained in 65% yield by exposing a CH_2Cl_2 solution (55 mL) of hexamethylbicyclo[2.2.0]hexa-2,5-diene (55 mL, 29.7 mmole) to a stream of dry gaseous hydrogen chloride over a period of 2 hr and stirred for a further 14 hr,[11] and iridium trichloride pentahydrate (4.15 g, 10.7 mmole) are refluxed in methanol (120 mL), under dinitrogen, to give 2.77 g of di-μ-chloro-bis[chloro(η^5-pentamethylcyclopentadienyl)iridium(III)] (3.48 mmole, 65%). Further reaction with triphenylphosophine (2.77 g, 10.57 mmole) in refluxing ethanol (140 mL), under dinitrogen, gives 3.97 g of $IrCl_2[\eta^5-C_5(CH_3)_5]$-$[P(C_6H_5)_3]$ (86%).

All the following reactions and operations are carried out under dinitrogen, and all solvents must be saturated with dinitrogen before use. A Schlenk tube (100 mL), equipped with a magnetic stirring bar, is charged with dichloro(η^5-pentamethylcyclopentadienyl)(triphenylphosphine)iridium(III) (0.3 g, 0.454 mmole) and tetrahydrofuran (70 mL). The suspension, maintained at room temperature, is treated dropwise over a period of 30 min, with 1.37 mmole (7.6 mL of 0.18 M) of μ-1,4-butanediyl-bis(bromomagnesium) in tetrahydrofuran prepared from 1,4-dibromobutane and magnesium turnings.[6] Stirring is continued for further 2.5 hr. The mixture is filtered through a sintered-glass filter, and the resulting yellow solution is evaporated to dryness under vacuum. The residue is extracted with a mixture of pentane (35 mL) and benzene (15 mL) for 2.5 hr, at room temperature. After filtration, a yellow solution is obtained which is dried under vacuum to leave a yellow-orange residue. This is dissolved in 3 mL of a solvent mixture of 1:1 pentane-benzene, applied to a 15 × 1.3-cm chromatographic column constructed with Merck alumina 90 (activity grade II-III) as the absorbent, and developed with a 1:1 pentane-benzene mixture as eluant. The product is collected as a very pale yellow band, as it elutes. After evaporation of the solvent an oil is obtained which is dissolved in pentane (60 mL); then the solution is reduced to 30 mL and cooled to $-30°$ to give overnight 69 mg of the title compound (23%).

Anal. Calcd. for $C_{32}H_{38}IrP$: C, 59.51; H, 5.93; P, 4.78; MW, 646.7. Found: C, 60.15; H, 6.04; P, 4.60; MW, 646 (mass spectroscopy).

Properties

(1,4-Butanediyl)(η^5-pentamethylcyclopentadienyl)(triphenylphosphine)iridium-(III) is a pale yellow crystalline product that remains unchanged when stored under an inert atmosphere at $-25°$. It is soluble in common organic solvents. The ^{1}H-NMR spectrum (100 MHz, benzene-d_6, δ in ppm downfield from TMS) shows two broad multiplets centered at 1.12 and at 1.96 (methylenes), a doublet at 1.45 (methyl protons, J_{PH} = 1.5 Hz), and two multiplets in the range 7.14-7.62 (phenyl protons). The ^{13}C-NMR spectrum (25 MHz, benzene-d_6, δ in ppm downfield from TMS) shows signals at: 5.68 (doublet, J_{PC} = 7.7 Hz, α-methylenes of the iridacyclopentane moiety), 8.92 (singlet, methyls), 35.21 (singlet, β-

methylenes of the iridacyclopentane moiety), 93.21 (singlet, carbon atoms of the cyclopentadienyl ring), 126.72–135.80 (multiplet, phenyl carbon atoms). Other chemical properties are given in the literature.[10]

References

1. G. Ingrosso, in *Reactions of Co-ordinated Ligands*, P. S. Braterman (ed.), Plenum Publishing Co. (in press).
2. R. B. King, *Inorg. Chem.*, **5**, 82 (1966).
3. T. S. Piper, F. A. Cotton, and G. Wilkinson, *J. Inorg. Nucl. Chem.*, **1**, 165 (1955).
4. R. B. King and F. G. A. Stone, *Inorg. Synth.*, **7**, 112 (1963).
5. M. D. Rausch and R. A. Genetti, *J. Org. Chem.*, **35**, 3888 (1970).
6. T. Yvernault, G. Casteignau, and M. Estrade, *C. R. Hebd. Sciences Acad. Sci., Ser. C*, **269**, 169 (1969).
7. J. W. Kang, K. Moseley, and P. M. Maitles, *J. Am. Chem. Soc.*, **91**, 5970 (1969).
8. J. X. McDermott, J. F. White, and G. M. Whitesides, *J. Am. Chem. Soc.*, **98**, 6521 (1976).
9. P. Diversi, G. Ingrosso, A. Lucherini, P. Martinelli, M. Benetti, and S. Pucci, *J. Organomet. Chem.*, **165**, 253 (1979).
10. P. Diversi, G. Ingrosso, A. Lucherini, W. Porzio, and M. Zocchi, *Inorg. Chem.*, **19**, 3590 (1980).
11. L. A. Paquette and G. R. Krow, *Tetrahedron Lett.*, 2139 (1968).

38. CYCLOOLEFIN COMPLEXES OF RUTHENIUM

Submitted by P. PERTICI* and G. VITULLI*
Checked by W. C. SPINK† and M. D. RAUSCH†

Cycloolefin–ruthenium complexes containing only olefins as ligands are normally difficult to prepare, and the yields are often very low. The classical methods involve Grignard reagents and UV irradiation or Ziegler catalysts.[1] The direct reaction of cycloolefins with hydrated ruthenium trichloride in the presence of metallic zinc offers a versatile and very useful reaction for the preparation of cycloolefin-ruthenium(0) and -ruthenium(II) complexes.[2]

The preparation by this route of the complexes (η^6-benzene)(η^4-1,3-cyclohexadiene)Ru(0), (η^4-1,5-cyclooctadiene)(η^6-1,3,5-cyclooctatriene)Ru(0), bis(η^5-2,4-cycloheptadienyl)Ru(II), and bis(η^5-cyclopentadienyl)Ru(II) is described here. These complexes are of interest in catalysis and in preparative chemis-

---

*Centro C.N.R. per le Macromolecole Stereoordinate ed Otticamente Attive, Istituto di Chimica Organica, Università di Pisa, via Risorgimento 35, 56100 Pisa, Italy.
†Department of Chemistry, University of Massachusetts, Amherst, MA 01003.

try,[3] and the availability of adequate amounts could stimulate studies of their chemistry.

All the operations described here must be carried out under a rigorously dry, oxygen-free nitrogen atmosphere, using conventional Schlenk-tube techniques.[4] Solvents should be purified by conventional methods, distilled and stored under nitrogen.

A. (η^6-BENZENE)(η^4-1,3-CYCLOHEXADIENE)RUTHENIUM(0)

$$2RuCl_3 + 6C_6H_8 + 3Zn \xrightarrow{\text{ethanol}} 2[Ru(C_6H_6)(C_6H_8)] + 3ZnCl_2 + 2C_6H_{10}$$

Procedure

In a 100-mL, two-necked, round-bottomed flask equipped with a nitrogen inlet and a magnetic stirring bar, 0.320 g (1.23 mmole) of hydrated ruthenium trichloride, obtained as $RuCl_3 \cdot 3H_2O$, is completely dissolved, under nitrogen, in 8 mL of absolute ethanol. 1,3-Cyclohexadiene previously distilled and stored under nitrogen (5 mL, 52 mmole) and zinc dust (3 g, 45 mmole) are added in that order, and the mixture is stirred for 3 hr at room temperature.

The resulting yellow-brown solution is filtered, and the solid residue is washed with benzene (50 mL). The collected solution is evaporated to dryness under reduced pressure and the solid residue obtained is extracted with pentane (2 × 60 mL). The light-yellow pentane solution is filtered. (As an alternative procedure, the pentane extract is concentrated and chromatographed with pentane on an alumina column, 1 × 20 cm, Merck product, activity II–III.) Reduction of the volume of the solvent, under vacuum, to about 5 mL and cooling to −78° give light-yellow crystals of $Ru(C_6H_6)(C_6H_8)$. The yield is 0.225 g (70% based on $RuCl_3 \cdot 3H_2O$). Mp, 118–119°.

Anal. Calcd. for $C_{12}H_{14}Ru$: C, 55.5; H, 5.4; M, 259.2. Found: C, 56.1; H, 5.5; M, 260.

Properties

(η^6-Benzene)(η^4-1,3-cyclohexadiene)ruthenium(0) is an air-sensitive light-yellow solid, soluble in the common organic solvents; it slowly decomposes in halogenated solvents. The complex can be purified by sublimation at 80°/0.1 torr and may be stored for a prolonged period of time under nitrogen.

The ^{1}H-NMR spectrum, in acetone-d_6 solution, shows resonances at 5.40(s, 6H), 4.75(dd, 2H), 3.02(m, 2H), and 1.39(m, 4H) ppm.

B. (η^4-1,5-CYCLOOCTADIENE)(η^6-1,3,5-CYCLOOCTATRIENE)-RUTHENIUM(0)

$$2RuCl_3 + 6C_8H_{12} + 3Zn \xrightarrow{\text{ethanol}} 2[Ru(C_8H_{10})(C_8H_{12})] + 3ZnCl_2 + 2C_8H_{14}$$

Procedure

In a 100-mL, three-necked, round-bottomed flask equipped with a magnetic stirring bar, a nitrogen inlet, and a water-jacketed condenser, 0.34 g (1.3 mmole) of hydrated ruthenium trichloride is completely dissolved in absolute ethanol (10 mL) under nitrogen. 1,5-Cyclooctadiene* (10 mL, 81 mmole) and zinc dust (3.0 g, 46 mmole) are added in that order, and the mixture is gently heated at 80° for 3 hr. Care should be taken as refluxing for a longer period results in significant isomerization to bis(η^5-1,5-cyclooctadienyl)ruthenium(II). The resulting yellow-brown solution is filtered and the residue is washed with benzene (50 mL). The filtrate is evaporated to dryness under reduced pressure at room temperature and the solid residue obtained extracted with pentane (2 × 60 mL).

The orange-yellow pentane solution is concentrated and chromatographed with pentane on an alumina column (1 × 20 cm; activity II-III). A yellow band is removed and collected, while traces of bis(η^5-1,5-cyclooctadienyl)ruthenium-(II) and small amounts of decomposition products remain on the top of the column. The yellow pentane solution is concentrated to 5 mL and cooled to -78° giving yellow crystals of [Ru(C$_8$H$_{10}$)(C$_8$H$_{12}$)]. The yield is 0.20 g (50% based on RuCl$_3$ · 3H$_2$O). Mp, 92-94°.

Anal. Calcd. for $C_{16}H_{22}Ru$: C, 60.1; H, 6.98; M, 315.4. Found: C, 59.53; H, 7.10; M, 316.

Properties

(η^4-1,5-Cyclooctadiene)(η^6-1,3-5-cyclooctatriene)ruthenium(0) is a yellow crystalline solid which is unstable toward atmospheric oxygen. When pure, the complex may be stored for a prolonged period of time under nitrogen at 0°. The compound is soluble in the common hydrocarbon solvents, ethanol, and tetrahydrofuran, but it decomposes in halogenated solvents. The ^{1}H-NMR spectrum, in benzene-d_6 solution, shows resonances at 5.22(dd, 2H), 4.78(m, 2H), 3.79(m, 2H), 2.92(m, 4H), 2.22(m, 8H), 1.64(m, 2H), and 0.90(m, 2H) ppm.

The complex is a useful starting material for the preparation of other ruthenium complexes and is of interest in catalysis.[3a,c] By heating at 100° in hydro-

*Merck or Aldrich, distilled and stored under nitrogen.

carbon solvents, the complex isomerizes to the corresponding bis(η^5-1,5-cyclooctadienyl)ruthenium(II).[2b]

C. BIS($\eta^5$2,4-CYCLOHEPTADIENYL)RUTHERNIUM(II)

$$2RuCl_3 + 6C_7H_{10} + 3Zn \xrightarrow{ethanol} 2[Ru(C_7H_9)_2] + 3ZnCl_2 + 2C_7H_{12}$$

Procedure

In a 100-mL three-necked round-bottomed flask equipped with a magnetic stirring bar, a nitrogen inlet, and a water-jacketed condenser, 0.34 g (1.3 mmole) of hydrated ruthenium trichloride is completely dissolved, under nitrogen, in absolute ethanol (8 mL). 1,3-Cycloheptadiene* (5.0 mL, 46 mmole) and zinc dust (3.0 g, 46 mmole) are added in that order, and the mixture is heated at 80° for 1 hr.

The resulting yellow-brown solution is filtered, and the solid residue is washed with benzene (50 mL). The collected solution is evaporated to dryness under reduced pressure, and the solid residue obtained is extracted with pentane (2 × 60 mL).† The yellow pentane solution is concentrated under vacuum to about 6 mL and cooled to −78° to give yellow crystals of [Ru(C$_7$H$_9$)$_2$]. The yield is 0.19 g (50% based on RuCl$_3$ · 3H$_2$O). Mp, 43–44°.

Anal. Calcd. for C$_{14}$H$_{18}$Ru: C, 58.55; H, 6.25; M, 287.30. Found: C, 57.95; H, 6.25; M, 288.

Properties

Bis(η^5-2,4-cycloheptadien-1-yl)ruthenium(II) is a yellow crystalline solid which is unstable toward atmospheric oxygen. The compound is soluble in common hydrocarbon solvents, ethanol, and tetrahydrofuran. It decomposes slowly in halogenated solvents.

The ^{1}H-NMR spectrum, in benzene-d_6 solution, shows resonances at 5.03 (t, 2H), 4.40 (dd, 4H), 3.90 (m, 4H), and 2.47–1.37 (bm, 8H) ppm.

---

*Can be obtained from Aldrich Chemical or may be prepared by bromination of cycloheptene with Br$_2$ in CCl$_4$ and dehydrobromination of the 1,2-dibromocycloheptane with KOH in triethylene glycol at 200°.[5,6]

†As an alternative procedure the pentane extract may be concentrated and chromatographed with benzene on an alumina column (1 × 20 cm; activity II–III). The yellow band is collected and the solvent completely removed under reduced pressure. The yellow solid obtained is recrystallized from pentane.

D. BIS(η^5-CYCLOPENTADIENYL)RUTHENIUM(II) (RUTHENOCENE)

$$2RuCl_3 + 6C_5H_6 + 3Zn \xrightarrow{\text{ethanol}} 2[Ru(C_5H_5)_2] + 3ZnCl_2 + 2C_5H_8$$

Procedure

In a 100-mL, two-necked round-bottomed flask equipped with a nitrogen inlet and a magnetic stirring bar, 0.25 g (0.95 mmole) of hydrated ruthenium trichloride is completely dissolved, under nitrogen, in 7 mL of absolute ethanol. Freshly distilled cyclopentadiene, prepared by literature methods,[8] (7.0 mL, 84.7 mmole) and zinc dust (2.5 g, 28.2 mmoles) are added in that order and the mixture stirred for 2 hr at room temperature. The resulting yellow-brown solution is filtered, and the solid residue is washed with benzene (50 mL). The filtrate is evaporated to dryness under reduced pressure, and the solid residue obtained is extracted with pentane (2 × 60 mL).

The light-yellow pentane solution is filtered. If the solution is not completely clear, it is concentrated and chromatographed with benzene on an alumina column. Reduction of the volume of the solvent, under vacuum, to about 5 mL and cooling at $-78°$ gives light-yellow crystals of $[Ru(C_5H_5)_2]$. The yield is 0.16 g (75% based on $RuCl_3 \cdot 3H_2O$). Mp, 199–200°.

Anal. Calcd. for $C_{10}H_{10}Ru$: C, 52.0; H, 4.35; M, 231.19. Found: C, 52.95; H, 4.60; M, 232.

Properties

Bis(η^5-cyclopentadienyl)ruthenium(II) is a pale-yellow solid, soluble in the common organic solvents.

The ^{1}H-NMR spectrum, in benzene-d_6 solution, shows a resonance at 4.6(s) ppm. Other physical properties of the complex are given in the literature.[7]

References

1. (a) D. Jones, L. Pratt, and G. Wilkinson, *J. Chem. Soc.*, 4458 (1962). (b) J. Muller, C. G. Kreiter, B. Mertschenk, and S. Schmitt, *Chem. Ber.*, **108**, 273 (1975) and refs. therein.
2. (a) G. Vitulli, P. Pertici, and L. Porri, Italian Patent 23725 A/75. (b) P. Pertici, G. Vitulli, M. Paci, and L. Porri, *J. Chem. Soc. Dalton Trans.*, 1961 (1980).
3. (a) P. Pertici, G. P. Simonelli, G. Vitulli, G. Deganello, P. L. Sandrini, and A. Mantovani, *J. Chem. Soc. Chem. Commun.*, 123 (1977). (b) A. Lucherini and L. Porri, *J. Organomet Chem.*, **155**, C45 (1978). (c) P. Pertici, G. Vitulli, W. Porzio, and M. Zocchi, *Inorg. Chim. Acta Lett.*, **37**, L521 (1979).

4. D. F. Shriver, *Manipulation of Air-Sensitive Compounds.* McGraw-Hill, New York, 1969.
5. P. W. Henninger, E. Wapenaar, and E. Havinga, *Rec. Trav. Chim.*, **81**, 1053 (1962).
6. D. Pocar, *Reazioni Organiche. Teoria e practica.* Casa Editrice Ambrosiana, Milano, 1966, p. 240.
7. D. E. Bublitz, W. E. McEwen, and J. Kleinberg, *Org. Synth.*, **5**, 1001 (1973).
8. R. B. Moffett, *Org. Synth.*, **4**, 238 (1963).

39. MONONUCLEAR PENTACARBONYL HYDRIDES OF CHROMIUM, MOLYBDENUM, AND TUNGSTEN

Submitted by MARCETTA Y. DARENSBOURG* and SYDNEY SLATER*
Checked by MARK A. PETRINOVIC† and JOHN E. ELLIS†

A. $(PNP)[BH_4] + [W(CO)_5NMe_3] \longrightarrow (PNP)[W(CO)_5(HBH_3)] + NMe_3 \longrightarrow$
$(PNP)[HW(CO)_5] + Me_3NBH_3$

B. $[Cr(CO)_5pip] + 2Na[C_{10}H_8] \longrightarrow Na_2[Cr(CO)_5] + pip + 2C_{10}H_8$

$Na_2[Cr(CO)_5] + (PNP)I + CH_3OH \longrightarrow (PNP)[HCr(CO)_5] + NaI + NaOCH_3$

The $[HM(CO)_5]^-$ (M = VIB metal) anions, isoelectronic with the neutral $HM(CO)_5$ (M = Mn and Re) and analogous to the easily prepared $[HFe(CO)_4]^-$,[1] aggregate with coordinatively unsaturated $M(CO)_5^0$ units with the greatest facility to yield the well known $[\mu\text{-}HM_2(CO)_{10}]^-$ anions.[2] Effective syntheses of the mononuclear hydrides must minimize the production of such $M(CO)_5^0$ fragments. The following two methods have proven successful. The first involves the interaction of $(PNP)[BH_4]$ ($(PNP)^+ = \mu$-nitrido-bis(triphenylphosphorous)-(1+)) with $M(CO)_5$(amine) complexes. It is limited to the more stable hydrides, M = Cr and W. The second involves protonation of $[M(CO)_5]^{2-}$ and is general for all three VIB metals. The two methods presented possess the following advantages over those available in the literature:[3,4]

1. The products are obtained in high yield (70-80%).
2. The starting materials are inexpensive and easily available.

*Department of Chemistry, Tulane University, New Orleans, LA 70118; current address (MYD), Department of Chemistry, Texas A & M University, College Station, TX 77843.
†Department of Chemistry, University of Minnesota, Minneapolis, MN 55455.

3. Although rigorous exclusion of air throughout all manipulations is imperative, common Schlenk apparatus and techniques can be used.
4. No unsavory by-products are produced.

■ **Caution.** *Volatile metal carbonyls are highly toxic and should be handled in the open only in a hood.*

A. µ-NITRIDO-BIS(TRIPHENYLPHOSPHORUS)(1+) PENTACARBONYLHYDRIDOTUNGSTEN(1−)

Procedure

All operations are carried out under nitrogen, with rigorous exclusion of air and moisture. All solvents used are dried and deoxygenated by distillation from sodium/benzophenone. The products are oxidized by air, consequently they should be handled under an inert gas environment at all times.

To a 24/40 ℑ 500-mL Schlenk flask fitted with magnetic stirring bar and a straight vacuum adapter (with a small rubber septum wired over the inlet) is added 2.13 g (5.56 mmole) of $[W(CO)_5(NMe_3)]$[5,6] and 3.1 g (5.56 mmole) of $(PNP)[BH_4]$.[7,8] The flask is then evacuated and backfilled with nitrogen. Three hundred mL of tetrahydrofuran (THF) is added via cannula or syringe through the vacuum adapter. The resulting mixture is stirred for 1.5 hr, after which the volume is reduced to 60 mL. Diethyl ether (100 mL) is then added to precipitate any unreacted $(PNP)[BH_4]$ and the mixture transferred by cannula into a Schlenk filter funnel containing Celite. The filtered diethyl ether/tetrahydrofuran mixture is concentrated to 50 mL, and 100 mL of hexane is added to effect precipitation of a fine yellow powder. The produce is allowed to settle and the supernatant removed by cannula. After being dried under vacuum at room temperature, the solid is redissolved in 50 mL of tetrahydrofuran, and the process described above (addition of diethyl ether, filtration, reduction of solvent volume, and addition of hexane) is repeated. The yellow powder isolated from this second purification is dried and weighed (4.0 g, 84% yield).

Acceptable elemental analyses on such samples are obtained; the minor contaminant of $[\mu\text{-}HW_2(CO)_{10}]^-$ is generally present in <2% amounts as determined by infrared (for $[\mu\text{-}HW_2(CO)_{10}]^-$, the major $\nu(CO)$ band is at 1945 cm^{-1} tetrahydrofuran solution). Well-developed crystals of the mononuclear hydride may be obtained by the diffusion of hexane vapor into a concentrated tetrahydrofuran solution of the hydride.

The chromium analog is prepared in precisely the same manner with similarly high yields of pure product. Attempts to use this synthetic approach for M = Mo however leads to a mixture of $[\mu\text{-}HMo_2(CO)_{10}]^-$ and $[(H_2BH_2)Mo(CO)_4]^-$.[8]

Two final points are of import: (1) similar results have been obtained by using $M(CO)_5(\text{piperidine})$[8] (M = Cr, W) as a starting material; and (2) the scaling up of the reaction must be accompanied by at least an equivalent increase in solvent, for solutions of greater concentration than used above lead to increased amounts of the $[\mu\text{-HM}_2(CO)_{10}]^-$ impurity.

B. μ-NITRIDO-BIS(TRIPHENYLPHOSPHORUS)(1+) PENTACARBONYLHYDRIDOCHROMIUM(1-)

The cannula and/or syringe techniques used to transfer solvents or solutions as described above are also employed in this synthesis. Sixty mL of tetrahydrofuran is added to an N_2-filled Schlenk flask equipped with vacuum adapter and stirring bar, and containing 1.6 g (5.8 mmole) of $[Cr(CO)_5pip]$,[9] (pip = piperidine). The resulting solution is cooled to $-78°$. A 0.2 M naphthalene/sodium solution (prepared by mixing stoichiometric amounts of *very clean* $Na°$ and naphthalene in the requisite amount of THF) is added by increments with stirring until the reaction solution maintains a dark green color.[6] With the reaction mixture still at $-78°$, a solution of (PNP)I (3.85 g, 5.8 mmole) in 20 mL MeOH is added. The color of the reaction solution is now yellow. After warming the solution to room temperature, the solvent is removed under vacuum. To the residue is added 40 mL of diethyl ether. The tetrahydrofuran/diethyl ether mixture is then filtered through Celite and concentrated to approximately 20 mL. Addition of 60 mL hexane gives the product as a yellow oil which solidifies on stirring. The solid is obtained by cannulation of the mother liquor away from the precipitate and is then dried under vacuum. Yield: 3.0 g (70%).* Crystals can be obtained by the diffusion of hexane vapors into a tetrahydrofuran solution of the hydride.

The pure tungsten analog is similarly prepared in this manner. Furthermore, a number of other ion-exchange reagents (e.g., $(Ph_4P)Br$, $(Ph_4As)Cl$, $(Et_4N)Cl$, (PNP)Cl) may be substituted for the (PNP)I. Alkali metal salts (Na^+, K^+) may be isolated by omitting the ion-exchange reagent in the protonation step. They are less stable salts and are difficult to obtain as solids.

The molybdenum analog is obtained pure by this method if extreme precaution is taken to exclude O_2 and moisture. In addition, the temperature of the solutions containing $[HMo(CO)_5]^-$ should never exceed $0°C$.

Properties

Spectroscopic constants of the hydrides are listed in Table I.[10] They are all yellow to yellow-orange solids which are air- and moisture-sensitive. They can

---

*The checkers also found that an alternative route[3] to $[HCr(CO)_5]^-$ involving the protonation of $Na_2Cr(CO)_5$, generated directly from the sodium reduction of $Cr(CO)_6$ in liquid ammonia, provides approximately the same yield of $(PNP)[HCr(CO)_5]$ reported herein.

TABLE I Spectral Properties of (PNP)[HM(CO)$_5$]

	Infrared[a] ν(CO)			^{1}H-NMR[b] δ(M—H)	^{13}C-NMR[b] δ(^{13}CO) (←TMS)	
M	A$_1^2$(vw)	E(s)	A$_1^1$(m)	(TMS→)	cis	trans
Cr	2016	1884	1856	−6.92	— 227.9[c]	228.7 231.5[c]
Mo	2030	1893	1858	−4.0	215.5	219.6
W	2029	1889	1859	−4.2[d]	205.3	209.6

[a] Tetrahydrofuran solution, measured on a Perkin-Elmer 283B infrared spectrophotometer.
[b] CD$_3$CN solution, Me$_4$Si reference, measured on a Varian EM-390 spectrometer (^{1}H-nmr) and a JEOL FX60 (^{13}C-nmr) at ambient temperatures except where noted.
[c] <0°C.
[d] J_{183_W-H} = 53.4 Hz.

be stored under N$_2$ or under vacuum apparently indefinitely; refrigeration is recommended. The tungsten and chromium hydrides are stable for days in solution at room temperature. However, the molybdenum analog decomposes at temperatures >0°, giving [μ-HMo$_2$(CO)$_{10}$]$^-$ (characteristic ν(CO) band at 1942 cm^{-1}; characteristic δ(hydride) = −12.0 ppm).[2]

The complexes are very soluble in tetrahydrofuran and acetonitrile, sparingly soluble in diethyl ether, and insoluble in hydrocarbon solvents. At elevated temperatures, decomposition occurs in acetonitrile.

References

1. M. Y. Darensbourg, D. J. Darensbourg, and H. L. C. Barros, *Inorg. Chem.*, **17**, 297 (1978).
2. R. G. Hayter, *J. Am. Chem. Soc.*, **88**, 4376 (1966).
3. H. Behrens and R. Weber, *Z. Anorg. Allgem. Chem.*, **241**, 122 (1957); H. Behrens and J. Vogl, *Z. Anorg. Allgem. Chem.*, **291**, 123 (1957); H. Behrens and J. Vogl, *Chem. Ber.*, **96**, 2220 (1963).
4. M. Y. Darensbourg and J. C. Deaton, *Inorg. Chem.*, **20**, 1644 (1981).
5. The W(CO)$_5$(NMe$_3$) was prepared by the reaction of Me$_3$NO with W(CO)$_6$ in the presence of excess NMe$_3$ as described in Reference 6.
6. N. J. Cooper, J. N. Maher, and R. P. Beatty, *Organometallics*, **1**, 215 (1982).
7. The (PPN)$^+$ [BH$_4$]$^-$ was prepared by the neutral ion exchange between (PPN)$^+$ Cl$^-$ and NaBH$_4$ as described in Reference 8.
8. S. W. Kirtley, M. A. Andrews, R. Bau, G. W. Grynkewich, T. J. Marks, D. L. Tipton, and B. R. Whittlesey, *J. Am. Chem. Soc.*, **99**, 7154 (1977).
9. D. J. Darensbourg and M. A. Murphy, *Inorg. Chem.*, **17**, 884 (1978). A nonphotochemical synthesis is described in G. Boxhoorn, G. C. Schoemker, D. J. Stufkens, A. Oskam, A. J. Rest, and D. J. Darensbourg, *Inorg. Chem.*, **19**, 3455 (1980).
10. M. Y. Darensbourg and S. Slater, *J. Am. Chem. Soc.*, **103**, 5914 (1981).

Chapter Five
COMPOUNDS OF BORON

40. [(2,2-DIMETHYLPROPANOYL)OXY]DIETHYLBORANE, ((PIVALOYLOXY)DIETHYLBORANE)

Submitted by R. KÖSTER* and G. SEIDEL*
Checked by JAMES E. MCGUIRE, Jr.† and SHELDON G. SHORE†

$$(C_2H_5)_3B + (CH_3)_3CC(OH)=O \longrightarrow (C_2H_5)_2BOCC(CH_3)_3=O + C_2H_6$$

The colorless, vacuum-distillable liquid [(2,2-dimethylpropanoyl)oxy]diethylborane (2,2-dimethylpropanoyl = pivaloyl) is a catalyst for the introduction of diethylboryl groups onto the oxygen atoms of hydroxy compounds using triethylborane. Diethylborylations of hydroxy compounds have both preparative[1-9] and analytical (determination of the ethane value EZ[6,9,10]) uses. [(2,2-Dimethylpropanoyl)oxy]diethylborane is itself a reagent for preparing aldol-condensation products[11] and for determining analytical pivalate values (PZ) of H-acidic compounds.[10]

*Max-Planck-Institut für Kohlenforschung, Kaiser-Wilhelm-Platz, D-4330 Mülheim/Ruhr, Germany.
†Department of Chemistry, Ohio State University, Columbus, OH 43210.

(Acyloxy)dialkylboranes are best prepared from carboxylic acids with trialkylboranes.[1,12-16] Other methods involve reactions of halodiorganoboranes with carboxylic acids[17] or with acetic anhydride.[18] Alternative syntheses from acetic anhydride using dialkyl(alkylthio)boranes or dialkylamino compounds have also been described.[19]

Pure [(2,2-dimethylpropanoyl)oxy]diethylborane is prepared from triethylborane and 2,2-dimethylpropanoic acid (pivalic acid) at room temperature.

The reactions can be carried out without solvent under cooling. The course of the reaction can easily be followed quantitatively by measuring the amount of liberated ethane. An excess of pivalic acid should be avoided.

■ **Caution.** *[(2,2-dimethylpropanoyl)oxy]diethylborane reacts vigorously with water. During the preparation an inert liquid (e.g., Aliphatin*) should be used for cooling (bath, reflux condenser).*

Procedure

A 2-L three-necked flask is fitted with a reflux condenser capped with an inert gas bypass, a 500-mL dropping funnel surrounded by a heating jacket, an inside thermometer, and a magnetic stirrer. The apparatus is evacuated, filled with argon and charged with 510 g† (5.2 mole) triethylborane.‡

■ **Caution.** *Triethylborane is pyrophoric. It should be handled with extreme caution and only by personnel familiar with its physical and chemical properties. Before using, the experimentalist should consult "Trialkylboranes. Technical and Handling Bulletin," Callery Chemical Company, April, 1978 or a similar document, which describes the properties of triethylborane and the precautions to be taken.*

Pivalic acid (warmed to 50°) is added dropwise in approximately 15 hr to the stirred triethylborane at 5-10° (cooling bath). After 111 L (99%, STP) ethane has been evolved, the reaction is complete. Vacuum distillation gives ~22 g forerun (bp ≤20°/13 torr, mainly triethylborane), 816 g (96%) colorless liquid product (^{1}H-NMR spectrum), bp = 62°/13 torr, and 18 g of yellow residue.

Properties

In contrast to most of the other low-molecular-weight (acyloxy)diethylboranes, [(2,2-dimethylpropanoyl)oxy]diethylborane is a liquid. The vacuum distillable liquid (bp = 52-54°/9 torr, bp = 62°/13 torr) must be stored under an inert gas (Ar, N_2) because of its high air and moisture sensitivity. [(2,2-Dimethylpropanoyl)oxy]diethylborane is thermally stable to ~100°.

*Aliphatin = mixture of saturated hydrocarbons with bp ≈190°; available from Esso A. G., D-2000 Hamburg, Germany.

†Available from Callery Chem. Co., Callery, PA 16024, or prepared by the method in Reference 21.

‡The checkers used 100 g of triethylborane and obtained a quantitative yield of product.

At room temperature, some of the compound is dimeric. Accordingly, the $\delta_{^{11}B}$ signal is concentration- and temperature-dependent. Undiluted [(2,2-dimethylpropanoyl)oxy]diethylborane has $\delta_{^{11}B}$ = 38 ppm (half width = 290 Hz) at 20° and $\delta_{^{11}B}$ = 54.6 ppm (half width = 75 Hz) at 84°. The characteristic ^{1}H-NMR signals (60 MHz, neat) are $\delta_{^1H}$ = 1.22 (s) and 0.87 (m) in the ratio 9:10; ^{13}C-NMR 15.2 (C-1), 7.75 (C-2), (C-3), 40.02 (C-4), and 27.18 (C-5) ppm.

$$H_3C^2-H_2C^1\diagdown\quad\quad\overset{O}{\overset{\|}{C^3}}\quad\overset{C^3H_3}{\underset{|}{\;}}$$
$$\quad\quad\quad\quad B-O\diagup\quad\diagdown\underset{|}{C^4}-C^5H_3$$
$$H_3C^2-H_2C^1\diagup\quad\quad\quad\quad C^5H_3$$

References

1. R. Köster, H. Bellut, and W. Fenzl, *Liebigs Ann. Chem.*, 54 (1974).
2. R. Köster, K.-L. Amen, H. Bellut, and W. Fenzl, *Angew. Chem.*, **83**, 805 (1971); *Angew. Chem. Int. Ed. Engl.*, **10**, 748 (1971).
3. R. Köster and W. Fenzl, *Liebigs Ann. Chem.*, 69 (1974).
4. R. Köster and W. Rothgery, *Liebigs Ann. Chem.*, 112 (1974).
5. R. Köster, W. Fenzl, and G. Seidel, *Liebigs Ann. Chem.*, 752 (1975).
6. R. Köster, K.-L. Amen, and W. V. Dahlhoff, *Liebigs Ann. Chem.*, 752 (1975).
7. W. V. Dahlhoff and R. Köster, *Liebigs Ann. Chem.*, 1625 (1975).
8. W. V. Dahlhoff and R. Köster, *Liebigs Ann. Chem.*, 1914 (1975).
9. R. Köster, *Pure Appl. Chem.*, **49**, 765 (1977).
10. R. Köster and A.-A. Pourzal, *Synthesis*, 674 (1973).
11. A.-A. Pourzal, Dissertation Univ. Bochum 1974; *Meth. Org. Chem.*, (Houben-Weyl) Bd. 7/2b, pp. 1456, 1482, 1496, 1497, 1500, 1501 (1976).
12. H. Meerwein and H. Sönke, *J. Prakt. Chem.*, **147**, 251 (1937).
13. J. Goubeau and H. Lehmann, *Z. Anorg. Allgem. Chem.*, **322**, 224 (1963).
14. *Gmelin Handbuch der Anorganischen Chemie* 8. Aufl. Ergänzungswerk Bd. 48/16, pp. 166–167 (1977).
15. L. H. Torporcer, R. E. Dessy, and S. I. E. Green, *J. Chem. Soc.*, 1236 (1965).
16. G. Seidel, Studienarbeit, Fachhochschule Aachen-Jülich, 1972.
17. W. Gerrard, M. F. Lappert, and R. Shafferman, *J. Chem. Soc.*, 3648 (1958).
18. B. M. Mikhailov and N. S. Fedotov, *Izv. Akad. Nauk SSSR Otdel. Khim Nauk*, 857 (1958); C. A. **53**, 1203 (1959).
19. V. A. Doroklov and B. M. Mikhailov, *Izv. Akad. Nauk SSSR*, 1804 (1970).
20. G. Baum, *J. Organomet. Chem.*, **22**, 269 (1970).
21. R. Köster, P. Binger, and W. V. Dahlhoff, *Synth. Inorg. Metal-Org. Chem.*, **3**, 359 (1973).

41. TETRAETHYLDIBOROXANE

Submitted by W. FENZL* and R. KÖSTER*
Checked by GOJI KODAMA† and MAMORU SHIMOI†

$$2(C_2H_5)_3B + H_2O \xrightarrow[-2C_2H_6]{\text{cat.}} [(C_2H_5)_2B]_2O$$

Tetraethyldiboroxane is a widely usable solvent, for example, for certain metal salts, in which reactions with aprotic compounds proceed smoothly. Tetraethyldiboroxane is also a reagent for O-diethylborylating hydroxy compounds and can be used for increasing the reactivity of hydroxy-element compounds.

Tetraalkyldiboroxanes[1] are anhydrides of dialkylhydroxyboranes (borinic acids), from which they can be easily prepared by dehydration.

$$2R_2BOH \longrightarrow R_2BOBR_2 + H_2O$$

Although tetraalkyldiboroxanes having long alkyl groups are readily prepared by dehydration of dialkylhydroxyboranes, this method is not suited for the preparation of tetraethyldiboroxane, as much product is lost during the removal of water and the distillative separation.

Pure tetraethyldiboroxane[2] is smoothly prepared from triethylborane and water in the presence of 2,2-dimethylpropanoic acid or [(2,2-dimethylpropanoyl)-oxy]diethylborane. Trialkylboranes with long alkyl groups react to give quantitative yields of tetraalkyldiboroxanes. However, the yield of tetraethyldiboroxane is relatively low because of the unavoidable losses of water and volatile ethylboranes. The reaction is therefore best carried out in two steps[2] with diethylhydroxyborane as an intermediate in this procedure. Tetraethyldiboroxane is then obtained in a total yield of 96% from triethylborane and water.

In the first stage, diethylhydroxyborane is obtained in quantitative yield by reaction of triethylborane with water (cf. Chapter 5, Section 43). The diethylhydroxyborane, which contains some 2,2-dimethylpropanoic acid, is then reacted with triethylborane at 50° to give tetraethyldiboroxane.

$$(C_2H_5)_2BOH + B(C_2H_5)_3 \xrightarrow[-C_2H_6]{\leqslant 50°} [(C_2H_5)_2B]_2O$$

*Max-Planck-Institut für Kohlenforschung, Kaiser-Wilhelm-Platz 1, D-4330, Mülheim/Ruhr, Germany.
†Department of Chemistry, University of Utah, Salt Lake City, UT 84112.

Vacuum distillation of the catalyst-containing tetraethyldiboroxane gives catalyst-free, pure product.

Procedure

■ **Caution.** *Triethylborane*[*3] *reacts vigorously with air and water when carboxylic acids are present. Thus, the conditions of the following preparations should be maintained strictly. Any liquid used for cooling (bath, reflux condenser) should be inert such as Aliphatin.*† *Before attempting to use triethylborane, the experimentalist should consult "Trialkylboranes, Technical and Handling Bulletin," Callery Chemical Company, April 1978.*

■ **Caution.** *[(2,2-Dimethylpropanoyl)oxy]diethylborane reacts violently with air or water. All manipulations must be handled under an inert atmosphere.*

To prepare tetraethyldiboroxane from diethylhydroxyborane using triethylborane, a 1-L three-necked flask is equipped with a magnetic stirrer, thermometer, 250-mL dropping funnel, and reflux condenser. The temperature of the reflux-condenser coolant should be 3-5°. Evaporation losses of ethylboranes are avoided under these conditions. The apparatus is evacuated (10^{-3} torr), filled with inert gas, then charged with 232 g (2.37 mole) triethylborane.

2,2-Dimethylpropanoic acid (~0.8 g) or 2 mL [(2,2-dimethylpropanoyl)oxy]diethylborane (cf. p. 185) (2,2-dimethylpropanoic acid or its derivatives are added to the triethylborane even if the diethylhydroxyborane already contains catalyst) is added to the triethylborane. A 171.9 g quantity (2.00 mole) of diethylhydroxyborane is added dropwise to the stirred solution at 55° in 5 hr. This results in an exothermic ($t \leqslant 70°$) reaction which evolves 48 L (~97%) ethane. On completion of the reaction, the flask contains a mixture of triethylborane and catalyst-containing tetraethyldiboroxane.

The catalyst-containing tetraethyldiboroxane is distilled under slightly reduced pressure through a 30-cm fractionating column filled with glass Raschig rings of 0.5-cm diameter and length. (The pressure should be adjusted so that the boiling point of tetraethyldiboroxane is ⩽60°. The pure compound boils at 40°/8 torr). After 27.2 g triethylborane (bp = 58-60°/240-250 torr) and 12.2 g forerun (bp ⩽64°/30 torr), 294.2 g (95.6%) pure tetraethyldiboroxane (found B_C = 9.36%) with bp = 64-65° (30 torr) distills and 4.6 g of residue containing [(2,2dimethylpropanoyl)oxy]diethylborane remains.

Properties

Tetraethyldiboroxane is a water-clear, mobile liquid which can be vacuum distilled without decomposition (bp = 47°/11 torr). The compound must be stored

*Callery Chemical Corp., Callery, PA 16024.
†Aliphatin is a mixture of saturated hydrocarbons with bp ≈ 190°; available from Esso A. G., D-2000, Hamburg, Germany.

with strict exclusion of air and moisture. In the presence of boranes, dismutation to triethylborane and triethylcyclotriboroxane occurs. Small amounts of BO_3 derivatives are also formed (^{11}B-NMR).

$$[(C_2H_5)_2B]_2O \;\overset{HB<}{\rightleftharpoons}\; B(C_2H_5)_3 + \tfrac{1}{3}(C_2H_5BO)_3$$

Analytical characterization: Mass spectrometry[2,4] is suitable for detecting triethylcyclotriboroxane (base peak m/z = 139) in tetraethyldiboroxane (base peak m/z = 125). 0.5% Triethylcyclotriboroxane ($\delta_{^{11}B}$ = 33 ppm) or triethylborane ($\delta_{^{11}B}$ = 86 ppm) in tetraethyldiboroxane ($\delta_{^{11}B}$ = 53.3 ppm) can be identified by ^{11}B-NMR spectroscopy.[5] Triethylcyclotriboroxane or triethylborane amounts can be determined quantitatively by finding the B_C-values using the trimethylamine N-oxide method[6] in conjunction with the combined pyridine N-oxide/trimethylamine N-oxide method and the total boron content.

Tetraethyldiboroxane: ^{1}H-NMR spectrum (60 MHz, $CDCl_3$), δ = 0.9 ppm (s): ^{11}B-NMR spectrum (32.1 MHz, $CDCl_3$), δ = 53.3 ppm ($h_{1/2}$ = 106 Hz). ^{17}O-NMR spectrum[7], δ = 223.0 ppm ($h_{1/2}$ = 70 Hz). ^{13}C-NMR spectrum (20.1 MHz, $CDCl_3$), δ = 14.3 (br), 7.5 ppm. Mass spectrum (70 eV), m/z = 154 ($M^{+\cdot}$), 125 (M-$C_2H_5^+$, base peak); no peak at m/z = 139.

References

1. *Gmelin Handbuch der Anorganischen Chemie*, 8. Aufl., Ergänzungswerk, Bd., **48/16**, p. 74–84, 94–95 (1977), and references cited therein.
2. R. Köster, H. Bellut, and W. Fenzl, *Liebigs Ann. Chem.*, 54 (1974).
3. R. Köster, P. Binger, and W. V. Dahlhoff, *Synth. Inorg. Metal-Org. Chem.*, **3**, 359 (1973).
4. D. Henneberg, Max-Planck-Institut für Kohlenforschung, Mülheim/Ruhr.
5. R. Mynott, Max-Planck-Institut für Kohlenforschung, Mülheim/Ruhr.
6. R. Köster and Y. Morita, *Liebigs Ann. Chem.*, **704**, 70 (1967).
7. W. Biffar, H. Nöth, H. Pommerening, and B. Wrackmeyer, *Chem. Ber.*, **113**, 333 (1980).

42. DIETHYLMETHOXYBORANE- (METHYL DIETHYLBORINATE)

Submitted by W. FENZL* and R. KÖSTER*
Checked by GOJI KODAMA† and MAMORU SHIMOI†

$$(C_2H_5)_3B + CH_3OH \xrightarrow{\text{cat.}} (C_2H_5)_2BOCH_3 + C_2H_6$$

*Max-Planck-Institut für Kohlenforschung, Kaiser-Wilhelm-Platz 1, D-4330, Mülheim/Ruhr, Germany.
†Department of Chemistry, University of Utah, Salt Lake City, UT 84112.

Diethylmethoxyborane (methyl diethylborinate) is a reagent useful for the preparation of O-diethylboryl derivatives of hydroxy compounds. (Alkoxy)diethyl- and (aryloxy)diethylboranes are formed in good to quantitative yields by transesterification. Di-, tri-, and oligohydroxy compounds can be per-diethylborylated with this reagent. The methanol liberated in these reactions forms an azeotrope with the diethylmethoxyborane. The O-diethylboryl derivatives, which can usually be vacuum distilled without decomposition, are easily deboronated with methanol under very mild conditions. Diethylmethoxyborane is therefore used as a reagent for the purification of thermolabile hydroxy compounds. The preparation of pure diethylmethoxyborane is of interest, for its use can be hampered by the presence of ethyldimethyloxyborane impurity.

Alkoxydialkylboranes[1] and dialkoxyalkylboranes are formed rapidly in the >BH-catalyzed redistribution reaction of trialkoxyboranes with trialkylboranes. If the difference in the chain length is sufficiently large, distillative separations are often possible.

(Alkoxy)dialkyboranes with longer alkyl groups are usually easily prepared by reaction of dialkylhaloboranes, dialkylaminoboranes, or in particular dialkylhydroxyboranes with alcohols. The side products are generally easily removed; water is best removed as the alcohol azeotrope. However, the yields of the lower-boiling alkoxydialkylboranes are low because they also form azeotropes with alcohols.

In contrast, the broadly applicable synthesis of alkoxydialkylboranes by catalyzed alcoholysis of trialkylboranes[2] enables diethylmethoxyborane to be prepared in ~93% yield.

Diethylmethoxyborane is easily prepared by reaction of triethylborane with methanol at room temperature. Small amounts (0.1 mole %) of 2,2-dimethylpropanoic acid are added to the reaction mixture.[2] (Triethylborane reacts irreversibly with 2,2-dimethylpropanoic acid at 0° to give [(2,2-dimethylpropanoyl)oxy]diethylborane and ethane. With methanol, 2,2-dimethylpropanoic acid and diethylmethoxyborane are formed).

Since the distillative separation of diethylmethoxyborane (bp = 88–89°) from triethylborane (bp = 95–96°) is laborious, a 1% excess of methanol is used. The loss of diethylmethoxyborane due to the formation of a 1:1 azeotrope (bp = 58°) with methanol is therefore minimal. Pure diethylmethoxyborane is obtained in 93% yield.

■ **Caution.** *Triethylborane is spontaneously flammable. Before using this material a technical bulletin such as "Trialkylboranes, Technical and Handling Bulletin," Callery Chemical Co., April, 1978, should be consulted. In the presence of 2,2-dimethylpropanoic acid, it reacts with water or alcohols at 0° with the liberation of ethane. All coolants must therefore be free of protonic components. Aliphatin* is well-suited as a coolant.*

*Mixture of saturated hydrocarbons with bp ≈ 190°, available from Esso A.G., D-2000 Hamburg, Germany.

Procedures

A 1-L two-necked flask is equipped with magnetic stirrer, thermometer, a 100-mL dropping funnel, and a reflux condenser which is connected to a gas clock. The apparatus is evacuated (10^{-3} torr) and then filled with inert gas (argon or nitrogen). Water-free methanol, 64.8 g (2.02 mole), is added dropwise at 10-15° in 5.5 hr to a stirred mixture of 196.1 g (2.0 mole) triethylborane* and 0.4 g 2,2-dimethylpropanoic acid.† Approximately 43 L ethane (STP) is evolved. After stirring for a further 5 hr at room temperature (cooling to 20°) a total of 48.4 L (99%) ethane has been evolved. The product mixture is fractionated (15-cm column) at atmospheric pressure to give 3.6 g forerun (bp = 54-83°) and 186.1 g (93%) pure product with bp = 88-89°.

Properties

Diethylmethoxyborane is a colorless, extremely moisture-sensitive liquid that can be distilled without decomposition, bp = 88°/760 torr. It forms a constant-boiling azeotrope (bp = 54°/760 torr) with an equimolar amount of methanol. The compound must be stored and handled under an inert gas.

Spectra of pure diethylmethoxyborane: ^{1}H-NMR spectrum (60 MHz, neat), δ = 3.62(s), 0.87(s) ppm in the ratio 3:10. ^{11}B-NMR spectrum (32.1 MHz, neat), δ = 54.0 ppm ($h_{1/2} \approx 80$ Hz). ^{13}C-NMR spectrum (25.2 MHz, neat), δ = 53.0 (CH$_3$-O), ~12 (br, $J_{BC} \approx 70$ Hz, CH$_2$), 7.8 (CH$_3$) ppm. Mass spectrum (70 eV), m/z = 100 (M$^+$), 71 (M-C$_2$H$_5$)$^+$.

References

1. *Gmelin Handbuch der Anorganischen Chemie*, 8 Aufl., Ergänzungswerk, 48/16, p. 134 (1977).
2. R. Köster, W. Fenzl, and G. Seidel, *Liebigs Ann. Chem.*, 352 (1975).
3. W. Fenzl, H. Kosfeld, and R. Köster, *Liebigs Ann. Chem.*, 1370 (1976).
4. R. Köster, P. Binger, and W. V. Dahlhoff, *Synth. Inorg. Metal-Org. Chem.*, 3, 359 (1973).

---

*Available from Callery Chem. Company, Callery, PA 16024, or prepared by the method in Reference 4.
†Available from Deutsche Shell, D-400 Düsseldorf, Germany.

43. DIETHYLHYDROXYBORANE (DIETHYLBORINIC ACID)

Submitted by W. FENZL* and R. KÖSTER*
Checked by GOJI KODAMA[†] and MAMORU SHIMOI[†]

A. $[(C_2H_5)_2B]_2O + H_2O \longrightarrow 2(C_2H_5)_2BOH$

B. $(C_2H_5)_3B + H_2O \xrightarrow{\text{cat.}} (C_2H_5)_2BOH + C_2H_6$

Pure diethylhydroxyborane (diethylborinic acid) is a colorless liquid which has hitherto not been characterized. It is a reagent for introducing diethylboryl protective groups into hydroxy and primary amino compounds. Diethylhydroxyborane is an intermediate in the efficient synthesis of tetraethyldiboroxane from triethylborane and water. It can also be used for preparing diethylorganooxyboranes.

In general, hydroxydiorganoboranes (borinic acids) can be prepared by hydrolysis of diorganoboranes R_2BX [R = organo group; X = R, H, halogens, OR′, SR′, and NR′$_2$].[1]

$$R_2BX + H_2O \longrightarrow R_2BOH + HX$$

The separation of the side products HX from the hydroxydiorganoboranes can lead to drastic losses of product in the case of the lower-boiling boranes.

Diethylhydroxyborane that is free of HX side products can be prepared only from either triethylborane-hydro-[HX = C_2H_6] or tetraethyldiboroxane [HX = HOB(C_2H_5)$_2$] and water. Diethyl-hydro-borane [HX = H_2] is not a suitable starting material as monoethylboranes are also formed by >BH-catalyzed ligand exchange.

The last step in the best synthesis of pure diethylhydroxyborane is the reaction of equimolar amounts of water and tetraethyldiboroxane. The latter compound is easily prepared from water and triethylborane via a catalyst-containing diethylhydroxyborane.[2]

The preparation of diethylhydroxyborane from triethylborane and water according to the equation

$$(C_2H_5)_3B + H_2O \xrightarrow[-C_2H_6]{\text{cat.}} (C_2H_5)_2BOH$$

---

*Max-Planck-Institut für Kohlenforschung, Kaiser-Wilhelm-Platz 1, D-4330 Mülheim/Ruhr, Germany.
[†]Department of Chemistry, University of Utah, Salt Lake City, UT 84112.

proceeds smoothly at ~20°. A quantitative yield of product containing small amounts of 2,2-dimethylpropanoic acid or [(2,2-dimethylpropanoyl)oxy]diethylborane is obtained if the temperature is kept below 30°; otherwise, yields decrease because of unavoidable loss of water during the evolution of ethane.

The catalyzed hydrolysis of triethylborane proceeds in two stages:

$$(C_2H_5)_3B + (CH_3)_3CCO_2H \xrightarrow[-C_2H_6]{} (C_2H_5)_2BO_2CC(CH_3)_3$$

$$(C_2H_5)_2BO_2CC(CH_3)_3 + H_2O \xrightarrow[-(CH_3)_3CCO_2H]{} (C_2H_5)_2BOH$$

The reaction is exothermic, so the reaction mixture is kept below ~30° by cooling, and the coolant in the reflux condenser is held at 3-4°. In this way only negligible amounts of water are lost.

Catalyst-free diethylhydroxyborane is formed in quantitative yield by reaction of tetraethyldiboroxane with water:

$$(C_5H_2)_2BOB(C_2H_5)_2 + H_2O \longrightarrow 2(C_5H_2)_2BOH$$

Although tetraethyldiboroxane and water are almost immiscible liquids, the exothermic reaction proceeds at room temperature when the mixture is slowly stirred. The temperature should be kept ~30°.

The preparation of pure diethylhydroxyborane is therefore closely coupled with the preparation of pure tetraethyldiboroxane.

Procedure

An apparatus consisting of a three-necked flask with a magnetic stirrer is fitted with a dropping funnel, a thermometer, and an argon bubbler. For the preparation by method B, a lead to a gasometer is attached to a reflux condenser whose temperature is maintained at 3-5°.

Method A. Pure Diethylhydroxyborane from Tetraethyldiboroxane and Water. Tetraethyldiboroxane (Chapter 5, Section 41), 153.9 g (1.00 mole), is added with stirring in 1.5 hr to 18 g (1.0 mole) water at 22°. Stirring is essential for a sufficiently fast reaction, as tetraethyldiboroxane and water are only poorly miscible, but the flask temperature should not be allowed to exceed 30°. The reaction is complete when the addition is finished.

Method B. Catalyst-Containing Diethylhydroxyborane from Triethylborane and Water. Triethylborane, 98.8 g (1.008 mole), is added dropwise at 20-25° (t_{max} 27°) in 9 hr to a stirred solution of 0.5 g (5 mmole) 2,2-dimethylpropanoic acid* in water, 18 g (1.0 mole), evolving 23.5 L (94%) of ethane (STP). The flask then contains 2 mole diethylhydroxyborane which is contaminated with ~0.5 mole % [(2,2-dimethylpropanoyl)oxy]diethylborane/2,2-dimethylpropanoic acid.

Properties

Diethylhydroxyborane is a colorless, oxygen-sensitive liquid which must be handled in an inert gas atmosphere (argon, nitrogen). The ^{1}H-NMR spectroscopically pure borane is associated through hydrogen bridges (^{1}H-NMR[4] at +30° to -70°: H_{OH} signals are temperature-dependent, but those for Et_2B remain constant), cf. the IR spectrum.[5] Only one signal for three-coordinate boron can be detected in the ^{11}B-NMR spectrum.

Diethylhydroxyborane cannot be vacuum distilled without decomposition. Tetraethyldiboroxane and water are formed which then recombine at room temperature. Diethylhydroxyborane forms 1:1 addition compounds with carboxylic acids and acyloxydialkylboranes as evidenced by ^{1}H-NMR spectroscopy below -30°.

Diethylhydroxyborane: IR spectrum (neat), ν_{OH} = 3620, 3600, 3400 cm^{-1}. ^{1}H-NMR spectrum (60 MHz, $CDCl_3$), δ = 5.85(s) and 0.9(s) ppm in the ratio 1:10. ^{11}B-NMR spectrum (32.1 MHz, $CDCl_3$), δ = 55.5 ppm ($h_{1/2}$ = 90 Hz). Mass spectrum (70 eV): m/z = 86 (M$^+$), 85 (M-H)$^+$, 69 (B_1), 57 (B_1, base peak). ^{13}C-NMR spectrum (20.1 MHz, $CDCl_3$), δ = 12.5(br), 7.2 ppm.

References

1. *Gmelin Handbuch der Anorganischen Chemie*, 8. Aufl. Ergänzungswerk 48/16, p. 126 ff (1977).
2. R. Köster, H. Bellut, and W. Fenzl, *Liebigs Ann. Chem.*, 54 (1974).
3. R. Köster, P. Binger, and W. V. Dahlhoff, *Synth. Inorg. Metal-Org. Chem.*, 3, 359 (1973).
4. E. G. Hoffmann and G. Schroth, Max-Planck-Institut für Kohlenforschung, Mülheim/Ruhr.
5. K. Seevogel, Max-Planck-Institut für Kohlenforschung, Mülheim/Ruhr.
6. R. Mynott, Max-Planck-Institut für Kohlenforschung, Mülheim/Ruhr.

---

*2,2-Dimethylpropanoic acid is supplied by Deutsche Shell, D-4000 Düsseldorf, Germany.

44. BIS[μ-(2,2-DIMETHYLPROPANOATO-O,O')] DIETHYL-μ-OXO-DIBORON

Submitted by R. KÖSTER* and W. FENZL*
Checked by JAMES E. MCGUIRE, Jr.† and SHELDON G. SHORE†

A. $2(CH_3)_3CCO_2B(C_2H_5)_2 \xrightarrow[-2C_2H_6]{+H_2O,\ 80°}$

B. $[(C_2H_5)_2B]_2O \xrightarrow[-2C_2H_6]{+2(CH_3)CCO_2H,\ 80°}$

C. $6(CH_3)_3CCO_2B(C_2H_5)_2 \xrightarrow[-3[(C_2H_5)_2B]_2O]{+2(C_2H_5BO)_3}$

$$H_5C_2-\overset{|}{\underset{|}{B}}-O-\overset{|}{\underset{|}{B}}-C_2H_5$$
(with bridging pivalate groups, C(CH$_3$)$_3$ substituents)

The title compound is a useful reagent for the introduction of O-(ethylboranediyl) groups into polyhydroxy compounds.[1-3] The compound is a side-product in aldol condensations of carbonyl compounds using [(2,2-dimethylpropanoyl)oxy]diethylborane.[4]

μ-Alkanoato-diorgano-μ-oxo-diboron compounds have been prepared from triorganoboranes and carboxylic acids[5-7] or from the reactions of several types of organo-μ-oxo-diboron compounds, such as tetraorgano-μ-oxo-diboron compounds,[6,8] diorganoboranes,[9] dihydroxyorganoboranes,[10] or organobis(organooxy)boranes[10] with carboxylic acids or their anhydrides. Other ducts that have been used are haloorgano-boranes[9] and alkylhalo(organooxy)boranes.[11]

Bis[μ-2,2-dimethylpropanoato-O,O')]diethyl-μ-oxo-diboron is best prepared in high yield and purity by using one of the three different methods A–C described below.

Elimination of ethane from [(2,2-dimethylpropanoyl)oxy]diethylborane[7] with water yields >98% of the μ-oxodiboron (method A).

The compound can also be prepared by the reaction of tetraethyldiboroxane[7] (Et$_2$B)$_2$O with two moles of 2,2-dimethylpropanoic acid in a ~98% yield (method B).

Finally, [2,2-dimethylpropanoyl)oxy]diethylborane[7] and triethylcyclotriboroxane can exchange ligands exothermally yielding equal amounts of bis[μ-(2,2-dimethylpropanoato-O,O')]diethyl-μ-oxo-diboron and tetraethyldiboroxane (method C).

*Max-Planck-Institut für Kohlenforschung, Kaiser-Wilhelm-Platz 1, D-4330 Mülheim/Ruhr, Germany.
†Department of Chemistry, Ohio State University, Columbus, OH 43210.

The latter reaction, which is particularly smooth, convenient, and fast, does not involve the loss of ethylboron groups. The isolation is extremely simple: bis[μ-2,2-dimethylpropanoato-O,O')] diethyl-μ-oxo-diboron is obtained quantitatively as a residue after distilling of the volatile tetraethyldiboroxane.

Procedures

Method A. A 9-mL quantity (0.5 mole) of water is added dropwise with stirring in ~70 min to 169 g (0.99 mole) of [(2,2-dimethylpropanoyl)oxy]diethylborane[7] at 95-105° (bath temp.). If the mode of addition is reversed, practically no gas is evolved even at 100°. On cooling to room temperature the mixture crystallizes. Distillation gives 145.7 g (98.4%) product, bp = 140-149°/15 torr (bath temp. 160-174°), mp = 54-58°, while 1.4 g of solid, off-white residue remains.

Method B. A 188 g quantity (1.22 mole) of tetraethyldiboroxane is added dropwise with stirring in ~2 hr to 257.6 g (2.52 mole) of 2,2-dimethylpropanoic acid at 80° (flask temp.). The ethane evolution (total 57.5 L) stops ~0.5 hr after the addition has ended. On cooling, the light-brown product crystallizes. Vacuum distillation gives 367 g colorless crystalline crude product (bp 103-128°/9 torr, bath temp. 150-164°) and 5.1 g dark brown residue. Renewed vacuum distillation (at 10^{-3} torr) gives 7.3 g forerun (bp 29-105°/14 torr, bath temp. ≤120°C) and 359.5 g (98.7%) colorless product with bp 56°/10^{-3} torr, mp 51°.

Anal. Calcd. for $C_{14}H_{28}B_2O_5$: B, 7.26; B_C (boron linked to carbon), 2.42. Found: B, 7.16; B_C, 2.30.[12]

Method C. A 12.5 g (223.7 mmole) quantity of triethylcyclotriboroxane (prepared from B_2O_3 and triethylborane, according to German Pat. 1,056,127 (1959)) is added dropwise with stirring in ~10 min to 33.9 g (0.199 mole) [(2,2-dimethylpropanoyl)oxy]diethylborane (exothermic to ~44°). The mixture is then heated to 60° for 10 min to give equimolar amounts (^{11}B-NMR) of product and tetraethyldiboroxane. The tetraethyldiboroxane (15.6 g) is removed under vacuum (bp ≤20°/0.05 torr) to leave 29.4 g (99%) colorless, crystalline product as residue, mp 50-52°.

Anal. Calcd. for $C_{14}H_{28}B_2O_5$: B_C, 2.42. Found: B_C, 2.42.

Properties

Bis[μ-(2,2-dimethylpropanoato-O,O')]diethyl-μ-oxo-diboron must be stored under an inert gas (Ar, N_2), because of its moisture sensitivity.

The boron atoms of the bicyclo[3.3.1]nonane-like[6]-structured bis[μ-(2,2-

Compounds of Boron

dimethylpropanoato-O,O')]diethyl-μ-oxo-diboron[7] are four-coordinate ($\delta_{11_B} \approx$ 8 ppm, half width 315 Hz in $CDCl_3$). That is the reason for the success of the method C, which involves the thermodynamically highly stable triethylcyclotriboroxane as a reagent.

References

1. R. Köster, *Pure Appl. Chem.*, **49**, 765 (1977).
2. R. Köster and W. V. Dahlhoff, *Synth. Methods Carbohydr.*, *ACS Symp. Ser.*, **39**, 1 (1976).
3. W. V. Dahlhoff and R. Köster, *J. Org. Chem.*, **41**, 2316 (1976).
4. R. Köster and A. A. Pourzal, *Synthesis*, 674 (1973).
5. B. M. Mikhailov and V. A. Vaver, *Z. Obshch Khim.*, **31**, 574 (1961); *C. A.* **55**, 23318 (1961).
6. J. Goubeau and H. Lehmann, *Z. Anorg. Allgem. Chem.*, **322**, 224 (1963).
7. R. Köster, H. Bellut, and W. Fenzl, *Liebigs Ann. Chem.*, **54**, 1974.
8. W. Gerrard, M. F. Lappert, and R. Shafferman, *Soc*, 3648 (1958).
9. B. M. Mikhailov and T. A. Shchegoleva, *Izv. Akad. Nauk SSSR*, 832 (1958).
10. B. M. Mikhailov and T. A. Shchegoleva, *Izv. Akad. Nauk SSSR*, 1345 (1959).
11. B. M. Mikhailov and T. A. Shchegoleva, *Izv. Akad. Nauk SSSR*, 835 (1958).
12. R. Köster and Y. Morita, *Liebigs Ann. Chem.*, **704**, 70 (1967).

45. ALKALI METAL (CYCLOOCTANE-1,5-DIYL)DIHYDROBORATES(1-)

Submitted by R. KÖSTER* and G. SEIDEL*
Checked by JAMES E. MCGUIRE† and SHELDON G. SHORE†

$$\left[\bigcirc\!\!\!\!\!\text{BH} \right]_2 \xrightarrow[(\leq 150°)]{+2MH \quad \Delta}{THF} 2M^\oplus \left[\bigcirc\!\!\!\!\!B\begin{matrix}H\\H\end{matrix} \right]^\ominus$$

M = Li, Na, K

$\bigcirc\!\!\!\!\!$ = Cyclooctane-1,5-diyl-group

Alkali metal organohydroborates(1-) are well suited for many preparative reductions and for the preparation of various organoborates(1-)[1,2] and organoboranes.[1] These organohydroborates(1-) are useful analytical nucleophilic

---

*Max-Planck-Institut für Kohlenforschung, Kaiser-Wilhelm-Platz, D-4330 Mülheim/Ruhr, Germany.
†Department of Chemistry, Ohio State University, Columbus, OH 43210.

hydro-reagents for determining reduction equivalents of organic and inorganic compounds.

Difficulties resulting from exchange of hydro and monofunctional organo groups can be encountered in some attempted preparations of alkali metal dihydrodiorganoborate(1-).[1,3] Alkanediyldihydroborates(1-) are however usually thermally stable compounds.[4]

Many metal (cyclooctane-1,5-diyl)dihydroborates(1-) can be prepared by reaction of the easily prepared 9-borabicyclo[3.3.1]nonane dimer (9-BBN)[5,6] with metal-organic compounds.[7] The reactions of 9-BBN with alkali metal hydrides in ethers are particularly smooth.

The preparations of lithium and sodium (cyclooctane-1,5-diyl)dihydroborates(1-) in tetrahydrofuran proceed via isolable, stable etherates. These can be made solvent-free simply by heating under vacuum. 9-Borabicyclo[3.3.1]-nonane dimer (9-BBN)* can easily be prepared from cycloocta-1,5-diene[2] by reaction with tetraethyldiborane(6), tetrahydrofuran-borane[8,9] or dimethyl sulfide-borane.[10] The synthesis of alkali metal (cyclooctane-1,5-diyl)dihydroborates is achieved by addition of 9-BBN to a suspension of the alkali metal hydride in tetrahydrofuran. Lithium hydride reacts more slowly than sodium or potassium hydride. The reactions are brought to completion by heating under reflux.

A. LITHIUM (CYCLOOCTANE-1,5-DIYL)DIHYDROBORATE(1-)

Procedure

■ **Caution.** *Sodium/potassium alloy is extremely moisture sensitive. Contact with water or moist air poses a significant explosion hazard.*

Tetrahydrofuran is dried by refluxing over Na/K and finally distilling from sodium tetraethylaluminate. The apparatus consists of a 500-mL three-necked flask equipped with an internal thermometer, a magnetic stirrer, and a reflux condenser capped with an inert gas (pure N_2, argon) bypass. The apparatus is evacuated and then filled with inert gas.

Two hundred mL of tetrahydrofuran and (3.4 g 430 mmole) LiH powder are transferred to the 500-mL flask. A total of (24.5 g, 100 mmole) 9-BBN-dimer is added in portions in ~1 hr to the well-stirred suspension by transferring it from a 100-mL flask via a 15-cm-long PVC tube (0.9 i.d.) by occasional shaking. The mixture is then heated ~3 hr under reflux. After cooling to room temperature, the excess LiH is separated by filtration.

■ **Caution.** *Filtration of LiH should be carried out under an inert atmosphere as contact with moist air can cause ignition.*

*Available from Aldrich Chemical Co.

B. SODIUM (CYCLOOCTANE-1,5-DIYL)DIHYDROBORATE(1-)

Procedure

To a stirred suspension of (6.7 g, 279 mmole) NaH in 200 mL THF at 50-60° is added (26.6 g, 109 mmole) 9-BBN-dimer as described in the previous experiment. After being heated for 4 hr under reflux, the mixture is filtered to remove excess NaH and the filtrate concentrated under vacuum (14 torr). The remaining tetrahydrofuran is removed by heating to 140-150° under vacuum to leave 31.5 g (99%) product as a colorless powder.

C. POTASSIUM (CYCLOOCTANE-1,5-DIYL)DIHYDROBORATE(1-)

Procedure

To a stirred suspension of (13.5 g, 337 mmole) KH* in 250 mL tetrahydrofuran at room temperature (30.6 g, 125 mmole), 9-BBN-dimer is added rapidly enough that the solvent starts refluxing. The same work-up as for the Na compound is used to give 39.9 g (98%) of a colorless, THF free powdery product.

Properties

The air-stable alkali metal (cyclooctane-1,5-diyl)dihydroborates(1-) are very moisture-sensitive. They must be stored and handled in an inert gas atmosphere. The salts are insoluble in aliphatic solvents and poorly soluble in aromatic solvents such as mesitylene.

Crystalline products with the following compositions

*Prepared from metallic potassium and hydrogen gas at 200° in aliphatin; cf. D. Simič, Dissertation University Bochum 1970, p. 62. Aliphatin is a mixture of saturated hydrocarbons boiling at about 190° and available from Esso A.-G., D-2000 Hamburg, Germany.

TABLE I Spectroscopic Parameters for Alkali Metal (Cyclooctane-1,5-diyl)dihydroborates(1−)

M	Li	Na	K
ν_{BH}(THF) (cm^{-1})	2110	2080	2085
δ_{11B}(THF) (ppm)	−17.4	−18.3	−16.3
t, J_{BH}	74 Hz	74 Hz	74 Hz

are obtained from tetrahydrofuran after drying under vacuum. The tetrahydrofuran can be quantitatively removed from the lithium and sodium salts by heating to ⩽150° under vacuum. Monosolvates are obtained from 1,1-oxybis(2-methoxyethane) (diglyme). The sodium salt crystallizes spontaneously on concentrating the diglyme solution, and the lithium and potassium salts are precipitated from diglyme solutions by addition of hexane and isolated by filtration.

The hydride content of both the ether-containing and solvent-free alkali metal (cyclooctane-1,5-diyl)dihydroborate is (1−) can be determined volumetrically by measuring the hydrogen evolved after treatment with an alcohol (e.g., 2-ethyl-1-hexanol) at room temperature.

The solvent-free salts do not melt below 250°, whereas the lithium salt slowly undergoes ligand exchange reactions at this temperature.

Characteristics for all these borates are the B-H stretching vibrations in the IR spectra and also the ^{11}B-triplet nmr signals (cf. Table I).

References

1. P. Binger, G. Benedikt, G. W. Rotermund, and R. Köster, *Liebigs Ann. Chem.*, 717, 21 (1968).
2. P. Binger and R. Köster, *Inorg. Synth.*, 15, 136 (1974).
3. P. C. Moews and R. W. Parry, *Inorg. Chem.*, 5, 1552 (1966).
4. H. C. Brown and W. Negishi, *J. Am. Chem. Soc.*, 93, 6682 (1971).
5. R. Köster, *Angew. Chem.*, 72, 626 (1960).
6. R. Köster and P. Binger, *Inorg. Synth.*, 15, 141 (1974).
7. J. L. Hubbard and G. W. Kramer, *J. Organomet. Chem.*, 156, 81 (1978).
8. E. F. Knights and H. C. Brown, *J. Am. Chem. Soc.*, 90, 5280 (1968).
9. H. C. Brown, E. F. Knights, and C. G. Scouten, *J. Am. Chem. Soc.*, 96, 7765 (1974).
10. H. C. Brown, A. K. Mandal, and S. U. Kulkarni, *J. Org. Chem.*, 42, 1392 (1977).

46. DECABORANE(14)

$$17\text{NaBH}_4 + 20\text{BF}_3 \cdot \text{O}(\text{C}_2\text{H}_5)_2 \longrightarrow$$
$$2\text{NaB}_{11}\text{H}_{14} + 15\text{NaBF}_4 + 20\text{H}_2 + 20\text{O}(\text{C}_2\text{H}_5)_2$$
$$2\text{NaB}_{11}\text{H}_{14} + 20\text{H}_2\text{O} + 8\text{H}_2\text{O}_2 + \text{H}_2\text{SO}_4 \longrightarrow$$
$$\text{B}_{10}\text{H}_{14} + 12\text{B}(\text{OH})_3 + 18\text{H}_2 + \text{Na}_2\text{SO}_4$$

Submitted by GARY B. DUNKS,* KATHY PALMER-ORDONEZ,† and EDDIE HEDAYA†
Checked by PHILIP KELLER‡ and PAUL WUNZ §

Previously, decaborane(14) was synthesized by pyrolysis of hazardous diborane(6) in elaborate apparatus which precluded the use of the method for general laboratory work.[1] The procedure given here is performed in standard glassware employing relatively safe materials and thus makes decaborane(14) generally available for the first time.

Procedure

■ **Caution.** *Decaborane(14) is a toxic material and care must be exercised to avoid breathing the vapor or allowing the solid or its solutions to contact skin. All operations in the following procedure should be conducted in a well ventilated hood.*

A 1-L, three-necked flask is equipped (Fig. 1) with a mechanical stirrer, a tube through which $\text{BF}_3 \cdot \text{O}(\text{C}_2\text{H}_5)_2$ can be pumped from a reservoir (a pressure-equalized addition funnel can be used), and a side tube connected to dry ice-cooled condenser. A long-stem thermometer is fitted through the side tube and allowed to extend almost to the stirring paddle. The condenser outlet is connected to a tube with T connections for the introduction of nitrogen gas and to equilibrate the pressure in the $\text{BF}_3 \cdot \text{O}(\text{C}_2\text{H}_5)_2$ reservoir (unnecessary if an addition funnel is used). The end of the tube is connected to a glass tube which extends into a 500-mL Erlenmeyer flask containing 200 mL of acetone. The apparatus is assembled (all joints should be securely joined using rubber bands) and swept thoroughly with nitrogen. With a slow stream of nitrogen passing through

---

*Rockwell International Corporation, Canoga Park, CA 91304.
†Union Carbide Corp., Tarrytown, NY 10591.
‡Department of Chemistry, University of Arizona, Tucson, AZ 85721.
§Department of Chemistry, Indiana University of Pennsylvania, Indiana, PA 15705.

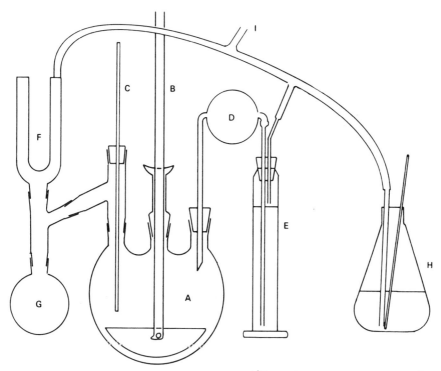

Fig. 1. *The apparatus used in the $[B_{11}H_{14}]^{1-}$ ion-forming step consists of a 1-L, three-necked reaction flask (A) fitted with a mechanical stirrer (B), a thermometer (C), and a pump (D) for the addition of $BF_3 \cdot O(C_2H_5)_2$ from reservoir (E). The reaction off-gas is passed through a dry-ice condenser (F) which condenses the diethyl ether into receiver (G). The noncondensible gas passes through the condenser and into the acetone scrubber (H); any diborane (6) formed as a by-product is destroyed by the acetone. The nitrogen inlet to the system is at (I).*

the system (approximately 40 mL/min), 60.0 g (1.59 mole) of sodium tetrahydroborate and 300 mL of dry diglyme* are charged to the flask through the neck containing the pump outlet tube (or the addition funnel).

■ **Caution.** *A positive nitrogen pressure must be maintained to avoid acetone suck-back.*

The pump outlet tube (or addition funnel) is put in place and secured. The ether receiver is cooled with an ice bath. The reservoir (or addition funnel) is

---

*Diglyme (bis(2-methoxyethyl) ether), Ansul-141, is heated over sodium and benzophenone until a dark blue color is obtained, then distilled at 60°/13 torr. Diglyme purified in this manner should be stored under nitrogen to minimize possible peroxide formation.

charged with 250 mL (2.04 mole) of boron trifluoride-diethyl ether which has been distilled from calcium hydride at 65° (41 torr) and the dry ice condenser is filled with dry ice and 2-propanol. The stirrer is started and the flask and contents are heated to 105 ± 2° with a heating mantle. The boron trifluoride-diethyl ether is added dropwise over 6 hr (45 mL/hr). Care should be taken to insure that a very slow stream of nitrogen passes through the system and that the dry ice is maintained in the condenser during the addition. When the addition is complete, the reaction mixture is maintained at 105° with stirring for 1 hr. The reaction mixture is allowed to cool overnight to ambient temperature, without stirring, under a nitrogen atmosphere. The condenser, ether receiver (containing approx. 211 mL of diethyl ether), side-arm, and pump outlet (or addition funnel) are removed and replaced with a concentrator condenser,* receiver, thermometer, and pressure-equalized addition funnel as shown in Fig. 2. Under slight positive pressure of nitrogen and with stirring, 105 mL of water is added *dropwise* to the yellow reaction mixture.

■ **Caution.** *Care must be exercised at this point to insure that the apparatus vent has sufficient capacity to allow the evolved gas to escape.*

The flask is heated to initiate distillation of the water/diglyme azeotrope (99.6°, 1 atm). As the distillation begins, water (983 mL) is added from the addition funnel at the same rate that distillate is collected to maintain the volume of the reaction mixture in the flask. The distillation is continued until a total of 1120 mL of distillate is collected or the temperature of the reaction mixture reaches 113° (3-5 hr are required). The flask and contents are allowed to cool to ambient temperature under nitrogen. The concentrator condenser and receiver are removed and the addition funnel attached to the reaction flask as shown in Fig. 3. Under nitrogen and with stirring, 4.8 g (0.02 mole) of $FeSO_4 \cdot 7H_2O$ and 250 mL of cyclohexane are added to the flask. A *cool* solution of 35 mL of conc. H_2SO_4 in 35 mL of water is added from the addition funnel. The addition funnel is charged with 72 mL (0.70 mole) of 30% hydrogen peroxide solution. The hydrogen peroxide solution is added dropwise over a 3-hr period. The temperature of the reaction mixture is maintained between 25-35° with an ice bath. When the addition is complete, the reaction mixture is filtered using a coarse glass frit. The reaction flask is washed with 2 × 45-mL portions of cyclohexane which are poured over the filter cake. The filter cake (95 g wet) is washed with 100 mL more of cyclohexane which is combined with the filtrate. The cyclohexane and water layers are separated and the cyclohexane fraction is washed with 5 × 50-mL portions of water. The resulting cyclohexane/decaborane(14) solution (approx. 305 mL) is transferred to a 500-mL three-necked flask equipped with an insulated Vigreux column topped with a distillation head and condenser, a mechanical stirrer, and a thermometer extend-

*Obtained from Kontes, Vineland, NJ (K-444500).

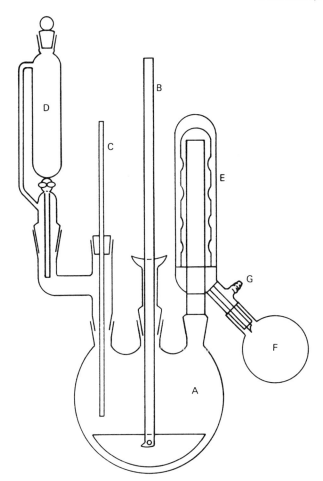

Fig. 2. The apparatus for the solvent exchange step (diglyme–water) consists of the same three-necked flask (A) that was used in the previous step fitted with an addition funnel (D), thermometer (C), a mechanical stirrer (B), a concentrator condenser (E), a receiver (F), and a nitrogen inlet (gas exit) (G).

ing into the flask to just above the stirrer blade. Under a blanket of nitrogen and with stirring, the flask is heated to distill the cyclohexane–water azeotrope (69°, 1 atm) then cyclohexane (81°, 1 atm). The distillation is continued until the temperature of the contents of the flask reaches 99° or 283 mL of distillate is collected.

■ **Caution.** *This step must be followed exactly to avoid distilling to dryness.*

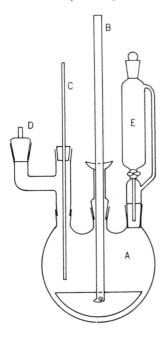

Fig. 3. The apparatus used in the oxidation step consists of the same three-necked flask (A), equipped with a mechanical stirrer (B), a thermometer (C), a nitrogen inlet (gas exit) (D), and a pressure-equalized addition funnel (E).

The heating mantle is immediately lowered (the nitrogen flow to the system must be sufficient to preclude suck-back of air), the stirrer stopped, and the flask allowed to cool to ambient temperature. The flask and contents are removed from the apparatus, stoppered, and cooled to −15° overnight. The decaborane(14) crystals formed are filtered out and air dried (yield 9.0 g, 0.07 mole, 79%).

Properties

Decaborane(14) is a toxic, volatile, white solid (mp 97-98°) that possesses a pungent, chocolate-like odor. Decaborane(14) is stable in air below 100°; however, it may explode in air above 100°. It is very soluble in aromatic hydrocarbons and less soluble in aliphatic hydrocarbons. Decaborane(14) solutions in certain halogenated solvents are shock sensitive. The ^{11}B-NMR spectrum measured in benzene at 28.9 MHz consists of three doublets of relative areas 4:4:2 at +10.5 ppm (J = 147 Hz), −1.17 ppm (J = 163 Hz), and −36.6 ppm (J = 158 Hz) (relative to $BF_3 \cdot O(C_2H_5)_2$ = 0.00).[2] The infrared spectrum (Nujol mull) consists of absorptions (cm^{-1}) at 2555(vs), 1920(m), 1875(m), 1370(m), 1100(w), 1040(w), 1010(s), 970(m), 964(m), 936(m), 920(s), 902(m), 858(s), 813(s), 765(s), 747(m), and 722(s).[3] The concentration of decaborane(14) in solution may be conveniently determined using a colorimetric method based upon the red complex formed with quinoline which absorbs at 490 nm.[4]

References

1. R. I. Holzmann (ed.), *Production of the Boranes and Related Research*, Academic Press, New York, 1967.
2. Chemical shifts up-field of the reference compound are negative, and downfield shifts are positive; *J. Organomet. Chem.*, 131, C43 (1977).
3. J. J. Miller and M. F. Hawthorne, *J. Am. Chem.*, 81, 4501 (1959).
4. W. H. Hill and M. S. Johnston, *Anal. Chem.*, 27, 1300 (1955).

47. DICHLOROPHENYLBORANE

$$Sn(C_6H_5)_4 + 4BCl_3 \longrightarrow SnCl_4 + 4C_6H_5BCl_2$$

Submitted by U. W. GERWARTH* and W. WEBER†
Checked by K. NIEDENZU‡

Various methods for the preparation of dichlorophenylborane, $C_6H_5BCl_2$, have been reported[1] but the interaction of tetraphenylstannane with boron trichloride is the most commonly adopted procedure. Although an almost complete phenyl transfer from tin to boron occurs in the absence of solvent,[2] the standard preparation[3,4] employs benzene as solvent but utilizes only two phenyl groups of the tetraphenylstannane. Since the reaction has been reported[2,5] to be very vigorous in the absence of solvent, this latter method is limited to the preparation of small quantities of dichlorophenylborane. However, under appropriate conditions the process based on the above equation may be applied to large-scale operations without difficulty.

Procedure

■ **Caution.** *Dichlorophenylborane is extremely sensitive to hydrolysis; furthermore, hot dichlorophenylborane is pyrophoric. Before disassembling any equipment it must cool to room temperature.*

Boron trichloride is condensed (acetone/dry ice) into a 500-mL two-necked flask equipped with a gas inlet and a $CaCl_2$ drying tube. In the meantime 480 g (1.12 mole) of tetraphenylstannane (technical grade) is placed in a 2-L three-necked flask equipped with a magnetic stirrer and water condenser topped with a $CaCl_2$ drying tube. The set-up is cooled in an ice bath.

---

*Institut f. Anorg. u. Analyt. Chemie, Universität Mainz, Mainz, West Germany. New address: Gmelin-Institut der Max-Planck-Gesellschaft, Varrentrappstr. 40/2, Frankfurt/Main, West Germany.

†Institut f. Anorg. u. Analyt. Chemie, Universität Mainz, Mainz, West Germany.

‡Department of Chemistry, University of Kentucky, Lexington, KY 40506.

When 560 g (4.78 mole) of boron trichloride is condensed, the entire quantity is quickly added to the cold tetraphenylstannane (in a well-ventilated hood). The mixture is stirred and the tin compound slowly dissolves. No vigorous reaction is observed, provided cooling with an ice-bath is maintained. After 7 hr of stirring, only a minor quantity of gray insoluble material is left, and the mixture is allowed to warm to room temperature. The reaction mixture is then heated to reflux for 4 hr. (Excess boron trichloride, which is gradually released, may be condensed in a trap for further use.) Distillation through a 1-m silver-mantle column yields 192 g of tin tetrachloride (bp 23–27° at 23 torr) and 612 g (87% yield) of dichlorophenylborane (bp 73–76° at 23 torr); 79 g of a mixture of tin tetrachloride and boron trichloride is collected in an acetone/dry ice trap, and 78 g of a brown residue crystallizes in the distillation flask at room temperature. Except for the first and last portions of the dichlorophenylborane distillate (each about 50 g) the obtained product remains colorless for at least 3 weeks, indicating the absence of tin-containing impurities.[5]

The use of a water aspirator vacuum is not recommended. Traps should be checked several times during the distillation and, if necessary to prevent blockage, any collected $SnCl_4$ should be discarded.

After the distillation is complete, the equipment should be flushed with dry nitrogen. It must be cooled to room temperature prior to disassembly. It is suggested that all equipment be rinsed with copious amounts of water immediately after use.

Since dichlorophenylborane tends to freeze joints, samples which are to be stored should be distilled into ampoules and sealed off.

The physical properties and reactions of dichlorophenylborane have been summarized in detail.[1]

References

1. *Gmelin Handbuch der Anorganischen Chemie*, Erg.-Werk (New Supplement Series), Vol. 34, "Borverindungen" Part 9, pp. 189–199.
2. J. E. Burch, W. Gerrard, M. Howarth, and E. F. Mooney, *J. Chem. Soc.*, **1960**, 4916.
3. K. Niedenzu and J. W. Dawson, *J. Am. Chem. Soc.*, **82**, 4223 (1964).
4. P. M. Treichel, J. Benedict, and R. G. Haines, *Inorg. Synth.*, **13**, 32 (1972).
5. J. C. Lockhart, *J. Chem. Soc. A*, **1966**, 1552.

48. (DIMETHYLAMINO)DIETHYLBORANE

$$2B(C_2H_5)_3 + B[N(CH_3)_2]_3 \longrightarrow 3(CH_3)_2NB(C_2H_5)_2$$

Submitted by F. ALAM* and K. NIEDENZU*
Checked by R. H. NEILSON†

(Dimethylamino)diethylborane, $(CH_3)_2NB(C_2H_5)_2$, was first prepared by the thermolysis of the dimethylamine adduct of triethylborane.[1] Later syntheses involved the dimethylaminolysis of chlorodiethylborane,[2] the B-alkylation of dichloro(dimethylamino)borane with a Grignard reagent,[3] and the ligand exchange between tris(dimethylamino)borane and triethylborane.[4] The preparation described here is a modified version of the last-named reaction. This procedure can also be used for the synthesis of a variety of (diorganylamino)-diorganylboranes.[5] The exchange reaction is quantitative within 10 hr at a reaction temperature of 120° or within 5 hr at 150° in a bomb tube.[4] Addition of a small quantity of material containing a B–H bond catalyzes the process and reduces the required reaction time and/or temperature.

Procedure

■ **Caution.** *Triethylborane is pyrophoric (a water-soaked towel should be kept available for immobilizing any spilled triethylborane) and tris(dimethylamino)borane as well as (dimethylamino)diethylborane are very sensitive to moisture. However, the following procedure does not require the use of a high-vacuum line.*

The apparatus consists of a 100-mL three-necked flask equipped with an argon inlet, a pressure-equalizing graduated 50-mL dropping funnel topped initially with an inlet tube, and a reflux condenser connected to a mercury bubbler (Fig. 1).

The equipment is flame-dried with all stopcocks (B, C, D) in an open position, while a slow stream of dry argon is passed through it, and inlet tube A is not yet connected to the triethylborane tank. A steady argon flow is maintained throughout all operations. After the apparatus cools to room temperature, stopcock B is briefly removed; a stirring bar is inserted into the flask, and the flask is charged with approximately 0.5 to 1 g of $NaBH_4$, and 14.3 g (0.1 mole) of tris(dimethylamino)borane.[6] After replacing stopcock B and waiting for 5 min,

---

*Department of Chemistry, University of Kentucky, Lexington, KY 40506.
†Department of Chemistry, Texas Christian University, Fort Worth, TX 76129.

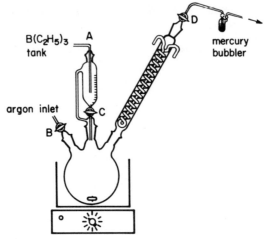

Fig. 1. *Apparatus for the synthesis of (dimethylamino)diethylborane.*

stopcock C is closed, and the inlet tube A is connected to the triethylborane tank by a short piece of Tygon tubing. Slightly less than the required quantity (28.1 mL, 19.6 g, 0.2 mole) of triethylborane* (approximately 26 to 27 mL) is transferred into the dropping funnel. (Employing a deficiency of triethylborane reduces the possibility of having some unreacted pyrophoric material left in the reaction product.) The triethylborane tank is disconnected from the inlet tube A, which is then quickly closed by a stopper. Traces of triethylborane adhering to A may ignite and are immobilized with the wet towel. The triethylborane is then slowly added with stirring to the reaction flask. After the addition is complete, the mixture is gently refluxed. The heating is gradually increased as the reflux temperature increases and, after about 2 hr, a bath temperature of approximately 135° is reached. This is sufficient to maintain reflux of the mixture. After a total 5 hr of reflux, the reaction is complete and the product is allowed to cool to room temperature. The product is then transferred under argon cover into a single-necked flask and is distilled under reduced pressure through a 25-cm silver-mantle column filled with stainless steel helices or, even better, through a small spinning band column. (Dimethylamino)diethylborane distills at 61–62°/120 torr (47°/48 torr, 121–122°/760 torr) and yields of 90% and better of pure product are obtained. Owing to the hydrolytic sensitivity of the compound, it is best to store it in sealed ampoules.

Properties

(Dimethylamino)diethylborane is a mobile liquid that is sensitive to moisture. The near-IR spectrum has major bands as follows (neat liquid between NaCl

*Callery Chemical Company, Callery, PA 16024.

plates, cm^{-1}): 3017(m), 2958(s), 2929(s, b), 2877(s), 2802(m), 1510(s), 1463(s), 1410(s), 1325(w), 1293(s), 1209(s), 1150(s), 1102(w), 1059(w), 1022(m), 926(m), 747(w). NMR data (neat liquid, chemical shifts δ in ppm with positive values indicating downfield from the given reference): δ ^{1}H (versus internal TMS) = 0.87 (broad singlet, 10 H), 2.66 (singlet, 6H); δ ^{11}B (versus external (C$_2$H$_5$)$_2$O · BF$_3$) = 45.7; δ ^{13}C (proton-decoupled; versus internal TMS) = 7.8 (singlet), 11.3 (broad singlet), 37.9 (singlet).

References

1. P. Buchheit, Dissertation, University of Munich, Germany, 1942.
2. H. Nöth, H. Schick, and W. Meister, *J. Organometal. Chem.*, **1**, 401 (1964).
3. K. E. Kalantar, cited in: C. E. Erickson and F. C. Gunderloy, *J. Org. Chem.*, **24**, 1161 (1959).
4. G. Abeler, H. Bayrhuber, and H. Nöth, *Chem. Ber.*, **102**, 2249 (1969).
5. K. Niedenzu, H. Beyer, J. W. Dawson, and H. Jenne, *Chem. Ber.*, **96**, 2653 (1963).
6. K. Niedenzu and J. W. Dawson, *Inorg. Synth.*, **10**, 135 (1967).

49. 2,3-DIETHYL-2,3-DICARBA-*nido*-HEXABORANE(8)

B$_5$H$_9$ + C$_2$H$_5$C≡CC$_2$H$_5$ + (C$_2$H$_5$)$_3$N ⟶ (C$_2$H$_5$)$_2$C$_2$B$_4$H$_6$ + (C$_2$H$_5$)$_3$NBH$_3$

Submitted by R. B. MAYNARD,* L. BORODINSKY,* and R. N. GRIMES*
Checked by N. HOSMANE†

Small carboranes of particular importance in syntheses are 2,3-dicarba-*nido*-hexaborane(8), (C$_2$B$_4$H$_8$), and its *C,C'*-substituted alkyl derivatives (Fig. 1).[1-3] These R$_2$C$_2$B$_4$H$_6$ compounds are highly versatile reagents which readily undergo complexation by main group and transition metals; η^1 (sigma), η^2 (bridged), and η^5 (face-bonded) complexes are well known and, in turn, have given rise to novel kinds of carborane stereochemistry. For example, bridged complexes such as η,η'-[(CH$_3$)$_2$C$_2$B$_4$H$_5$]$_2$Hg undergo oxidative linkage with loss of the metal to give B,B,-linked carboranes,[4] whereas bis(carborane) sandwich complexes containing the [R$_2$C$_2$B$_4$H$_4$]$^{2-}$ ligands exhibit oxidative fusion to produce four-carbon carboranes,[5] for example, [R$_2$C$_2$B$_4$H$_4$]$_2$FeH$_2$ → R$_4$C$_4$B$_8$H$_8$.

Until recently, the standard preparative route to 2,3-R$_2$C$_2$B$_4$H$_6$ compounds (R = H or alkyl) was the gas-phase reaction of alkyne with pentaborane(9), B$_5$H$_9$, at elevated temperature.[6,7] The scale of this reaction is limited by considerations of safety and practicality which center on (1) the utilization of large-

*Department of Chemistry, University of Virginia, Charlottesville, VA 22901.
†Department of Chemistry, Virginia Polytechnic Institute, Blacksburg, VA 24061.

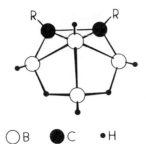

○ B ● C • H   Fig. 1. Structure of $2,3\text{-}R_2C_2B_4H_6$.

volume (1-2-L) glass bulbs to maintain the reactants in the gas phase, (2) the requirement of a large (~9:1) excess of alkyne, most of which is not recovered, (3) the dangers of handling pentaborane(9)-alkyne mixtures at high temperatures, and (4) the fact that the pure carbaborane product must be separated from the product mixture by preparative gas-liquid chromatography (GLC), which is of necessity a small-scale operation. Hence only a few millimoles of $2,3\text{-}R_2C_2B_4H_6$ product are obtainable per run. An alternate approach[7,8] utilizes the alkyne-B_5H_9 reaction at room temperature in the presence of 2,6-dimethylpyridine, but again the product must be purified by GLC on a millimole scale, and the removal of the dimethylpyridine is difficult.

A recently described procedure[9] which employs a strong Lewis base, triethylamine, generates 2,3-dialkyl derivatives of $C_2B_4H_8$ under mild conditions in good yield, and the product can be purified on a multigram scale by standard vacuum-line fractionation. This method has not, so far, proved successful for parent $C_2B_4H_8$ or for the 2,3-diphenyl derivative, but is the best available synthesis for $2,3\text{-}R_2C_2B_4H_6$ derivatives in which R = CH_3, C_2H_5, or C_3H_7. The most consistently successful results are obtained for the 2,3-diethyl derivative, the preparation of which is described here.

Procedure

■ **Caution.** *Pentaborane(9) is toxic and spontaneously inflames or explodes on exposure to air. This procedure should be carried out only by chemists experienced in vacuum-line techniques.*[10,11]

Triethylamine* is freed from primary and secondary amine impurities by refluxing and distillation from phthalic anhydride, after which it is dried by distillation from BaO under N_2. 3-Hexyne† is purified by twice stirring over metallic sodium and distilling under N_2, and pentaborane(9)‡ is used as received. All the

*Eastman Kodak Company, Rochester, NY. 14608.
†Albany International, Farchan Labs, Willoughby, OH. 44094.
‡Callery Chemical Company, Callery, PA 16024.

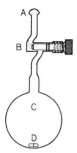

Fig. 2. Apparatus for the preparation of $2,3\text{-}(C_2H_5)_2C_2B_4H_6$. (A) Ball-and-socket connection to vacuum line; (B) high-vacuum valve with Teflon plug, 0–5-mm orifice; (C) 500-ml round-bottomed flask; (D) magnetic stirring bar.

above reagents are stored under vacuum in 100-mL Pyrex bulbs equipped with greaseless, O-ring, high-vacuum Teflon valves.

A 500-mL Pyrex reactor fitted with a Teflon high-vacuum valve (Fig. 2) is attached to a conventional vacuum line along with the bulbs containing the three reactants. Into the reactor, immersed in liquid nitrogen, are condensed 50 mmole of B_5H_9, 20 mmole of 3-hexyne, and 8.5 mmole of $(C_2H_5)_3N$. The liquid nitrogen bath is removed. When the contents of the reactor have melted, the reactor is placed in an ice bath, and the mixture is stirred for approximately 2 hr. After again cooling the reactor in liquid nitrogen, the stopcock is opened to the vacuum line and the trace of noncondensable gas is evacuated. At this point, the material in the reactor has begun to turn yellow. A second addition of 20 mmole of 3-hexyne and 8.5 mmole of $(C_2H_5)_3N$ is made, and the mixture is again stirred in an ice bath for approximately 2 hr. The above procedure is repeated until a total of 100 mmoles of 3-hexyne and 42.5 mmoles of $(C_2H_5)_3N$ have been added over the course of approximately 10 hr. At this point, the mixture is stirred overnight in an ice bath. The following day, the reactor is allowed to warm to room temperature and then stirred continuously for 3 days.

On the fourth day following the additions, the volatile reaction products are fractionated at room temperature over a period of 4 hr through a trap at 0° (ice bath) and a trap at −196° (liquid N_2). The reaction vessel is then heated in a water bath to 70° for an additional 2 hr. The $[(C_2H_5)_3N]BH_3$ in a trap at 0° and the orange-red nonvolatile polymer remaining in the reactor are discarded. The material in the trap at −196° is warmed to room temperature and allowed to stand overnight, at which point the color is pale yellow. The material is fractionated through traps at 0°, −63° (chloroform slush), and −196°; the small amount of pentaborane(9) in the trap at −196° can either be saved for subsequent reactions or destroyed with an excess of $(C_2H_5)_3N$. Again, the material in the trap at 0° remaining nonvolatile materials are discarded. The $(C_2H_5)_2C_2B_4H_6$ in the trap at −63° is twice fractionated at −63° and −196° to remove any traces of residual B_5H_9. The trap containing the pure $(C_2H_5)_2C_2B_4H_6$ is flooded with nitrogen gas, and the carborane is transferred to a storage container under nitrogen or vacuum. Yields of $(C_2H_5)_2C_2B_4H_6$ obtained by this procedure have ranged from 1.82 to 1.97 g (14-15 mmole), 33-35% based on $(C_2H_5)_3N$ employed. The reaction has been scaled up to twice the above quantity with no decrease in product yield or purity.

Properties

2,3-Diethyl-2,3-dicarba-*nido*-hexaborane(8) is a colorless, moderately air-sensitive liquid with a vapor pressure of 14 torr at 24°. The spectrum (neat sample between NaCl plates) exhibits major absorption bands at 2975(vs), 2940(s), 2883(s), 2600(vs), 1952(w), 1919(w), 1468(s), 1382(m), 1073(w), 998(w), 946(m), 924(m), 848(m), 799(w), 770(w), 724(w), 661(m) cm^{-1}. The unit-resolution mass spectrum exhibits an abundant molecular ion with a cutoff at m/z 133 corresponding to $^{13}C^{12}C_5{}^{11}B_4{}^1H_{16}^+$.

References

1. T. Onak, in *Boron Hydride Chemistry*, E. L. Muetterties (ed.), Academic Press, New York, 1975, Chapter 10.
2. R. N. Grimes, "Metallacarboranes and Metallaboranes," in *Comprehensive Organometallic Chemistry*, G. Wilkinson, F. G. A. Stone, and E. Abel (eds.), Pergamon Press, Oxford, 1982, Chapter 5.5.
3. R. N. Grimes, *Acc. Chem. Res.*, **11**, 420 (1978); Ibid **16**, 22 (1983).
4. N. S. Hosmane and R. N. Grimes, *Inorg. Chem.*, **18**, 2886 (1979).
5. W. M. Maxwell, V. R. Miller, and R. N. Grimes, *Inorg. Chem.*, **15**, 1343 (1976).
6. T. Onak, R. P. Drake, and G. B. Dunks, *Inorg. Chem.*, **3**, 1686 (1964).
7. T. Onak, R. E. Williams, and H. G. Weiss, *J. Am. Chem. Soc.*, **84**, 2830 (1962).
8. T. Onak, F. J. Gerhart, and R. E. Williams, *J. Am. Chem. Soc.*, **85**, 3378 (1963).
9. N. S. Hosmane and R. N. Grimes, *Inorg. Chem.*, **18**, 3294 (1979).
10. D. F. Shriver, *The Manipulation of Air Sensitive Compounds*, McGraw-Hill Book Co., New York, 1969, Chapter 1.
11. W. L. Jolly, *The Synthesis and Characterization of Inorganic Compounds*, Prentice-Hall, Inc., Englewood Cliffs, New Jersey, 1970, Chapters 7 and 8.

50. 2,2′,3,3′-TETRAETHYL-1,1-DIHYDRO-[1,1′-commo-BIS(2,3-DICARBA-1-FERRA-closo-HEPTABORANE)](12), $[2,3-(C_2H_5)_2C_2B_4H_4]_2FeH_2$ AND 2,3,7,8-TETRAETHYL-2,3,7,8-TETRACARBADODECABORANE(12), $(C_2H_5)_4C_4B_8H_8$

Submitted by RICHARD B. MAYNARD* and RUSSELL N. GRIMES*
Checked by NARAYAN S. HOSMANE†

The red diamagnetic complexes $(2,3-R_2-2,3-C_2B_4H_4)_2FeH_2$ (R = CH_3, C_2H_5, or C_3H_7) are highly reactive species which are starting materials for several types of syntheses.[1] One of the more important reactions exhibited by these complexes is the oxidative fusion of the formal $[R_2C_2B_4H_4]^{2-}$ carbaborane ligands to give neutral $R_4C_4B_8H_8$ tetracarbon systems (Fig. 1). This process occurs in tetrahydrofuran (THF) solution on exposure to air at room temperature and is essentially quantitative.[1,2] The $R_4C_4B_8H_8$ products (R = CH_3, C_2H_5, or C_3H_7) are also of unusual interest in their own right. These compounds are air- and heat-stable colorless crystalline solids which undergo reversible fluxional cage rearrangement in solution and are known to have distorted icosahedral cage structures in the solid state.[3] Insertion of transition metal groups such as $Fe(\eta^5-C_5H_5)$, $Co(\eta^5-C_5H_5)$, and $Ni[(C_6H_5)_2PCH_2]_2$ into neutral $R_4C_4B_8H_8$ or the $[R_4C_4B_8H_8]^{2-}$ dianion generates tetracarbon metallacarbaboranes having 11- to 14-vertex cage systems, many of which are structurally novel.[1,4] Additionally, $(CH_3)_4C_4B_8H_8$ can be degraded to give $(CH_3)_4C_4B_7H_9$, an 11-vertex arachno system.[5]

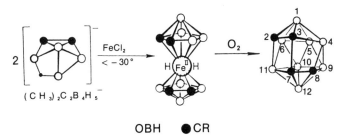

Fig. 1. Conversion of $[R_2C_2B_4H_5]^-$ carborane ions (R = CH_3, C_2H_5, or C_3H_7) to $[R_2C_2B_4H_4]_2FeH_2$ and oxidative fusion of the latter complex to $R_4C_4B_8H_8$ carboranes.

*Department of Chemistry, University of Virginia, Charlottesville, VA 22901.
†Department of Chemistry, Virginia Polytechnic Institute and State University, Blacksburg, VA 24061.

216 Compounds of Boron

This article describes two syntheses: the bis(carbaborane)iron(III) complex and the conversion of this compound to the neutral tetracarbon carborane. Both procedures generate high yields of product and entail normal vacuum-line synthetic techniques. The starting material, 2,3-diethyl-2,3-dicarba-*nido*-hexaborane(8) can be obtained by the procedure described in the preceding article.

A. 2,2',3,3'-TETRAETHYL-1,1-DIHYDRO-[1,1'-*commo*-BIS(2,3-DICARBA-1-FERRA-*closo*-HEPTABORANE)](12)

$$(C_2H_5)_2C_2B_4H_6 + NaH \longrightarrow Na^+[(C_2H_5)_2C_2B_4H_5]^- + H_2$$

$$2Na^+[(C_2H_5)_2C_2B_4H_5]^- + FeCl_2 \longrightarrow [(C_2H_5)_2C_2B_4H_4]_2FeH_2 + 2NaCl$$

■ **Caution.** *When free of mineral oil, sodium hydride will inflame in air or on exposure to moisture. Sodium hydride may be destroyed by careful addition of dry ethyl acetate until gas evolution ceases, followed by addition of ethanol and finally water.*

The preparation and properties of $(C_2H_5)_2C_2B_4H_6$ are described in the preceding synthesis.

A 100-mL, round-bottom flask equipped with a magnetic stirring bar and a glass stopcock (Fig. 2) is charged with 0.5 g (4 mmole) of anhydrous $FeCl_2$. This flask is attached to a coarse sintered-glass filter with a side-arm connected to a 100-mL, two-necked flask. The two-necked flask is charged with 0.24 g (10 mmole) of NaH (50% in mineral oil) which has been rinsed with dry pentane. To the remaining neck is attached a tipper equipped with a greaseless stopcock containing 0.9 g (7 mmole) of $(C_2H_5)_2C_2B_4H_6$. The entire apparatus is attached to a vacuum line and evacuated for several hours.

Following this, the stopcock leading to the $FeCl_2$ is closed, and ~50 mL of

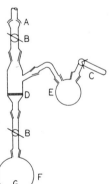

Fig. 2. Apparatus for the preparation of $[(C_2H_5)_2C_2B_4H_4]_2FeH_2$. (A) Ball-and-socket connection to vacuum line; (B) high-vacuum greaseless stopcock; (C) side-arm tipper fitted with high-vacuum greaseless stopcock; (D) coarse sintered-glass filter; (E) two-necked 100-mL round-bottom flask; (F) single-neck 100-mL round-bottom flask; (G) magnetic stirring bar.

THF is condensed into the two-necked flask immersed in liquid nitrogen, after which the contents of the tipper are added to the flask. The mixture is allowed to warm to ~-30°, at which point hydrogen evolution begins. The flask is briefly cooled with liquid nitrogen whenever the bubbling becomes too vigorous, to ensure that the material remains in the flask. Following the cessation of hydrogen evolution, the contents of the flask are refrozen and the noncondensables are pumped away. The contents of the flask are then allowed to warm to room temperature.

This solution is filtered onto the $FeCl_2$ in the lower (100-mL) flask which is cooled in liquid nitrogen. The solution is allowed to warm to ~-15°, where reaction with the $FeCl_2$ occurs. Stirring is continued at -15° for 2 hr, at which point the solution is a deep wine-red color. The THF is then removed under vacuum to leave a dark red residue. This material is transferred into a dry box where it is washed with hexane through a 1-cm layer of silica gel on a sintered glass filter until the washings are colorless. The hexane is removed under vacuum to give 0.83 g (2.6 mmole, 75% yield) of $[2,3\text{-}(C_2H_5)_2C_2B_4H_4]_2FeH_2$. The product may be used without further purification in the procedure in Part B.

Properties

2,2',3,3'-Tetraethyl-1,1-dihydro-[1,1'-*commo*-bis(2,3-dicarba-1-ferra-*closo*-heptaborane)] (12) is a moderately air-sensitive dark red oil. It can be easily vacuum-distilled (10^{-5} torr) from a water bath heated to 70° and is soluble in most common organic solvents.

The ^{11}B-NMR spectrum (115 MHz) exhibits three doublets of relative area 1:2:1 at (δ, ppm(Hz)) -0.96(163), -8.26(157), and -19.84(149). The unit-resolution mass spectrum contains a strong molecular ion signal with a high-mass cut-off (exclusive of ^{13}C peaks) at m/z 318 corresponding to the $^{56}Fe^{11}B_8\text{-}^{12}C_{12}{}^1H_{30}^{+\cdot}$ ion.

B. 2,3,7,8-TETRAETHYL-2,3,7,8-TETRACARBADODECABORANE(12)

$$[(C_2H_5)_2C_2B_4H_4]_2FeH_2 \longrightarrow (C_2H_5)_4C_4B_8H_8 + H_2 + \text{solids}$$

An 0.80-g sample (2.6 mmole) of $[2,3\text{-}(C_2H_5)_2C_2B_4H_4]_2FeH_2$ is dissolved under vacuum in 100 mL of hexane or dichloromethane, and O_2 gas is bubbled through the solution with stirring until there results a clear supernatant and a dark precipitate. The mixture is filtered in air, and the filtrate is evaporated on the rotary evaporator to near dryness. This material is eluted in hexane on preparative silica-gel TLC plates to give 0.59 g (2.3 mmole, 87% yield) of 2,3,7,8-$(C_2H_5)_4C_4B_8H_8$, R_f = 0.34 in *n*-hexane.

Properties

2,3,7,8-Tetraethyl-2,3,7,8-tetracarbadodecaborane(12) is a colorless, air-stable solid, melting at 57 ± 1° (uncorrected) and soluble in most common organic solvents. The infrared spectrum (KBr pellet) exhibits major absorption bands at 2970(s), 2936(s), 2874(s), 2515(vs), 1442(s), 1377(s), 1278(w), 1118(w), 1059(m), 1024(m), 1007(m), 950(s), 938(s), 870(m), 794(w), 766(w), 732(w), 713(w), 682(w), 645(w) cm^{-1}. The unit-resolution mass spectrum shows an abundant molecular ion, exhibiting a cutoff at m/z 261 corresponding to $^{13}C^{12}C_{11}{}^{11}B_8{}^1H_{28}^+$ ion.

The proton-decoupled 115-MHz ^{11}B-NMR spectrum of the pure carborane in CDCl$_3$, observed within a few minutes of preparation of the solution, consists of two resonances in a 3:1 area ratio at δ -3.11 and -12.08. (Gradual appearance of additional peaks as the solution stands at room temperature is due to reversible formation of an isomer.[2,3c])

References

1. R. N. Grimes, *Acc. Chem. Res.*, **11**, 420 (1978) and references therein; Ibid, **16**, 22 (1983).
2. W. M. Maxwell, V. R. Miller, and R. N. Grimes, *Inorg. Chem.*, **15**, 1343 (1976).
3. (a) D. P. Freyberg, R. Weiss, E. Sinn, and R. N. Grimes, *Inorg. Chem.*, **16**, 1847 (1977); (b) R. N. Grimes, W. M. Maxwell, R. B. Maynard, and E. Sinn, *Inorg. Chem.*, **19**, 2981 (1980); (c) R. B. Maynard, T. L. Venable, and R. N. Grimes, Abstr. Papers, 183rd National Meeting of the American Chemical Society, Las Vegas, Nevada, March 1982, Abstract INOR-140.
4. R. B. Maynard, E. Sinn, and R. N. Grimes, *Inorg. Chem.*, **20**, 1201 (1981) and references therein.
5. D. C. Finster and R. N. Grimes, *J. Am. Chem. Soc.*, **103**, 2675 (1981).

51. ^{10}B-LABELED BORON COMPOUNDS

Submitted by H. NÖTH* and R. STAUDIGL*
Checked by DONALD F. GAINES† and WILLIAM R. BEARD†

The course of chemical reactions in boron chemistry can be readily followed using ^{10}B-labeled compounds. For instance, using isotopically labeled boron compounds will show whether the exchange of substituents between two boron compounds is accompanied by boron transfer or not. Only a few such studies

*Institut für Anorganische Chemie der Universität München, Meiserstr. 1, D-8000 München 2, Germany.
†Department of Chemistry, University of Wisconsin-Madison, Madison, WI 53706.

have so far been published,[1] but these will become increasingly important if a better understanding of reactivity patterns in boron chemistry is to be achieved.

Boron-10 isotopic labeling of boron compounds is more useful than ^{11}B labeling since the latter isotope is more abundant in nature (80.39%). Commercially available starting materials for ^{10}B labeling are ^{10}B-enriched elemental boron or boric acid.* The ^{10}B content may be higher than 90%, and it is convenient for mechanistic studies to use a material containing 93.7% ^{10}B and 6.3% ^{11}B.

The compound $[^{10}B](OH)_3$† proved a convenient starting material for the preparation of more reactive boron compounds. The procedures follow classical reactions, though some modifications were introduced in order to achieve high yields of the ^{10}B-labeled species.

A. [^{10}B] BORON TRIBROMIDE

$$K[[^{10}B]F_4] + AlBr_3 \longrightarrow \text{``}KAlF_4\text{''} + [^{10}B]Br_3$$

The metathesis outlined in the equation produces $[^{10}B]Br_3$ in unsatisfactory yield[2] unless a large excess of $AlBr_3$ is used. A ratio of $K[[^{10}B]F_4]$: $AlBr_3$ = 1:10 is the most suitable.[3]

The starting material, $K[[^{10}B]F_4]$, is readily prepared by dissolving $[^{10}B](OH)_3$ in hydrofluoric acid, and neutralizing the solution with aqueous KOH^4 or K_2CO_3. Although any pure, freshly sublimed $AlBr_3$[5] can be used in the preparation of $[^{10}B]Br_3$, it is advisable to synthesize aluminum bromide shortly before it is used.

■ **Caution.** *Due to the high sensitivity of all compounds to moisture, care must be taken in these synthesis to provide strictly anhydrous conditions. Therefore, all parts of the apparatus used are to be dried carefully and are best kept under an atmosphere of dry, oxygen-free nitrogen. All reactions are carried out in an efficiently ventilated hood, and provisions must be made for absorbing or freezing out effluent gases. Care must also be exercised when cleaning the apparatus. 2-Propanol admitted under nitrogen gas is a good agent for destroying and (partially) dissolving residual compounds before disassembling the apparatus.*

Procedure

A 500-mL round-bottomed flask is charged with 35.0 g (1.3 mole) of aluminum turnings and a Teflon-coated magnetic stirring bar (length ~2.5 cm). The flask

*Oak Ridge National Laboratories, Oak Ridge, TN 37830; Sonneborn Refractories and Chemicals Inc., Aiken, SC 29801.

†The formula $[^{10}B](OH)_3$ does not represent boric acid containing only the ^{10}B isotope but stands for a compound highly enriched in ^{10}B. This holds also for all the other ^{10}B-labeled compounds described here.

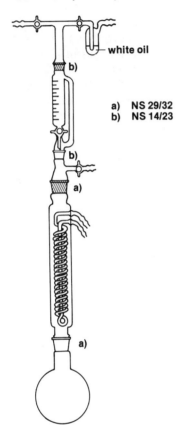

a) NS 29/32
b) NS 14/23

Fig. 1. *Reaction apparatus for synthesis of $AlBr_3$.*

is attached to a Dimroth reflux condenser (see Fig. 1) which carries a dropping funnel with pressure equalizer. Vacuum is then obtained by means of an oil pump. The apparatus is flamed to remove adsorbed water and finally filled with dry, oxygen-free nitrogen. Dry bromine, (77 mL, 1.5 mole), is transferred into the dropping funnel, and its stopper replaced by a bubbler filled with Nujol. The flask which is immersed in an oil bath to a level no higher than the turnings, is heated to 120° using a hot plate-stirrer. Bromine is then carefully added dropwise onto the stirred aluminum turnings.

■ **Caution.** *The reaction is highly exothermic and forms white clouds of $AlBr_3$. Sparks can sometimes be observed when a bromine drop hits the metal.*

Part of the aluminum bromide deposits on the cooler parts of the flask, forming colorless crystals. The dropping rate is adjusted so that the $AlBr_3$ vapor does not reach the coil of the reflux condenser. A pressure drop observed towards the end of the reaction can be readily equalized by slowly admitting nitrogen

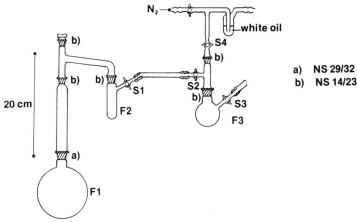

Fig. 2. Vacuum set-up for synthesis of $[^{10}B]Br_3$, part 1. (F1) flask; (F2) Schlenck or U-tube; (F3) flask; (S1, S2, S3, S4) stopcocks.

into the apparatus via the bubbler. Two to 2.5 hr are required to complete the reaction. The brownish liquid is then stirred for an additional 20 min at 120°. During this time, nitrogen gas is passed through the bubbler, carrying away most of any free bromine which may be present. The contents of the flask solidify on cooling to room temperature. The product has a dark appearance due to excess finely divided aluminum metal. This material can be used without further purification for the reaction with $K[[^{10}B]F_4]$.

The apparatus shown in Fig. 2 is evacuated,* thoroughly flamed, and filled with dry nitrogen gas. Finely ground, dry $K[[^{10}B]F_4]$, (12.5 g, 0.1 mole) is introduced into the flask containing the freshly prepared $AlBr_3$ under a countercurrent flow of nitrogen. This flask is then rapidly exchanged for the empty flask attached to the apparatus of Fig. 2 (F1), while a slow flow of nitrogen through the bubbler is maintained to prevent any air from diffusing into the apparatus.

■ **Caution.** *Stopcocks S1, S2, and S4 must be opened while S3, leading to a high-vacuum system, remains closed. Safety traps should be used in the gas lines and wherever else deemed appropriate.*

The Schlenck tube† (F2) is cooled by liquid nitrogen while the flask (F1) containing $AlBr_3$ and $K[[^{10}B]F_4]$ is slowly heated in an oil bath sitting on a hot plate-stirrer. As the $AlBr_3$ melts, rapid stirring becomes essential since the solid tetrafluoroborate must mix quickly and effectively with the $AlBr_3$. This generates some $[^{10}B]Br_3$, which condenses in F2. When the oil bath reaches 160°,

---

*Kel-F or a similar stopcock grease should be employed.
†A pump trap or U tube will also suffice.

the stopcocks S2 and S4 are closed. Vacuum is then applied to flask F3 by opening S3. The stopcock S3 is closed after a pressure of ~1 torr is attained, and stopcock S2 is slowly and cautiously opened. More $[^{10}B]Br_3$ condenses quickly into F2. Careful control at this stage is necessary since the reduction of pressure causes frothing in the reaction flask. After about 5 min, S2 is closed again and S3 opened. This procedure, by which the pressure in the apparatus is intermittently reduced, is repeated several times until the $AlBr_3$ starts volatilizing. It is sufficient to maintain this condition for 15 min. It is important at this stage to control the temperature so that no $AlBr_3$ sublimes into F2. If, however, significant quantities of $AlBr_3$ condense into F2, on warming to room temperature its contents may appear solid. In such a case, work-up for $[^{10}B]Br_3$ may proceed as described; however, it is preferable to repeat the purification step.

After removal of the oil bath, nitrogen gas is admitted through S4 until the pressure has equalized and the flask F1 has cooled to room temperature. This flask contains "$KAlF_4$," Al, and excess $AlBr_3$. Part of the excess $AlBr_3$ can be recovered by sublimation in a vacuum. The residue may then be removed from the flask by slowly and carefully hydrolyzing its contents over a period of days by adding small quantities of water. This procedure requires an efficient hood.

The Dewar flask containing liquid nitrogen is removed from F2, allowing its contents to warm up slowly to room temperature. Some gas (HBr, BF_3) may escape through the bubbler during this process, but it can be absorbed in NaOH solution. The material in F2 is a slightly reddish liquid, the color of which derives from small amounts of free bromine. Purification of the $[^{10}B]Br_3$ is carried out using either a conventional high vacuum system or the apparatus shown in Fig. 3.

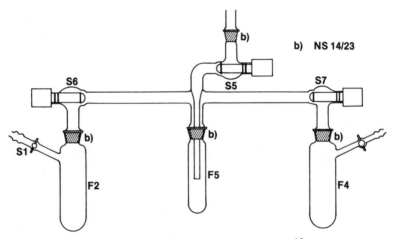

Fig. 3. Vacuum system for purification of $[^{10}B]Br_3$.

The contents of F2 are cooled with liquid nitrogen and F2 is attached to the apparatus (Fig. 3) while a slow nitrogen flow is passed through S1. Trap F4 contains a few drops of mercury and a small stirring bar, while F5 is an empty trap. High vacuum is applied to the system, and the valve S5 is closed. The Dewar flask is then removed from S2 and used to cool F4. Both [^{10}B]Br$_3$ and Br$_2$ condense in F4. The temperature in F2 must not exceed 0° to prevent the vaporization of AlBr$_3$. After all the volatile material has accumulated in F4, the valves S6 and S7 are closed and the contents in F4 warmed to room temperature. The liquid in F4 is now stirred for 30 min, and valve S7 is opened. Trap F5 is then cooled with dry ice/2-propanol, and valve S7 is opened. The impure [^{10}B]Br$_3$ is distilled at 0° from F4 into F5. This product may be further fractionated in a train of traps to remove traces of HBr. However, the [^{10}B]Br$_3$ obtained contains very little HBr, and another purification step is not necessary in most cases. The product is transferred into a vacuum storage flask. Yield: 19.0 g (76%).*

Properties

[^{10}B]Boron tribromide is a colorless liquid which fumes copiously on contact with air, bp 90.8°/735 torr, mp −46°. δ ^{11}B (^{10}B) 39 ppm (versus BF$_3$ · OR$_2$).

The ^{10}B content of the product can be readily and accurately determined by mass spectrometry using the peaks resulting from BBr$_3^+$ and/or BBr$_2^+$.

■ **Caution.** *Boron tribromide vapors are harmful to the respiratory system. Contact with the skin should also be avoided.*

B. DIBROMOMETHYL[^{10}B]BORANE

$$2[^{10}B]Br_3 + Sn(CH_3)_4 \longrightarrow 2CH_3[^{10}B]Br_2 + Br_2Sn(CH_3)_2$$

A large variety of methylated metals can be used selectively to methylate boron bromide to dibromomethylborane.[6] Tetramethylstannane is the most convenient to use since it does not require a solvent, and this allows an easy separation of the highly volatile CH$_3$[^{10}B]Br$_2$.

■ **Caution.** *Dibromomethylborane is not only readily hydrolyzed; it may ignite spontaneously in air if it is warm or if the heat liberated on oxidation or hydrolysis cannot be dissipated. It must therefore be handled with care in an atmosphere of dry nitrogen or argon. Use of syringe techniques when transferring is recommended. Also, all methylstannane compounds are poisonous, and any contact with the skin must be prevented.*

---

*The yield depends greatly on the efficiency of mixing AlBr$_3$ and K[[^{10}B]F$_4$] and on the particle size of the tetrafluoroborate. Yields between 65 and 85% have been obtained consistently.

Procedure

A 25-mL, two-necked, round-bottomed flask is fitted with a magnetic stirring bar, a dropping funnel, and a 25-cm Vigreux column with a column head and a receiver. The column head must allow a complete reflux of the distilling product.* The outlet of the receiver is attached to a bubbler.

The entire apparatus is filled with dry nitrogen and the flask charged with [^{10}B]Br$_3$ (8.40 mL, 22.1 g, 88.6 mmole). It is then cooled to -20° to -30° by means of a dry ice/2-propanol bath. Magnetic stirring is commenced, and Sn(CH$_3$)$_4$ (6.0 mL, 7.9 g, 44 mmole) is added dropwise within 20-25 min.

■ **Caution.** *The addition of BBr$_3$ to Sn(CH$_3$)$_4$ must be avoided. This leads to a spontaneous reaction which cannot be readily controlled, producing (CH$_3$)$_2$BBr and even B(CH$_3$)$_3$ in addition to CH$_3$BBr$_2$.*

The mixture is then allowed to warm slowly to room temperature by no longer adding dry ice to the cooling bath. At approximately -15°, a white solid forms which redissolves later. The cooling bath is removed when the temperature reaches 0°. Further stirring for 20 min is followed by heating the flask for 30 min to 90° (oil bath). The mixture is heated at reflux for an additional 30 min, then fractionally distilled by raising the oil bath temperature to ~180°. This yields 14.7 g of a colorless liquid, bp 57-61°. Cooling the receiving flask with liquid nitrogen and applying vacuum (oil pump) increases the yield of the impure product to 15.8 g (~95%). It contains about 3% (CH$_3$)$_2$[^{10}B]Br and [^{10}B]Br$_3$. Both impurities can be readily removed by fractional redistillation, using a 50-cm spinning band column (yield 12-14 g), bp 59-60°/735 torr. This purification is not necessary if the product is to be used for the preparation of (CH$_3^{10}$B)$_2$S$_3$.

The colorless residue of the distillation is primarily (CH$_3$)$_2$SnBr$_2$. This can be purified by distillation, bp 208-213°.

Properties

Dibromomethyl [^{10}B]borane is a colorless liquid, which fumes strongly in contact with moisture. Larger quantities when exposed to air may ignite; bp 59°/736 torr, δ ^{11}B(^{10}B) 65.2 ppm, δ ^{1}H 1.42 ppm (5% in CCl$_4$). The compound dissolves readily in various hydrocarbons. It forms stable addition products with tertiary nitrogen bases and is readily solvolyzed by alcohols as well as primary and secondary amines. Boronium salts may form with the amines. Reaction with ethers ruptures the C-O bond.

---

*The model 9089 Vigreux column made by NORMAG-Otto Fritz GmbH, Hofheim/Taunus, Germany, was used.

C. 3,5-DIMETHYL[$^{10}B_2$]-1,2,4,3,5-TRITHIADIBOROLANE

$$2CH_3[^{10}B]Br_2 + 2H_2S_x \longrightarrow$$

$$+ 4HBr + \frac{2x-3}{8} S_8$$

Trithiadiborolanes have been obtained by various procedures, and many derivatives are known.[7] Many of these can be prepared by the reaction of a dimethyl derivative with BX_3 via a CH_3/X exchange reaction.[1c] The procedure developed for the ^{10}B-labeled compound follows the original preparation by Schmidt and Siebert.[7c]

■ **Caution.** *Although 3,5-dimethyl-1,2,4,3,5-trithiadiborolane has a low vapor pressure, it hydrolyzes readily with formation of H_2S. It must therefore be regarded as poisonous and should be handled with care.*

Procedure

A 50-mL, two-necked, round-bottomed flask fitted with a dropping funnel, a reflux condenser, and a stirring bar is purged with dry nitrogen gas and charged with 14.5 g $CH_3[^{10}B]Br_2$ (80 mmole). A mercury-filled bubbler is attached to the condenser, and its outlet is connected through Tygon tubing to an Erlenmeyer flask containing a KOH solution for absorbing the HBr liberated during the reaction.

The temperature of the reflux condenser is kept at $-50°$ by circulating cold acetone cooled by a dry ice/acetone bath.

The dropping funnel is then charged with 14 mL of dry hydrogen polysulfide[8] which is added within 30 min to the rapidly stirred $CH_3[^{10}B]Br_2$. Further stirring for 1 hr at $40°$ completes the reaction. After that time, the reflux condenser is replaced by a 15-cm Vigreux column carrying a column head and a receiver through which a current of nitrogen is being passed. The flask is then slowly heated to $160°$ at 8.5 torr to yield a yellowish distillate which on redistillation, using a 10-cm column filled with glass beads, gives 3.2 g (22 mmole, 55% yield) of nearly colorless, pure $(CH_3[^{10}B])_2S_3$, bp $52°/5.6$ torr.

Properties

The product is a colorless or slightly yellowish, malodorous liquid, by $52°/5.6$ torr; δ ^{11}B 70.5 ppm, δ ^1H 1.20 ppm; IR spectrum, bands (cm^{-1}) at 2980(s),

2900(w), 2800(w), 2120(w), 1495(w), 1300(vs), 1105(s), 1040(vs), 980(m), 870(m), 835(vs), 770(vs), 750(sh), 525(vs), 486(w), 390(w), 355(s).

The compound hydrolyzes readily and dissolves easily in hydrocarbons and ethers. It is solvolyzed by primary and secondary amines and by alcohols. It is quickly destroyed either by a hypochlorite solution or by bromine dissolved in methanol.

References

1. (a) M. Nadler and R. F. Porter, *Inorg. Chem.*, 6, 1192 (1966); (b) H. Nöth and T. Taeger, *Z. Naturforsch.*, **B34**, 135 (1979); (c) H. Nöth, R. Staudigl, and T. Taeger, *Chem. Ber.*, 114, 1157 (1981).
2. E. Lee Gamble, *Inorg. Synth.*, 3, 29 (1950).
3. H. Nöth and R. Staudigl, *Chem. Ber.*, 111, 3280 (1978).
4. P. A. van der Meulen and H. C. van Mater, *Inorg. Synth.*, 1, 24 (1939).
5. D. G. Nicholson, P. K. Winter, and G. Fineberg, *Inorg. Synth.*, 3, 31 (1950).
6. For literature see: H. Nöth and W. Storch, *Synth. Inorg. Metalorg. Chem.*, 1, 197 (1971); H. Nöth and P. Fritz, *Z. Anorg. Allgem. Chem.*, 322, 297 (1963).
7. M. Schmidt and W. Siebert, (a) *Angew. Chem.*, 76, 607, 687 (1964); (b) *Z. Anorg. Allgem. Chem.*, 345, 87 (1966); (c) *Chem. Ber.*, 102, 2752 (1969).
8. R. Brauer, *Handbuch der präparativen anorganischen Chemie*, 2nd ed., vol. 1, Ferdinand Enke Verlag, Stuttgart, 1960, p. 315.

52. THE THIADECABORANES
$arachno$-$[6$-$SB_9H_{12}]^-$, $nido$-6-SB_9H_{11}, AND $closo$-1-SB_9H_9

$$B_{10}H_{14} + S^{2-} + 4H_2O \longrightarrow [SB_9H_{12}]^- + [B(OH)_4]^- + 3H_2$$

$$[SB_9H_{12}]^- + 1/2 I_2 \longrightarrow SB_9H_{11} + 1/2 H_2 + I^-$$

$$SB_9H_{11} \longrightarrow SB_9H_9 + H_2$$

Submitted by R. W. RUDOLPH* and W. R. PRETZER†
Checked by A. ARAFAT,‡ D. HERNANDEZ-SZCZUREK,‡ T. HANUSA,‡ and L. J. TODD‡

Systematic approaches to the chemistry of heteroboranes suggest that the *closo*, *nido*, and *arachno* clusters of a given size heteroborane are related by a simple

---

*Deceased-Department of Chemistry, University of Michigan, Ann Arbor, MI 48019.
†Gulf Research and Development, Pittsburgh, PA 15230.
‡Department of Chemistry, Indiana University, Bloomington, IN 47405.

redox chemistry.[1] A prime example of this is found in the case of the thiadecaboranes discussed here. The procedures described here employ the readily available starting material decaborane(14), $B_{10}H_{14}$. Preparation of *arachno*-$[6\text{-}SB_9H_{12}]^-$ presents a slight modification of the originally reported procedure.[2] However, the iodine oxidation of the latter anion described herein is superior to the original protolytic preparation. Finally, 1-SB_9H_9 is readily prepared in good yield by a simple pyrolytic disproportionation of 6-SB_9H_{11}.

These thiadecaboranes have a rich chemistry which includes their conversion to other thiaboranes,[2,3] the preparation of metallothiaboranes[2,4] and homogeneous catalysis,[5] Friedel-Crafts substitution,[6,7] hydroboration reactions,[8] and acid-base reactions.[2,9]

A. CESIUM DODECAHYDRO-6-THIA-*arachno*-DECABORATE(1-)

Procedure

All preparations are conducted under an atmosphere of prepurified nitrogen unless specified otherwise. Solvents are vacuum distilled from reservoirs containing appropriate drying agents. Decaborane(14) (See Chapter 5, Section 46) is freshly sublimed prior to use.

■ **Caution.** *Decaborane(14) is purified by sublimation at 60° and 10 torr. The apparatus should be situated in a hood which is well shielded. If air should accidentally contact decaborane at elevated temperatures, a violent oxidation might result. Decaborane(14) is toxic; therefore, contact of the solid with the skin and exposure to the vapor should be avoided. Also, decaborane(14) and halogenated hydrocarbons can form extremely shock-sensitive solutions.*

The procedures are confined to a well-ventilated hood because of the sulfide odors. Waste sulfide solutions should not be mixed with acids because of H_2S danger.

The apparatus consists of a 500-mL, three-necked flask equipped with a heating mantle, a magnetic stirrer, an air-cooled condenser, a nitrogen inlet, a bubbler, and a powder addition funnel which can be placed through one of the necks. The apparatus is purged with nitrogen, 200 mL of concentrated dark polysulfide [ammonium sulfide, assay 23% $(NH_4)_2S$] is added,* and the solution is warmed to about 45° under the nitrogen atmosphere. One neck of the flask is opened, and the powder addition funnel is placed there while a steady stream of nitrogen is maintained through the open neck. Decaborane(14) (12 g, 0.1 mole) is added cautiously through the powder addition funnel to the magnetically stirred solution. The rate of addition is maintained so that the temperature of the reaction flask does not rise much above 60°. Occasionally,

*The checkers used potassium polysulfide K_2S_n prepared from KOH and sulfur.

the decaborane may stick in the powder funnel and must be cautiously pushed into the flask with a spatula. The reaction is evidenced by a vigorous evolution of gas. After approximately 1/2 hr at 60°, there is no more evidence of gas evolution, and the reaction is complete. At this point the reaction mixture is poured into a 500-mL beaker containing a solution of CsF (60 g, 0.4 mole in 100 mL). The solution is cooled in an ice bath for 1 hr to assure precipitation of $Cs[6\text{-}SB_9H_{12}]$. The crude product is collected on a filter in the air and washed with 100 mL of cold water. The mustard-yellow precipitate is purified by recrystallization from 150 mL of water heated to 90°. Upon dissolution of the crude product in the hot water, an olive-green solution is obtained which should be filtered through a hot funnel to remove small amounts of insoluble material. As the filtrate cools, crystals of the pure cesium salt are obtained. The pure product is dried under vacuum (21 g, 90% yield). Other salts such as the rubidium and tetramethylammonium salts are prepared similarly and recrystallized from water and acetone, respectively.

Properties

The salts of dodecahydro-6-thia-*arachno*-decaborate(1-) are air-stable white solids. The infrared spectrum of $Cs[6\text{-}SB_9H_{12}]$ shows prominent absorptions at 2525(vs), 2470(s), 2380(s), 2355(m), 1189(m), 1049(s), 1010(vs), 990(sh), 966(w), 938(w), 920(m), 905(sh), 784(m), 758(m), 726(m), 690(w), and 657(w) cm^{-1} in a KBr disk. The ^{11}B-NMR spectrum of $Cs[SB_9H_{12}]$ in acetonitrile consists of six multiplets of relative intensities 1:2:1:1:2:2 centered at 4.0, -7.9, -11.3, -15.0, -33.4, and -36.6 ppm from $BF_3 \cdot OEt_2$, respectively.[10]

B. 6-THIA-*nido*-DECABORANE(11)

Procedure

A 200-mL volume of dry benzene and 11.82 g (46.6 mmole) of I_2 are placed in a 500-mL, three-necked flask equipped with a condenser, a nitrogen inlet, an outlet and bubbler, a magnetic stirrer, and a heating mantle. The flask is thoroughly purged with nitrogen, and 25.6 g (93.1 mmole) of $Cs[6\text{-}SB_9H_{12}]$ is added quickly to the stirred iodine solution through one of the side necks. The neck is stoppered, and the flask contents are brought to reflux. Gas evolution is seen, even at room temperature, and the intense iodine color slowly fades to give a yellow solution and a white solid. Best results are obtained when reflux is continued about 3 hr beyond the initial fading in color. The reaction mixture is cooled and filtered quickly in air. The filtrate is promptly taken to dryness on a rotary evaporator to give a yellow solid. The crude product is purified by sublimation at 40° under vacuum. Exposure of the material to air, especially humid air, should be minimized. It is suggested that the sublimator be loaded in a nitrogen-filled glove bag. This procedure typically gives 11.3 g (86%) of 6-SB_9H_{11}.

Properties

Pure 6-thia-*nido*-decaborane(11) is a clear, crystalline solid with a foul sulfur odor. It does not survive prolonged exposure to air. However, it is indefinitely stable at 25° in evacuated ampules. It melts with decomposition at 76.5–77° in a sealed capillary and is very soluble in a wide variety of organic solvents. The infrared spectrum displays bands at 2530(vs), 1950(w), 1900(w), 1049(s), 1028(sh), 952(w), 936(m), 903(s), 882(m), 834(s), 806(s), 790(w), 749(s), 730(sh), 721(s), and 702(s) cm^{-1}. The ^{11}B-NMR spectrum in benzene has six doublets of relative intensity 2:1:2:2:1:1 centered at 24.9, 17.3, 6.8, −10.1, −21.5, and −30.7 ppm from $BF_3 \cdot OEt_2$, respectively.

C. 1-THIA-*closo*-DECABORANE(9)

Procedure

In a typical pyrolysis, 7.91 g (56.3 mmol) of 6-SB_9H_{11} is placed in the bottom of a 24-mm o.d. borosilicate glass tube 70-cm long, round at the bottom and fitted with an outer 24/40 standard taper joint at the open end. The open end of this tube is attached to a vacuum sublimator, the bottom of which is fitted with an inner 24/40 standard taper joint to permit coupling to the pyrolysis tube. Prior to connection of the tube to the sublimator, glass wool, which has been degreased by washing with CCl_4, is loosely packed in the tube, starting 25 cm from the bottom and extending upwards for 15 cm (see Fig. 1).

The entire apparatus is evacuated through the sublimator. A furnace is placed around the tube in the region of the glass wool packing, and heating tape is wrapped around the bottom of the tube so as gently to sublime the SB_9H_{11} into the furnace hot zone. The furnace is heated to 450°, and the system is continuously evacuated during the entire pyrolysis while the sublimator cold finger is maintained at −78° (dry ice). After about 9 hr, no further hydrogen evolution is observed as monitored by a vacuum gauge. Then heating is stopped and the system back-filled with nitrogen. Crude 1-SB_9H_9 (4.62 g, 59%) is recovered from the cold finger, and small amounts of mixed isomers of $(SB_9H_8)_2$ can be recovered from the glass tube just above the pyrolysis zone (0.63 g, 8%).[3] Impurities imparting a slight yellow color to the 1-SB_9H_9 are removed by washing with acetonitrile. Removal of excess CH_3CN by rotary evaporation and vacuum sublimation at room temperature affords 4.35 g (56%) of pure white crystals.

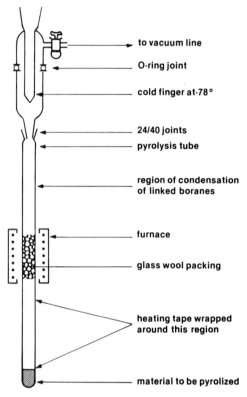

Fig. 1. Apparatus used for pyrolysis.

Properties

Pure 1-thia-*closo*-decaborane(9) is a clear crystalline solid with a distinctively "sweet" odor with no repulsive "sulfur" smell. It is quite resistant to air and moisture but is best stored for extended periods under N_2 or in evacuated ampoules. It melts sharply at 216.5–217.0° in a sealed capillary and is soluble in a wide variety of organic solvents. The IR spectrum shows absorptions at 2600(vs), 2560(vs), 1011(w), 972(m, sh), 958(s), 910(w), 902(m), 810(s), 750(w), 738(m), 721(m), 685(s), 660(w), 610(s), and 603(s) cm^{-1}. In benzene, the ^{11}B-NMR spectrum contains three doublets of 1:4:4 relative intensity at 74.5, −4.8, and −17.6 ppm from $BF_3 \cdot OEt_2$, respectively.[3]

References

1. R. W. Rudolph, *Acc. Chem. Res.*, **9**, 446 (1976).
2. W. R. Hertler, F. Klanberg, and E. L. Muetterties, *Inorg. Chem.*, **6**, 1696 (1967).

3. W. R. Pretzer and R. W. Rudolph, *J. Am. Chem. Soc.*, 98, 1441 (1976); *Inorg. Chem.*, 15, 1779 (1976).
4. T. K. Hilty, D. A. Thompson, W. M. Butler, and R. W. Rudolph, *Inorg. Chem.*, 18, 2642 (1979).
5. D. A. Thompson and R. W. Rudolph, *J. Chem. Soc. Chem. Commun.*, 770 (1976).
6. W. L. Smith, B. J. Meneghelli, D. A. Thompson, P. Klymko, N. McClure, M. Bower, and R. W. Rudolph, *Inorg. Chem.*, 16, 3008 (1977).
7. B. J. Meneghelli and R. W. Rudolph, *J. Organomet. Chem.*, 133, 139 (1977).
8. B. J. Meneghelli, M. Bower, N. Canter, and R. W. Rudolph, *J. Am. Chem. Soc.*, 102, 4355 (1980).
9. T. K. Hilty and R. W. Rudolph, *Inorg. Chem.*, 18, 1106 (1979).
10. A. R. Siedle, G. M. Bodner, A. R. Garber, and L. J. Todd, *Inorg. Chem.*, 13, 1756 (1974).

53. POTASSIUM DODECAHYDRO-7,8-DICARBA-*nido*-UNDECABORATE(1−), K[7,8-$C_2B_9H_{12}$], INTERMEDIATES, STOCK SOLUTION, AND ANHYDROUS SALT

$$1,2\text{-}C_2B_{10}H_{12} + 3CH_3OH + KOH \xrightarrow{\text{boil}}$$

$$K[7,8\text{-}C_2B_9H_{12}] + B(OCH_3)_3 + H_2O + H_2$$

Submitted by J. PLEŠEK,* S. HEŘMÁNEK,* and B. ŠTÍBR*
Checked by L. WAKSMAN† and L. G. SNEDDON†

The very important anion [7,8-$C_2B_9H_{12}$]$^-$ was first obtained by alkaline solvolysis of 1,2-dicarba-*closo*-dodecaborane(12), 1,2-$C_2B_{10}H_{12}$.[1] Later, various modifications of the original procedure were described.[2,3] This anion represents a basic substance for preparing a large number of metallocarbaborane sandwich compounds[3-6] as well as for the syntheses of several *nido*- and *closo*-carbaboranes[7-9,13] and heteroboranes.[3,9-12]

The present procedure is a modification of the original method,[1] using potassium hydroxide instead of sodium hydroxide and methanol instead of ethanol. This combination leads to the desired product much more quickly than the original one and makes the isolation easier. The yields are practically quantitative. Moreover, the same basic procedure and a slightly modified working up lead to optimum intermediate solutions, directly suitable for further syntheses.

*Institute of Inorganic Chemistry, Czechoslovak Academy of Sciences, 250 68 Řež near Prague, Czechoslovakia.
†Department of Chemistry, University of Pennsylvania, Philadelphia, PA 19104.

Reagents

Commercially available 1,2-dicarba-*closo*-dodecaborane(12), technical purity of about 95%, can be used without further purification. Technical-grade methanol is distilled prior to use. Technical-grade solid potassium hydroxide pellets are used. Any additional water should be avoided, because the solvolysis with methanol is practically stopped upon an addition of several volume percent of water to the reaction mixture.

Apparatus

A 1-L round-bottomed flask fitted with a water reflux condenser and a mechanical stirrer is used.

■ **Caution.** *Provision must be made to vent, into a hood, after dilution with an inert gas, hydrogen which is generated in this reaction.*

Prior to adding methanol to both solid components, the content of the apparatus is evacuated, and the apparatus is filled with nitrogen.

Procedure

To the mixture of solid 1,2-dicarba-*closo*-dodecaborane(12) (43.4 g, 0.3 mole) and potassium hydroxide (56 g, 1.0 mole) in the reaction flask, 300 mL of methanol is added in one stream through the reflux condenser. Upon slow stirring, a vigorous reaction starts, the solid substances dissolve, and much heat and hydrogen are evolved. The reaction is moderated by occasional cooling of the reaction flask by immersion into cold water. After about 30 min, the spontaneous reaction subsides. The reaction flask is then heated in a water bath to reflux for about 1.5 hr. The end of the reaction is signaled by cessation of the hydrogen evolution.

At this point, the solution contains the $K[7,8-C_2B_9H_{12}]$ salt in essentially quantitative yield, along with excess potassium hydroxide and some potassium borate in anhydrous methanol. This solution is ideally suited for synthesis of the mixed cyclopentadienyl-metallocarbaborane sandwich compounds of the type $(C_5H_5)M^{III}(C_2B_9H_{11})$ according to the "alcoholic route",[6] cf. this volume, Chapter 5, Section 54.

The volatile solvents are then distilled off at 50° and 10 torr. When approximately 100 mL of the distillate (containing essentially methanol and some trimethyl borate) have been collected, the remainder is diluted with 400 mL of water and the vacuum distillation is continued until about 300 mL of distillate are collected. A viscous water solution of the $K[7,8-C_2B_9H_{12}]$ with excess potassium hydroxide and potassium borate remains.

This solution is a useful intermediate for syntheses of the metallocarbaborane sandwich compounds of the $[(C_2B_9H_{12})_2M^{III}]^-$ type according to the "wet route."[5] It is noteworthy that any residual methanol in this solution dramatically lowers the yield of the sandwich anions. However, for mixed cyclopentadienyl-metallocarbaborane sandwiches, alcohol as a solvent is mandatory.

For the isolation of the dry potassium salt, $K[7,8-C_2B_9H_{12}]$, the aqueous solution is diluted with an additional 200 mL of water, saturated with carbon dioxide, and an essentially pure product is extracted with three successive 150-mL portions of diethyl ether. Much water is coextracted. The combined extracts are dried with solid potassium carbonate until a water layer separates and is discarded. Then an equal volume of benzene is added to the ether layer, and the drying is continued as long as a water layer is formed. The clear diethyl etherbenzene solution is filtered, and the solvents are evaporated at 50° and 10 torr. The solid residue is then dried under vacuum (100°, 1 torr). After cooling the solid to room temperature, nitrogen is introduced into the evacuated flask. The white solid product (48.3 g, 94.3%) is pure enough for any subsequent synthetic procedure. At ambient temperature in an inert atmosphere, the salt is stable indefinitely. It can be handled briefly in air without any risk, but it is quite hygroscopic.

For the preparation of the standard 1.0 M stock solution of the potassium salt, the drying operation is superfluous. The original diethyl ether extract is washed with 100 mL of water, and the diethyl ether is evaporated at ambient temperatures and 10 torr. At the end of this operation the temperature is raised for a short time to 40°. When practically all the diethyl ether has been stripped off, the remaining viscous solution is diluted with 200 mL of water, and approximately 50 mL of additional distillate are collected under vacuum to ensure that the residual solution is not contaminated with diethyl ether. The residue of the distillation is then transferred to a calibrated flask which is filled with water to the 280-mL mark. The solution is filtered after the addition of about 3.0 g of charcoal. The resulting filtrate is very close to the 1.0 M concentration, as can be verified by weighing the almost insoluble tetramethylammonium salt precipitated from an aliquot. The stock solution is indefinitely stable in an inert atmosphere in brown glass containers. It serves as a standard starting material for the syntheses of various intermediate *nido-* and *closo-*carbaboranes[7,9] and heteroboranes.[3,10,11]

Properties

Because of the very extensive chemistry of the $[7,8-C_2B_9H_{12}]^-$ anion, its properties are thoroughly characterized and have been described in the literature.[1-3,12] Its salts with bulky organic cations, such as tertiary and quaternary ammonium ions, are virtually insoluble in water, but they easily dissolve in wet ethers, alcohols, ketones, nitriles, nitro compounds, etc. The cesium salt, which is suitable for characterization of the compound, is moderately soluble in water. On silica gel thin layers, such as Silufol (aluminum foil bearing a 10-μm layer of silica gel bound with 5% of starch), with a mixture of CH_3CN-$CHCl_3$ (1:3) as the eluant, a single spot of R_f 0.15 is obtained. The spot appears as a white one on a lilac background when the Silufol foil is immersed in iodine vapors. After spraying with 1% $AgNO_3$ solution, the spot slowly darkens and turns black after several hours.

For easy identification, it is convenient to measure the ^{1}H-NMR spectrum of the tetramethylammonium salt in acetone-d_6. The CH carbaborane signal is found at 1.72 ppm (int 2), and the tetramethylammonium group signal appears at 3.43 ppm (int 12). The ^{11}B-NMR spectrum in acetone-d_6 shows five doublets of relative intensities 2:3:2:1:1 centered at -11.2, -17.5, -22.4, -33.4, and -38.0 ppm related to $BF_3 \cdot O(C_2H_5)_2$ and read upfield.

References

1. R. A. Wiesboeck and M. F. Hawthorne, *J. Am. Chem. Soc.*, **86**, 1642 (1964).
2. P. M. Garrett, F. N. Tebbe, and M. F. Hawthorne, *J. Am. Chem. Soc.*, **86**, 5016 (1964).
3. R. N. Grimes, *Carboranes*, Academic Press, New York, 1970.
4. M. F. Hawthorne, D. C. Young, and P. A. Wegner, *J. Am. Chem. Soc.*, **87**, 1818 (1965).
5. M. F. Hawthorne, D. C. Young, T. D. Andrews, D. V. Howe, R. L. Pilling, A. D. Pitts, M. Reintjes, L. F. Warren, Jr., and P. A. Wegner, *J. Am. Chem. Soc.*, **90**, 879 (1968).
6. J. Plešek, B. Štíbr, and S. Heřmánek, *Synth. Inorg. Metallorg. Chem.*, **3**, 291 (1973).
7. J. Plešek and S. Heřmánek, *Chem. Ind.* (London), **1971**, 1267.
8. M. F. Hawthorne and P. A. Wegner, *J. Am. Chem. Soc.*, **80**, 896 (1968).
9. J. Plešek, Б. Štíbr, and S. Heřmánek, *Chem. Ind. (London)*, 626 (1980).
10. J. Plešek, B. Štíbr, and S. Heřmánek, *Chem. Ind. (London)*, 662 (1974).
11. J. Plešek and S. Heřmánek, *Pure Appl. Chem.*, **39**, 431 (1974).
12. M. F. Hawthorne, *Pure Appl. Chem.*, **29**, 547 (1972).
13. F. N. Tebbe, P. M. Garrett, and M. F. Hawthorne, *J. Am. Chem. Soc.*, **90**, 869 (1968).

54. 3-(η^5-CYCLOPENTADIENYL)-1,2-DICARBA-3-COBALTA-closo-DODECABORANE(11), 3-(η^5-C_5H_5)-3-Co-1,2-$C_2B_9H_{11}$

$$2[C_2B_9H_{11}]^{2-} + 2[C_5H_5]^- + 3Co^{2+} \longrightarrow 2[\eta^5\text{-}(C_5H_5)Co^{III}C_2B_9H_{11}] + Co$$

Submitted by J. PLEŠEK,* S. HEŘMÁNEK,* and B. ŠTÍBR*
Checked by T. K. DUTTA† and T. P. FEHLNER†

Mixed cyclopentadienyl-metallocarbaborane sandwiches were first prepared by Hawthorne and coworkers[1,2] under strictly aprotic conditions. The title compound was obtained in 15% yield. The "alcoholic route" described here for the example compound 3-(η^5-cyclopentadienyl)-1,2-dicarba-3-cobalta-closo-dodecaborane(11) was found to be suitable for synthesis of essentially any compound of this kind.[3-6] This procedure is far more convenient than the aprotic route, gives high yields, and can be easily modified for other ligand and metal combinations. The title compound is the best known representative of the mixed η^5-(cyclopentadienyl)metallocarbaboranes. It was used as a starting material for various physical[7] and chemical transformations.[8,10]

Reagents

Starting reagents are methanol, potassium hydroxide pellets of commercial grade, freshly distilled cyclopentadiene, and $CoCl_2 \cdot 6H_2O$ of analytical purity. A methanolic solution (150 mL) of $K[7,8\text{-}C_2B_9H_{12}]$ containing excess KOH prepared from 0.1 mole of $1,2\text{-}C_2B_{10}H_{12}$ as described in this volume (Chapter 5, Sec. 53) is used as a source of the carbollide ligand (undecahydro-7,8-nido-undecaborate(2−)).

Apparatus

A 1-L three-necked, round-bottomed flask is fitted with a strong mechanical stirrer, pressure-equalizing funnel, and a stopcock, attached to a rubber "breathing bag." This rubber container serves as a portable nitrogen source and as a pressure compensator.

---

*Institute of Inorganic Chemistry, Czechoslovak Academy of Sciences, 250 68 Řež near Prague, Czechoslovakia.
†Department of Chemistry, University of Notre Dame, Notre Dame, IN 46556.

Procedure

To the stirred solution of $CoCl_2 \cdot 6H_2O$ (35.6 g, 0.15 mole) in 100 mL of warm methanol are gradually added potassium hydroxide pellets (56 g, 1.0 mole) during 30 min with occasional cooling by immersion of the reaction flask into cold water. Potassium hydroxide is added through the neck designed for holding the dropping funnel in the next step. The flask, which now contains a gray-blue slurry of cobalt(III) hydroxide, is cooled in an ice water bath, the dropping funnel is attached, and the apparatus is evacuated and filled with nitrogen. A mixture of freshly distilled cyclopentadiene (20 g, 0.3 mole) and 150 mL of the methanolic solution containing 0.1 mole of $K[7,8-C_2B_9H_{12}]$ is introduced into the dropping funnel. The mixture is added dropwise during 1 hr into the stirred cobalt(II) hydroxide slurry at 0°, and the stirring is continued for an additional 5 hr at ambient temperature. The resulting mixture is allowed to stand overnight.

The flask, which now contains a very thick gray-brown suspension, is evacuated carefully to 20 torr at ambient temperature to evaporate the excess cyclopentadiene. This operation lasts about 30 min and is attended by some foaming. The contents are diluted with 400 mL of water, and the solids are collected by suction filtration through a fritted-glass filter. The filter cake is thoroughly washed with three 100-mL portions of water, then rinsed with three successive 100-mL volumes of 10% hydrochloric acid to dissolve the excess cobalt(II) hydroxide. The remaining yellow platelets contaminated with cobalt powder, are washed with water and dissolved in 200 mL of acetone. The filtered yellow solution is added to 200 mL of benzene, and acetone is stripped off under vacuum at ambient temperature until the residual volume amounts to about 200 mL. This residue is warmed in a water bath at 60° to dissolve all solids. Then 200 mL of hexane is carefully poured on the surface of the yellow benzene solution. The two-layer mixture is set aside overnight. Bright yellow platelets separate on standing. They are filtered out and dried at 50°/10 torr for 3 hr; 20.5 g (78.2%) of the title compound is obtained.

■ **Caution.** *Benzene is carcinogenic and should be handled in a hood.*

Properties

The 3-(η^5-cyclopentadienyl)-1,2-dicarba-3-cobalta-*closo*-dodecaborane(11) is a yellow solid of mp 238–239°. The melting point is, however, dependent on the heating rate owing to easy isomerization. The mass spectrum of the compound exhibits a cut-off at m/z 258 and by TLC on Silufol silica gel sheet (see p. 234) with benzene as an eluant the compound shows an R_f value of 0.35. Two absorption bands in the electronic spectrum are recorded at 277 nm (ϵ = 58,720) and 423 nm (ϵ = 487) in dichloromethane as a solvent. The ^{1}H-NMR spectrum in acetone-d_6 shows a distinct signal of cyclopentadienyl protons at δ 5.87(5) and a broad C–H carbaborane signal at 4.49 ppm. The ^{11}B-NMR spectrum in the same solvent is composed of five doublets of relative intensities 1:1:4:2:1 centered at 6.6, 2.7, -4.8, -16.1, and -21.8 ppm (chemical shifts are relative to $BF_3 \cdot O(C_2H_5)$ and are ordered according to the increasing magnetic field).

References

1. M. F. Hawthorne and R. L. Pilling, *J. Am. Chem. Soc.*, 87, 3987 (1965).
2. M. F. Hawthorne, D. C. Young, T. D. Andrews, D. V. Howe, R. L. Pilling, A. D. Pitts, M. Reintjes, L. F. Warren, Jr., and P. A. Wegner, *J. Am. Chem. Soc.*, 90, 879 (1968).
3. J. Plešek, B. Štíbr, and S. Heřmánek, Czech. Pat. 175,600 (1976).
4. J. Plešek, B. Štíbr, and S. Heřmánek, *Synth. Inorg. Metallorg. Chem.*, 3, 291 (1973).
5. J. Dolanský, K. Baše, and B. Štíbr, *Chem. Ind.* (London), 1976, 853.
6. J. Plešek, B. Štíbr, and S. Heřmánek, *Chem. Ind.* (London), 1980.
7. M. F. Hawthorne, *Pure. Appl. Chem.*, 29, 557 (1972).
8. J. Plešek and S. Heřmánek, *Coll. Czech. Chem. Commun.*, 43, 1325 (1978).
9. L. I. Zakhakin, I. V. Pisareva, and R. Ch. Bikkineev, *Izv. Akad. Nauk SSSR*, 26, 641 (1977).
10. J. Plešek, K. Baše, and S. Heřmánek, 2nd Int. Meeting on Boron Chemistry, Leeds, 25–29th March 1974, Abstr. No. 46.

55. 2,6-DICARBA-*nido*-NONABORANE(11),* 2,6-$C_2B_7H_{11}$

$$7,8\text{-}C_2B_9H_{12}^- + 2CH_2O + 5H_2O + H_3O^+ \longrightarrow$$

$$2,6\text{-}C_2B_7H_{11} + 2CH_3OH + 2B(OH)_3 + 2H_2$$

Submitted by B. ŠTÍBR,† J. PLEŠEK,† and S. HEŘMÁNEK†
Checked by L. WAKSMAN‡ and L. G. SNEDDON‡

2,6-Dicarba-*nido*-nonaborane(11) is a representative example of a nine-vertex molecular structure with a five-membered open face containing one hydrogen bridge and one BH_2 group.[1,2]

The title compound was first prepared as one of the products of the reaction of octaborane(12) with acetylene.[1] Subsequently, 2,6-$C_2B_7H_{11}$ was obtained in

---

*Numbering is based on the closed bicapped square antiprism with the missing vertex assigned position 9 (highest possible); this numbering is in accord with Rule 3.221 in the IUPAC Nomenclature of Inorganic Boron Compounds (*Pure Appl. Chem.*, 30, 683–710 (1972) and is consistent with Rule 2.321 in the ACS Nomenclature of Boron Compounds (*Inorg. Chem.* 1968, 7, 1945–1964). In the literature, the carbon atoms of this polyboron have been numbered 4,5-, in agreement with the rules for numbering an earlier proposed structure.

†Institute of Inorganic Chemistry, Czechoslovak Academy of Sciences, 250 68 Řež, near Prague, Czechoslovakia.

‡Department of Chemistry, University of Pennsylvania, Philadelphia, PA 19104.

a moderate yield as a by-product in the oxidation of the $[7,8\text{-}C_2B_9H_{12}]^-$ anion with ferric chloride.[2] Both these methods are rather inconvenient because of either the relative unavailability of B_8H_{12} or a complicated work up of the reaction mixture. These complications can be avoided by using the readily available $[7,8\text{-}C_2B_9H_{12}]^-$ ion which, on treatment with aqueous formaldehyde in the presence of dilute hydrochloric acid, produces $2,6\text{-}C_2B_7H_{11}$ as the sole product in good yield.[3] This preparation makes 2,6-dicarba-*nido*-nonaborane(11) one of the most available of the intermediate carboranes.

The chemistry of $2,6\text{-}C_2B_7H_{11}$ has not been studied to any great extent due to the difficulty in obtaining it. Recently, however, it has been used to prepare the $(C_5H_5Co)_2C_2B_7H_9$ and $Pt[P(C_6H_5)_3]_2C_2B_7H_{11}$ transition metal complexes.[3,4]

Procedure

■ **Caution.** *The reaction should be carried out under nitrogen, and the excess of explosive hydrogen must be vented properly.*

A 500-mL, two-necked, round-bottomed flask fitted with a reflux condenser, a nitrogen inlet, a dropping funnel, and a magnetic stirring bar is charged with 84 mL of 0.313 M aqueous $K[C_2B_9H_{12}]$ (0.026 mole; for preparation see Chapter 5, Sec. 53), 10.5 mL of 38% aqueous formaldehyde (approx. 0.12 mole), and 100 mL of hexane. The reaction apparatus is purged with nitrogen. The reaction flask is immersed in a water bath at 20° and 26 mL of concentrated (12M) hydrochloric acid is added dropwise with stirring over a period of 10 min. Stirring is continued for 6 hr while the mixture is maintained at 20°. The hexane layer is separated and the water solution is extracted with two 25-mL portions of hexane. The combined hexane fractions are dried over anhydrous magnesium sulfate and filtered. The solvent is then removed under vacuum at room temperature to give white crystals of pure $2,6\text{-}C_2B_7H_{11}$.* The yields vary from 1.65 to 2.15 g (57–75%, based on starting $K[C_2B_9H_{12}]$). Further purification is accomplished by subliming the product at $50°/10^{-2}$ torr onto a cooling finger kept at 0°.

■ **Caution.** *The product is obtained as white needles extremely sensitive to oxygen and inflammable in air. The compound should be considered to be toxic.*

Properties

2,6-Dicarba-*nido*-nonaborane(11) is a white, crystalline solid with mp 90°, and is soluble in all aprotic solvents. With alcohols and water it decomposes with

*The checkers found that appreciable carbaborane was lost in this step unless a trap (e.g., −46°) was employed during this operation.

evolution of hydrogen. Thin-layer chromatography in benzene (Silufol, i.e., silica gel on aluminum foil, starch as binder; Kavalier n.p., Votice, Czechoslovakia; detection I_2 vapor–$AgNO_3$ spray) shows an R_f 0.51. The mass spectrum at 70 eV exhibits a molecular ion at m/z 112 corresponding to $^{12}C_2{}^{11}B_7H_{11}^+$.[2]

The IR spectrum[1] (thin film on Irtran-2 window at $-196°$) consists of following characteristic vibrations: 3048(w), 2982(w) (cage CH stretch); 2585(s, sh), 2505(s), 2460(s) (terminal BH stretch), 1880(vw) (symmetric B–H–B stretch), 1535(m, b) (asymmetric B–H–B stretch).

The 70.6-MHz ^{11}B-NMR spectrum[1] is composed of five doublets at 10.4, 4.7, -3.0, -5.2, and -55.4 ppm and one triplet at -28.3 ppm (chemical shifts relative to external $BF_3 \cdot O(C_2H_5)_2$, signals ordered according to increasing shielding).

The 220-MHz ^{1}H-NMR spectrum in CS_2 shows two sharp 1:1 CH singlets at 2.81 and 1.95 ppm (relative to external TMS) and a broad singlet of a hydrogen bridge at -0.87, besides overlapping signals of terminal B–H protons in the region of +3.72 to -0.15 ppm.

References

1. R. R. Rietz and R. Schaeffer, *J. Am. Chem. Soc.*, 95, 6254 (1973).
2. H. M. Colquhoun, T. J. Greenhough, M. G. H. Wallbridge, S. Heřmánek, and J. Plešek, *J. Chem. Soc., Dalton Trans.*, 1978, 944.
3. J. Plešek, B. Štíbr, and S. Heřmánek, *Chem. Ind.* (London), 1980, 626–627.
4. B. Štíbr, S. Heřmánek, J. Plešek, and B. Baše, *Chem. Ind.* (London), 1980, 468.

56. 9-(DIMETHYL SULFIDE)-7,8-DICARBA-*nido*-UNDECABORANE(11), 9-[(CH$_3$)$_2$S]-7,8-C$_2$B$_9$H$_{11}$

$$7,8\text{-}[C_2B_9H_{12}]^- + (CH_3)_2SO + H^+ \longrightarrow 9\text{-}[(CH_3)_2S]\text{-}7,8\text{-}C_2B_9H_{11} + H_2O$$

Submitted by J. PLEŠEK,* Z. JANOUŠEK,* and S. HEŘMÁNEK*
Checked by J. SMITH† and H. D. SMITH, JR.†

The title compound is a typical representative of the $L \cdot C_2B_9H_{11}$ series, containing up to eleven isomers for each family of the same formula. All these compounds may be useful as precursors of sandwich complexes, forming $[L \cdot C_2B_9H_{10}]^-$

*Institute of Inorganic Chemistry, Czechoslovak Academy of Sciences, 250 68 Řež near Prague, Czechoslovakia.

†Department of Chemistry, Virginia Polytechnic Institute and State University, Blacksburg, VA 24061.

anions which are isoelectronic with the $C_5H_5^-$ ion in respect to sandwich chemistry.[1]

Analogs with other ligands are obtainable by an oxidation of the $[C_2B_9H_{12}]^-$ ion in the presence of a base L, namely with iron(III) chloride[2] or copper(II) chloride[3] under aprotic conditions. These far less convenient routes are, however, further complicated by a simultaneous formation of isomers with the ligand L in the 9- and 10-positions. Four more isomers of the title compound were prepared by specific methods.[1]

The procedure detailed here makes the title compound the most easily accessible representative of the class of sandwich-forming ligands.

Apparatus

A 500-mL Erlenmeyer flask fitted with a pressure-equalizing dropping funnel and a magnetic stirring bar is immersed in a 2-L cold water bath. The entire apparatus is placed on a magnetic stirring device.

Procedure

The Erlenmeyer flask is charged with 100 mL of the 1.0 M stock solution of $K[C_2B_9H_{12}]$ (see this volume, Chapter 5, Section 53) and 31.2 g (0.4 mole) of dimethyl sulfoxide. While the solution is being stirred, 100 mL of concentrated sulfuric acid is added dropwise over a 1-hr period. Stirring is continued for an additional 5 hr. The resultant two-layer mixture is then set aside overnight. After dilution with 500 mL of cold water, the mixture is vacuum filtered, to isolate the solid residue, and this is washed with two 100-mL portions of water and recrystallized from 80% methanol. White needles separate which are isolated by vaccum filtration and dried in air at ambient temperature for 8 hr. The yield is 10.8 g (55.6%).

Properties

The 9-(dimethyl sulfide)-7,8-dicarba-*nido*-undecaborane(11) is a solid of mp 147–148°, giving a molecular ion at m/z 196. It is soluble in polar organic solvents and in benzene, insoluble in water and paraffinic hydrocarbons. By TLC on Silufol (see this volume, p. 234) and with benzene as an eluant, it shows an R_f value of 0.21. The ^{1}H-NMR spectrum in $CDCl_3$ shows two C-H carbaborane signals at 2.22 and 2.70 ppm respectively. The signal of the B-H-B bridge hydrogen is found at -3.37 ppm, and both signals of the CH_3 groups appear at 2.62 and 2.80 ppm respectively. All values are related to TMS. The ^{11}B-NMR spectrum in acetone-d_6 is composed of one singlet at -3.2 ppm and of eight doublets, centered at -2.6, -9.7, -13.4, -14.8, -19.3, -22.1, -25.8, and -34.2

ppm (chemical shifts are related to $BF_3 \cdot O(C_2H_5)_2$ and signals are ordered according to the increasing magnetic shielding).

References

1. J. Plešek, Z. Janoušek, and S. Heřmánek, *Coll. Czech. Chem. Commun.*, **43**, 2862 (1978).
2. D. C. Young, D. V. Howe, and M. F. Hawthorne, *J. Am. Chem. Soc.*, **91**, 859 (1969).
3. V. A. Brattsev and V. I. Stanko, *Zh. Obshch. Khim.*, **38**, 1657 (1968).

57. DIRECT SULFHYDRYLATION OF BORANES AND HETEROBORANES, 1,2-DICARBA-*closo*-DODECABORANE(12)-9-THIOL, 9-HS-1,2-$C_2B_{10}H_{11}$

$$1,2\text{-}C_2B_{10}H_{12} + S \xrightarrow[5 \text{ hr}]{AlCl_3, 120°} 9\text{-HS-}1,2\text{-}C_2B_{10}H_{11}$$

Submitted by J. PLEŠEK,* Z. JANOUŠEK,* and S. HEŘMÁNEK*
Checked by H. D. SMITH, JR.†

The B-substituted borane- and heteroboranethiols represent an almost unexplored class of compounds, because until recently, no general method for their syntheses was known. A few examples of mercapto HS derivatives mentioned in the literature were prepared by rather specific methods. The $[B_{12}H_{11}SH]^{2-}$ anion has been obtained by the reaction of the conjugate acid of $[B_{12}H_{12}]$ with hydrogen sulfide[1,2] or with 3-methylbenzothiazole-2(3H)-thione compound, hydrolyzable to the mercapto derivative.[2] The [1,10-$(HS)_2B_{10}Cl_8]^{2-}$ anion has been prepared by the exchange of diazo groups in the [1,10-$(N_2)_2B_{10}Cl_8]^{2-}$ anion by treatment with hydrogen sulfide.[3] The [(8-(HS)$C_2B_9H_{10})_2$Co]$^-$ sandwich anion results from the reaction of the parent anion with carbon disulfide in the presence of aluminum chloride, followed by hydrolysis of the cyclic dithio product.[4] Each of these methods is a specific one and is unsuited for sulfhydrylation of another type of skeleton.

The recent finding[5] that *B*-thiols can be prepared in high yields by heating the appropriate borane compounds with sulfur and $AlCl_3$ has made these com-

*Institute of Inorganic Chemistry, Czechoslovak Academy of Sciences, 250 68 Řež near Prague, Czechoslovakia.
†Department of Chemistry, Virginia Polytechnic Institute and State University, Blacksburg, VA 24061.

pounds easily accessible, and, consequently, the chemistry of this class of compounds can be developed.

The general preparation of borane- and heteroborane-B-thiols is demonstrated for the example of 9-mercapto-1,2-dicarba-*closo*-dodecaborane(12). About 20 analogs have been prepared in a similar way starting with all three isomeric icosahedral dicarbaboranes[5,6] and their substitution derivatives,[7] cyclopentadienylcobaltacarbaborane,[5] decaborane(14),[8] and several *nido*-heteroboranes.[8] Moreover, slight modification of this procedure leads to particular vicinal B,B'-dithiols.[9]

Apparatus

A 500-mL round-bottomed flask is fitted with a 30-cm long tube for use as an air-cooled reflux condenser. This tube is closed by a stopcock fitted with a rubber "breathing bag," allowing for thermal expansion of gases but maintaining the inertness of the atmosphere inside.

Procedure

A thorough mixture of powdered reagents, for example, 14.4 g (0.1 mole) of 1,2-dicarba-*closo*-dodecaborane(12) (technical grade, about 95% purity), 8.0 g (0.25 mole) of sulfur (reagent grade), and 16.0 g (0.12 mole) of anhydrous aluminum chloride (reagent grade), is placed in the reaction flask. The apparatus is evacuated and filled with dry nitrogen. The flask is placed in a hood and immersed in an oil bath kept at 100°. During 1 hr the temperature is raised to 120° and held at that value for an additional 5 hr. A clear, almost homogeneous melt is obtained, and a small portion of aluminum chloride sublimes into the air-cooled condenser. Some hydrogen sulfide is liberated during the reaction, owing to partial oxidation of the product by excess sulfur.

After cooling to room temperature, the melt is dissolved in 200 mL of ethanol, the mixture being cooled occasionally in tap water. The viscous yellowish solution is filtered and poured into 500 mL of water, resulting in a precipitate. The mixture is then filtered and the filter cake washed with three 50-mL portions of 10% hydrochloric acid and three 50-mL portions of water. It is then dried and dissolved in 200 mL of benzene. This solution is filtered, and 6.0 g (0.1 mole) of Zn powder and 20 mL (0.3 mole) of acetic acid are added to the filtrate with stirring. The mixture is heated to the boiling point and then allowed to cool to ambient temperature. The product is filtered, and the filtrate (which contains the carbaboranethiol) is washed with three 50-mL portions of 100% hydrochloric acid and one 50-mL portion of water. The carbaboranethiol is then extracted with three 100-mL portions of 10% aqueous potassium hydroxide, and the solution is filtered. The slightly yellowish filtrate is saturated with carbon dioxide and allowed to stand overnight. The white precipitate is collected by filtration, washed with three 50-mL portions of water, dried at 50°/10 torr for 3 hr and sublimed at 130°/1 torr. The sublimate is recrystallized from boiling cyclohexane, yielding a snow-white crystalline substance which, after drying at ambient temperature and 10 torr for 3 hr, weighs 9.5 g (54.3%).

The benzene layer remaining after the extraction of the crude carbaboranethiol with potassium hydroxide is evaporated under vacuum, and the residue is sublimed at 80°/1 torr, producing 5.1 g (35.4%) of unreacted carbaborane. This increases the yield of the carbaboranethiol to 84.4%, calculated with respect to reacted carbaborane.

Properties

1,2-Dicarba-*closo*-dodecaborane-(12)-9-thiol is a white solid substance of mp 203–205° and giving a molecular ion at m/z 178. It is insoluble in water but readily soluble in organic solvents with the exception of cold hexane and cyclohexane. It is easily sublimable under vacuum and can be handled in air. The thiol is slightly acidic, with a pK_a value of 10.08 in 50% ethanol.

The ^{1}H-NMR (200 MHz) spectrum of 9-HS-1,2-$C_2B_{10}H_{11}$ in $CDCl_3$ exhibits two CH signals at δ 3.44 and 3.58 ppm, and the SH signal at 0.42 ppm (±0.3 ppm). The ^{11}B-NMR spectrum in $CDCl_3$ consists of one singlet of intensity one at 4.5 ppm and of five doublets of following ppm (int.)· −1.6(1), −8.2(2), −13.6(2), −14.6(2), −15.7(2). Some chemical and physical properties of 9-HS-1,2-$C_2B_{10}H_{11}$ are described in the literature.[10,11]

References

1. W. H. Knoth, J. C. Sauer, D. C. England, W. R. Hertler, and E. L. Muetterties, *J. Am. Chem. Soc.*, **86**, 3973 (1964).
2. E. T. Tolpin, G. R. Wellum, and S. A. Berley, *Inorg. Chem.*, **17**, 2867 (1978).
3. W. H. Knoth, *J. Am. Chem. Soc.*, **88**, 935 (1966).
4. M. R. Churchill, K. Gold, J. N. Francis, and M. F. Hawthorne, *J. Am. Chem. Soc.*, **91**, 1222 (1969).
5. J. Plešek and S. Heřmánek, *Chem. Ind.* (London), **1977**, 360.
6. J. Plešek and S. Heřmánek, *Coll. Czech. Chem. Commun.*, **43**, 1325 (1978).
7. J. Plešek and S. Heřmánek, *Coll. Czech. Chem. Commun.*, **46**, 687 (1981).
8. Z. Janoušek, J. Plešek, and Z. Plzák, *Coll. Czech. Chem. Commun.*, **44**, 2904 (1979).
9. J. Plešek, Z. Janoušek, and S. Heřmánek, *Coll. Czech. Chem. Commun.*, **45**, 1775 (1980).
10. J. Plešek, Z. Janoušek, and S. Heřmánek, *Coll. Czech. Chem. Commun.*, **45**, 2862 (1978).
11. J. Plešek and S. Heřmánek, *Coll. Czech. Chem. Commun.*, **44**, 24 (1979).

INDEX OF CONTRIBUTORS

Abu Salam, O. M., 21:107
Ackerman, J. F., 22:56
Ahuja, H. S., 21:187
Alam, F., 22:209
Andersen, R. A., 21:116
Anderson, D. M., 21:6
Andrews, Mark, 22:116
Ang, K. P., 21:114
Angelici, Robert J., 20:1; 22:126, 128
Arafat, A., 22:226
Aragon, R., 22:43
Armit, P. W., 21:28
Aufdembrink, Brent A., 21:16

Bagnall, K. W., 21:187
Bailar, John C., Jr., 22:124, 126, 128
Balasubramaniam, A., 21:33
Balch, Alan L., 21:4*l*
Basil, John D., 21:47
Beard, William R., 22:218
Beaulieu, Roland, 22:80
Begbie, C. M., 21:119
Benner, Linda S., 21:47
Bennett, M. A., 21:74
Bercaw, John, 21:181
Bisset, Graham, 22:156
Blake, D. M., 21:97
Booth, Carlye, 22:73
Borodinsky, L., 22:211
Bradley, John S., 21:66
Branch, J. W., 21:112
Brewer, Leo, 21:180
Broomhead, John A., 21:127
Brown, C., 22:131
Bruce, M. I., 21:78, 107
Burkhardt, E., 21:57
Butler, I. S., 21:28
Butler, Jan S., 21:1

Cahen, D., 22:80
Case, Christopher, 22:73
Chakravorti, M. C., 21:116, 170

Chandler, T., 21:170
Chen, M. G., 21:97
Chisholm, Malcolm H., 21:51
Chivers, T., 21:172
Christou, George, 21:33
Cisar, A., 22:151
Coffey, Christopher C., 21:142, 145, 146
Combs, Gerald L., Jr., 21:185
Cooper, C. B., III, 21:66
Corbett, J. D., 22:1, 15, 23, 26, 31, 36, 39, 151
Cote, W. J., 21:66
Cotton, F. A., 21:51
Coucouvanis, D., 21:23
Cronin, James L., 21:12
Cuenca, R., 21:28
Cushing, M. A., Jr., 21:90

Daake, Richard L., 22:26
Daniel, Cary S., 21:31
Darensbourg, Marcetta Y., 22:181
Dash, A. C., 21:119
Dassanayake, N. L., 21:185
Deaton, Joe, 21:135
Delmas, C., 22:56
Del Pilar De Neira, Rosario, 21:12, 23
De Renzi, A., 21:86
Deutsch, Edward, 21:19
Diversi, P., 22:167, 171
Dixon, N. E., 22:103
Dolcetti, Giuliano, 21:104
Draganjac, M., 21:23
DuBois, M. Rakowski, 21:37
Dudis, D. S., 21:19
Dunks, Gary B., 22:202
Durfee, Wm. S., 21:107
Dutta, T. K., 22:235
Dwight, A. E., 22:96
Dye, J. L., 22:151

Eisenberg, R., 21:90
Elder, R. C., 21:31

245

Eller, P. Gary, 21:162
Ellis, John F., 22:181
Enemark, John, 21:170
English, Ann M., 21:1
Enright, W. F., 22:163
Erner, K. A., 22:167, 171

Fackler, John P., Jr., 21:6
Fagan, Paul J., 21:181
Fang, Lawrence Y., 22:101
Fehlner, T. P., 22:235
Fenzl, W., 22:188, 190, 193, 196
Fjare, D. E., 21:57
Foley, H. C., 21:57
Fouassier, C., 22:56
Fox, J. R., 21:57
Foxman, Bruce M., 22:113
Frajerman, C., 22:133
Francis, Colin G., 22:116
Freeman, Wade A., 21:153
Furuya, F. R., 21:57

Gaines, D., 21:167
Gaines, Donald F., 22:218
Garito, A. F., 22:143
Geller, S., 22:76
Geoffroy, G. L., 21:57
Gewarth U. W., 22:207
Ghedini, Mauro, 21:104
Ginsberg, A. P., 21:153
Gladfelter, W. L., 21:57, 163
Goedken, V. L., 21:112
Goli, U., 22:103
Gould, E. S., 22:103
Grashens, Thomas, 22:116
Gray, William M., 21:104
Grims, R. N., 22:211, 215
Gruen, D. M., 22:96

Hagenmuller, P., 22:56
Haitko, Deborah A., 21:51
Halstead, Gordon W., 21:162
Hameister, C., 21:78
Han, Scott, 21:51
Haneister, C., 21:107
Hanusa, T., 22:226
Harley, A. D., 21:57
Harris, G. M., 21:119
Harris, Loren J., 22:107
Harrison, H. R., 22:43

Harrod, J. F., 21:28
Heck, Richard F., 21:86
Hedaya, Eddie, 22:200
Heeger, A. J., 22:143
Henderson, S. G. D., 21:6
Herkovitz, T., 21:99
Hermanek, S., 22:231, 235, 237, 239, 241
Hernandez-Szczurek, D., 22:226
Hodes, G., 22:80
Holida, Myra D., 21:192
Holt, M. S., 22:131
Honig, J. M., 22:43
Hood, Pam, 21:127
Hoots, John E., 21:175
Horowitz, H. S., 22:69, 73
Hosmane, N., 22:211, 215
Huang, T. N., 21:74
Hubert-Pfalzgraf, Lillian G., 21:16
Hudson, R. F., 22:131
Hull, J. W., Jr., 21:57
Hwu, S.-J., 22:1, 10

Ingrosso, G., 22:167, 171
Ittel, S. D., 21:74, 78, 90

Jackels, Susan C., 22:107
Jackson, W. G., 22:103, 119
Janousek, Z., 22:239, 241
Jaselskis, Bruno J., 22:135
Jaufmann, Judy, 22:113
Jensen, C. M., 22:163
Johnson, J. N., 21:112
Jones, Sharon A., 21:192

Kaesz, H. D., 21:99, 163
Kampe, C., 21:99
Kauffman, George B., 22:101, 149
Keem, J. E., 22:43
Keller, Philip, 22:200
Khan, I. A., 21:187
Kim, Leo, 22:149
Klabunde, Kenneth, 22:116
Knachel, Howard C., 21:175
Kodama, Goji, 22:188, 190, 193
Kohler, A., 22:48
Korenke, William J., 21:157
Koster, R., 22:181, 185, 188, 190, 193, 196, 198
Kozlowski, Adrienne W., 21:12
Krause, Ronald A., 21:12

Kreevoy, M. M., 21:167
Krusic, P. J., 21:66
Kubas, G. J., 21:37
Kwik, W. L., 21:114

Laguna, Antonio, 21:71
Lau, C., 21:172
Lawrance, G. A., 22:103
Lehr, S. D., 21:97
Lewandowski, J. T., 22:69
Liang, T. M., 21:167
Lin, S. M., 21:97
Lintvedt, Richard L., 22:113
Longo, J. M., 22:69, 73
Lucherini, A., 22:167, 171
Lund, G., 22:76

McCarly, R. E., 21:16
Macdiarid, Alan G., 22:143
Macdougall, J. Jeffrey, 22:124
McGrady, Nancy, 21:181
McGuire, James E., Jr., 22:185, 196, 198
Manassen, J., 22:80
Manriquez, Juan M., 21:181
Manzer, L. E., 21:84, 135
Marino, Dean F., 22:149
Marks, Tobin J., 21:84, 181
Masuo, Steven T., 21:142, 149, 151
Matheson, T. W., 21:74
Maynard, R. B., 22:211, 215
Mendelsohn, M. H., 22:96
Meunier, B., 22:133
Meyer, Gerd, 22:1, 10, 23
Mikulski, Chester M., 22:143
Miller, Douglas J., 21:37
Miller, Joel S., 21:142, 149, 151
Mintz, E. A., 21:84
Mondal, J. U., 21:97
Morita, Hideyoshi, 22:124, 126, 128
Murillo, Carlos A., 21:51
Murphy, C. N., 21:23
Murphy, Donald W., 22:26
Murphy, J., 22:69

Neilson, R. H., 22:209
Nelson, J. H., 22:131
Nickerson, William, 21:74
Niedenzu, K., 22:207, 209
Nielsen, Kj., Fl., 22:48
Nosco, Dennis L., 21:19

Noth, H., 22:218

Olah, George, 21:185
Ozin, Geoffrey A., 22:116

Paine, Robert T., 21:162
Palmer-Ordonez, Kathy, 22:200
Pan, Wie-Hin, 21:6, 33
Pandit, S. C., 21:170
Panuzi, A., 21:86
Pertici, P., 22:176
Petrinovic, Mark A., 22:181
Pfrommer, G., 22:48
Pickering, Ruth A., 21:1
Plesek, J., 22:231, 235, 237, 239, 241
Plowman, Keith R., 21:1
Poeppelmeier, K. R., 22:23
Pomerantz, Martin, 21:185
Pragovich, Anthony F., Jr., 21:142
Pretzer, W. R., 22:226

Rauchfuss, Thomas B., 21:175
Rausch, M. D., 22:176
Reilly, J. J., 22:90
Ridge, Brian, 21:33
Rollmann, L. D., 22:61
Rose, N. J., 21:112
Rosen, R., 21:57
Ross, J., 21:66
Rudolph, R. W., 22:226
Ruse, G. F., 22:76
Rydon, H. N., 21:33

Sandow, Torsten, 21:172
Sandrock, G. D., 22:90, 96
Sargeson, A. M., 22:103
Scalone, M., 21:86
Schertz, Larry D., 21:181
Schmitz, Gerard P., 21:31
Schoenfelner, Barry A., 22:113
Schonherr, E., 22:48
Schrock, R. R., 21:135
Schroeder, Norman C., 22:126, 128
Seidel, G., 22:185, 198
Seidel, W. M., 21:99
Shannon, R. D., 22:61
Sharp, Paul, 21:135
Shaw, C. Frak, III, 21:31
Sheridan, Richard, 21:157
Shimoi, Mamoru, 22:188, 190, 193

Index of Contributors

Shore, Sheldon G., 22:185, 196, 198
Shriver, D. F., 21:66
Siedle, A. R., 21:192
Simhon, E., 21:23
Simon, Arndt, 20:15; 22:31, 36
Sinf, Lee, 21:187
Slater, Sidney, 22:181
Smith, A. K., 21:74
Smith, H. D., Jr., 22:239, 241
Smith, J., 22:239
Smith, L. R., 21:97
Sneddon, L. G., 22:231, 237
Solomon, E. I., 21:114
Spencer, John L., 21:71
Spink, W. C., 22:176
Spira, D., 21:114
Sprinkel, C. R., 21:153
Staudigl, R., 22:218
Steehler, G., 21:167
Stephenson, T. A., 21:6, 28
Steudel, Rolf, 21:172
Stevens, R. E., 21:57; 22:163
Stibr, B., 22:231, 235, 237
Stiefel, Edward, I, 21:33
Strem, M. E., 22:133
Stremple, P., 21:23
Sullivan, Brian, 22:113
Surya Praksh, G. K., 21:185
Sutton, Lori J., 21:192
Svoboda, John, 21:180
Swincer, A. G., 21:78
Switkes, E. S., 21:28

Taylor, Michael J., 22:135
Taylor, William H., 21:192
Thomas, Rudolf, 22:107
Thorn, David L., 21:104
Todd, L. J., 22:226
Townsend, G., 22:101
Trenkle, A., 21:180

Truitt, Leah E., 21:142, 153
Tsunoda, Mitsukimi, 21:16
Tuck, Dennis G., 22:135
Tucka, A., 21:28
Tulip, T. H., 22:167, 171
Turner, David G., 21:71
Turney, John H., 22:149

Urbach, F. L.,
Uson, Rafale, 21:71

Vahrenkamp, H., 21:180
Valyocsik, E. W., 22:61
Van Eck, B., 22:141
Vergamini, P. J., 21:37
Vidusek, David A., 21:149, 151
Viswanathan, N., 22:101
Vitulli, G., 22:176

Waksman, L., 22:231, 237
Wallis, R. C., 21:78
Wartew, G. A., 22:131
Weber, David C., 22:143
Weber, W., 22:207
Weir, John R., 21:86
Wenckus, J. F., 22:43
White, Curtiss L., Jr., 21:146
Whitmire, K., 21:66
Wilber, S. A., 22:76
Williams, Jack M., 21:141, 142, 145, 146, 149, 151, 153, 192
Winkler, M., 21:114
Witmer, William B., 22:149
Wold, Aaron, 22:80
Wong, Ching-Ping, 22:156
Wrobleski, Debra A., 21:175
Wunz, Paul, 22:200

Yanta, T. J., 22:163
Young, Charles G., 21:127

SUBJECT INDEX

Names used in this Subject Index for Volumes 21–25 are based upon IUPAC *Nomenclature of Inorganic Chemistry,* Second Edition (1970), Butterworths, London; IUPAC *Nomenclature of Organic Chemistry,* Sections A, B, C, D, E, F, and H (1979), Pergamon Press, Oxford, U.K.; and the Chemical Abstracts Service *Chemical Substance Name Selection Manual* (1978), Columbus, Ohio. For compounds whose nomenclature is not adequately treated in the above references, American Chemical Society journal editorial practices are followed as applicable.

Inverted forms of the chemical names (parent index headings) are used for most entries in the alphabetically ordered index. Organic names are listed at the "parent" based on Rule C-10, Nomenclature of Organic Chemistry, 1979 Edition. Coordination compounds, salts and ions are listed once at each metal or central atom "parent" index heading. Simple salts and binary compounds are entered in the usual uninverted way, e.g., *Sulfur oxide* (S_8O), *Uranium (IV) chloride* (UCl_4).

All ligands receive a separate subject entry, e.g., *2,4-Pentanedione,* iron complex. The headings *Ammines, Carbonyl complexes, Hydride complexes,* and *Nitrosyl complexes* are used for the NH_3, CO, H, and NO ligands.

Acetic acid, 2-mercapto-cobalt complex, 21:21
Acetone:
  compd. with tri-μ-chloro-chloro-(thiocarbonyl)tetrakis(triphenylphosphine)-diruthenium (1:1), 21:29
  compd. with carbonyltri-μ-chloro-chloro-tetrakis(triphenylphosphine)diruthenium (2:1), 21:30
Acetonitrile:
  copper, iron, and zinc complexes, 22:108, 110, 111
  iridium complex, 21:104
  iron complex, 21:39
Actinides:
  5,10,15,20-tetraphenylporphyrin complexes, 22:156
Alkali metal alkyldihydroborates, 22:198
Alkali metal rare earth bromides and chlorides, 22:1, 10
Alkali metal transition metal oxides, 22:56
Aluminosilicates:
  mol. sieves, 22:61
Aluminum lanthanum nickel hydride ($AlLaNi_4H_4$), 22:96
Aluminum potassium sodium tetramethyl-ammonium silicate hydrate
  $[K_2Na[(CH_3)_4N]Al_4(Si_{14}O_{36})]\cdot 7H_2O$, 22:66
Aluminum sodium silicate hydrate ($NaAlSiO_4 \cdot 2.25H_2O$), 22:61
Aluminum sodium silicate hydrate ($Na_2Al_2Si_5O_{14} \cdot XH_2O$), 22:64
Aluminum sodium tetrapropylammonium silicate hydrate
  $(Na_{2.4}[(C_3H_7)_4N]_{3.6}Al_{2.6}(Si_{100}O_{207}))$, 22:67
Amine:
  cobalt(III) trifluoromethylsulfonate complexes, 22:103
Ammines:
  cobalt(III) trifluoromethanesulfonate complexes, 22:104
  platinum, 22:124
Ammonia, intercalate with hydrogen pentaoxoniobatetitanate(1−), 22:89
Ammonium, tetrabutyl-:
  tetrachlorooxotechnetate(V) (1:1), 21:160
  tetrakis(benzenethiolato)tetra-μ-seleno-tetraferrate(2−) (2:1), 21:36
  tetrakis(benzenethiolato)tetra-μ-thio-tetraferrate(2−) (2:1), 21:35
  tetrakis(1,1-dimethylethanethiolato)tetra-μ-seleno-tetraferrate(2−) (2:1), 21:37

249

———, tetramethyl-:
 potassium sodium aluminum silicate hydrate
 [K$_2$Na[(CH$_3$)$_4$N]Al$_4$(Si$_{14}$O$_{36}$)]·7H$_2$O, 22:66
 tetrakis(1,1-dimethylethanethiolato)tetra-μ-thio-tetraferrate(2−) (2:1), 21:36
———, tetrapropyl-:
 bis(pentasulfido)platinate(II) (2:1), 21:13
 sodium aluminum silicate
  (Na$_{2.4}$[(C$_3$H$_7$)$_4$N]$_{3.6}$Al$_{2.6}$(Si$_{100}$O$_{207}$)), 22:67
Ammonium bis(hexasulfido)palladate(II) (2:1): nonstoichiometric, 21:14
Ammonium diphosphate ((NH$_4$)$_4$(P$_2$O$_7$)), 21:157
Ammonium pentasulfide((NH$_4$)$_2$S$_5$, 21:12
Ammonium tris(pentasulfido)platinate(IV) (2:1), 21:12, 13
Ammonium tris(pentasulfido)rhodate(III) (3:1), 21:15
Arsine, [2-[(dimethylarsino)methyl]-2-methyl-1,3-propanediyl]bis(dimethyl-:
 niobium complex, 21:18
———, o-phenylenebis(dimethyl-:
 niobium complex, 21:18
 rhodium complex, 21:101
Azide, cesium tetracyanoplatinate (0.25:2:1): hydrate, 21:149

Benzene:
 chromium complex, 21:1, 2
———, ethynyl-:
 ruthenium complex, 21:82, 22:177
———, hexamethyl-:
 ruthenium complex, 21:74–77
———, 1-isopropyl-4-methyl-:
 ruthenium complex, 21:75
———, pentafluoro-:
 lithium and thallium complex, 21:71, 72
———, 1,2,3,5-tetrafluoro-:
 thallium complex, 21:73
———, 1,2,4,5-tetrafluoro-:
 thallium complex, 21:73
———, 1,3,5-trifluoro-:
 thallium complex, 21:73
Benzaldehyde, 2-(diphenylphosphino)-, 21:176
Benzenethiol:
 cadmium, cobalt, iron, manganese, and zinc complexes, 21:24–27
 iron complex, 21:35
Benzoic acid, 2-(diphenylphosphino)-, 21:178
2,2'-Bipyridine:
 cobalt complex, 21:127

palladium complex, 22:170
[1,1'-*commo*-Bis(2,3-dicarba-1-ferra-*closo*-heptaborane)](12):
 2,2',3,3'-tetraethyl-1,1-dihydro-, 22:215
Bismuthide(2−), tetra-:
 bis[(4,7,13,16,21,24-hexaoxa-1,10-diazabicyclo[8.8.8]hexacosane)potassium](1+), 22:151
Borane, dichlorophenyl-, 22:207
———, diethylhydroxy-, 22:193
———, diethylmethoxy-, 22:190
———, (dimethylamino)diethyl-, 22:209
———, [(2,2-dimethylpropanoyl)oxy]diethyl-, 22:185
———, (pivaloyloxy)diethyl, *see*—Borane, [(2,2-dimethylpropanoyl)oxy]diethyl-, 22:185
[^{10}B]Borane, dibromomethyl-, 22:223
Borate(1−), cyanotri[(^{2}H)hydro]sodium, 21:167
———, (cyclooctane-1,5-diyl)dihydro-:
 lithium, 22:199
 potassium, 22:200
 sodium, 22:200
———, dodecahydro-7,8-dicarba-*nido*-undeca-:
 potassium, 22:231
———, dodecahydro-6-thia-*arachno*-deca-:
 cesium, 22:227
———, hydrotris(pyrazolato)-:
 copper complex, 21:108
———, tetrakis(pyrazolato)-:
 copper complex, 21:110
———, tetraphenyl-:
 tetrakis(1-isocyanobutane)-bismethylenebis(diphenylphosphine)]-dihodium(I), 21:49
 tetrakis(1-isocyanobutane)rhodium(I), 21:50
———, tris(3,5-dimethylpyrazolato)hydro-:
 boron-copper complex, 21:109
Borate(2−), tris[μ-[(1,2-cyclohexanedione dioximato)-O:O']diphenyldi-:
 iron complex, 21:112
Borinic acid, diethyl-, *see*—Borane, diethylhydroxy-, 22:193
———, diethyl, methyl ester, *see*—Borane, diethylmethoxy-, 22:190
Boron, bis-μ-(2,2-dimethylpropanoato-O,O')-diethyl-μ-oxo-di-, 22:196
[^{10}B]Boron bromide (^{10}BBr$_3$), 22:219
Boron compounds:
 labeling of, with boron-10, 22:218

Bromides:
  of rare earths and alkali metals, 22:1, 10
1-Butanamine:
  intercalate with hydrogen pentaoxoniobatetitanate(1−), 22:89
Butane:
  iridium and rhodium cobalt, complexes, 22:171, 173, 174
  palladium complex, 22:167, 168, 169, 170
Butane, isocyano-:
  rhodium complex, 21:49

Cadmate(II), tetrakis(benzenetholato)-:
  bis(tetraphenylphosphonium), 21:26
Cadmium chalcogenides:
  on metallic substrates, 22:80
Cadmium selenide (CdSe):
  on titanium, 22:82
Cadmium selenide telluride:
  on molybdenum, 22:84
Cadmium selenide telluride ($CdSe_{0.65}Te_{0.35}$), 22:81
Calcium manganese oxide ($Ca_2Mn_3O_8$), 22:73
Carbon dioxide:
  iridium complex, 21:100
Carbon diselenide, 21:6, 7
Carbonic acid:
  cobalt complex, 21:120
  platinum chain complex, 21:153, 154
Carbonyl complexes:
  chromium, 21:1, 2
  chromium, molybdenum and tungsten, 22:81
  cobalt, iron, osmium, and ruthenium, 21:58–65
  copper, 21:107–110
  iridium, 21:97
  iron, 21:66, 68
  iron and ruthenium, 22:163
  palladium, 21:49
  ruthenium, 21:30
Cerium:
  porphyrin complexes, 22:156
  ———, bis(2,4-pentanedionato)[5,10,15,20-tetraphenylporphyrinato(2−)]-, 22:160
Cesium azide tetracyanoplatinate (2:0.25:1): hydrate, 21:149
Cerium chloride ($CeCl_3$), 22:39
Cesium chloride tetracyanoplatinate (2:0.30:1), 21:142
Cesium [hydrogen bis(sulfate)] tetracyanoplatinate (3:0.46:1), 21:151

Cesium lithium thulium chloride ($Cs_2LiTmCl_6$), 20:10
Cesium lutetium chloride (Cs_2LuCl_5), 22:6
Cesium lutetium chloride (Cs_3LuCl_6), 22:6
Cesium lutetium chloride ($Cs_3Lu_2Cl_9$), 22:6
Cesium praseodymium chloride ($CsPr_2Cl_7$), 22:2
Cesium scandium chloride ($CsScCl_3$), 22:23
Cesium scandium chloride ($Cs_3Sc_2Cl_9$), 22:25
Chlorides:
  of rare earths and alkali metals, 22:1, 10
Chromate(1−), pentacarbonylhydrido-μ-nitrido-bis(triphenylphosphorus)(1+), 22:183
Chromium(O), ($η^6$-benzene)-dicarbonyl(selenocarbonyl), 21:1, 2
———, pentacarbonyl(selenocarbonyl)-, 21:1, 4
Chromium potassium oxide ($KCrO_2$), 22:59
Chromium potassium oxide ($K_{0.5}CrO_2$): bronze, 22:59
Chromium potassium oxide ($K_{0.6}CrO_2$): bronze, 22:59
Chromium potassium oxide ($K_{0.7}CrO_2$): bronze, 22:59
Chromium potassium oxide ($K_{0.77}CrO_2$): bronze, 22:59
Cluster compounds:
  cobalt, iron, molybdenum, ruthenium, and osmium, 21:51–68
Cobalt, (1,4-butanediyl)($η^5$-cyclopentadienyl)(triphenylphosphine)-, 22:171
Cobalt(III), (2-aminoethanethiolato-N,S)bis(1,2-ethanediamine)-:
  diperchlorate, 21:19
———, [N-(2-aminoethyl)-1,2-ethanediamine]tris(trifluoromethylsulfonato)-:
  fac-, 22:106
———, aquabromobis(1,2-ethanediamine)-:
  dibromide, cis-, monohydrate, 21:123
  dithionate, $trans$-, monohydrate, 21:124
———, aquachlorobis(1,2-ethanediamine)-:
  dithionate, $trans$-, monohydrate, 21:125
———, bis(1,2-ethanediamine)-bis(trifluoromethanesulfonato)-:
  cis-, trifluoromethanesulfonate, 22:105
———, bis(1,2-ethanediamine)(2-mercaptoacetato(2−)-O,S)-:
  perchlorate, 21:21

——, (carbonato)bis(1,2-ethanediamine)-:
    bromide, 21:120
——, dibromobis(1,2-ethanediamine)-:
    bromide, cis-, monohydrate, 21:121
    bromide, trans-, 21:120
——, pentaammine(trifluoromethane-
    sulfonato)-:
    trifluoromethanesulfonate, 22:104
Cobaltate(1 −), tridecacarbonyltriruthenium-:
    μ-nitrido-bis(triphenylphosphorus)(1 +),
    21:61
Cobaltate(II), tetrakis(benzenethiolato)-:
    bis(tetraphenylphosphonium), 21:24
Cobalt potassium oxide ($KCoO_2$), 22:58
Cobalt potassium oxide ($K_{0.5}CoO_2$):
    bronze, 22:57
Cobalt potassium oxide ($K_{0.67}CrO_2$):
    bronze, 22:57
Cobalt sodium oxide ($NaCoO_2$), 22:56
Cobalt sodium oxide ($Na_{0.6}CoO_2$), 22:56
Cobalt sodium oxide ($Na_{0.64}CoO_2$), 22:56
Cobalt sodium oxide ($Na_{0.74}CoO_2$), 22:56
Cobalt sodium oxide ($Na_{0.77}CoO_2$), 22:56
Containers:
    tantalum, as high-temp., for reduced
    halides, 20:15
Copper(I), carbonyl[hydrotris(pyrazolato)-
    borato]-, 21:108
——, carbonyl[tetrakis(pyrazolato)-borato]-,
    21:110
——, carbonyl[tris(3,5-dimethylpyr-
    azolato)hydroborato]-, 21:109
Copper(II), (2,9-dimethyl-3,10-diphenyl-
    1,4,8,11-tetraazacyclotetradeca-1,3,8,10-
    tetraene)-:
    bis[hexafluorophosphate(1 −)], 22:110
——, (1,10-phenanthroline)[serinato(1 −)]-:
    sulfate (2:1), 21:115
Copper iodide (CuI), 22:101
Crystal growth:
    of Li_3N, 22:51
    of oxides by skull melting, 22:43
    of silver tungstate $Ag_8(W_4O_{16})$, 22:78
Cyano complexes:
    boron, 21:167
    platinum chain complexes, 21:142–156
1,3-Cycloheptadiene:
    ruthenium complex, 22:179
1,3-Cyclohexadiene:
    ruthenium complex, 21:77; 22:177

1,2-Cyclohexanedione, dioxime:
    boron-iron complex, 21:112
1,5-Cyclooctadiene:
    ruthenium complex, 22:178
Cyclooctane:
    boron complex, 22:199
1,3,5-Cyclooctatriene:
    ruthenium complex, 22:178
1,3-Cyclopentadiene:
    cobalt complex, 22:171, 235
    iron complexes, 21:37–46
    ruthenium complex, 21:78; 22:180
    titanium and vanadium complexes, 21:84,
    85
——, 1,2,3,4,5-pentamethyl-, 21:181
L-Cysteine:
    gold complex, 21:31

DC510150, see—Poly(dimethylsiloxane-co-
    methylphenylsiloxane), 22:116
Decaborane(14), 22:202
diars, see—Arsine, o-phenylenebis(dimethyl-,
    21:18
Diboroxane, tetraethyl-, 22:188
1,2-Dicarba-3-cobalta-closo-dodecaborane(11):
    3-(η⁵-cyclopentadienyl)-, 22:235
1,2-Dicarba-closo-dodecaborane(12)-9-thiol,
    22:241
2,3-Dicarba-nido-hexaborane(8):
    2,3-diethyl-, 22:211
2,6-Dicarba-nido-nonaborane(11), 22:237
7,8-Dicarba-nido-undecaborane(11):
    9-(dimethyl sulfide)-, 22:239
Dimethyl sulfide:
    boron complex, 22:239
    niobium complex, 21:16
    platinum(II) complexes, 22:126, 128
diphos, see—Phosphine, 1,2-ethanediyl-
    bis(diphenyl-, 22:167
Diphosphate, tetraammonium (($NH_4)_4(P_2O_7)$),
    21:157
Diselenocarbamic acid, N,N-diethyl-:
    nickel complex, 21:9
Divanadium:
    polymer-stabilized, 22:116
dppe, see—Phosphine, 1,2-ethanediyl-
    bis(diphenyl-, 21:18
Dysprosium:
    porphyrin complexes, 22:156

———, (2,4-pentanedionato)[5,10,15,20-tetraphenylporphyrinato(2 −)]-, 22:160
———, (2,2,6,6-tetramethyl-3,5-heptanedionato)[5,10,15,20-tetraphenylporphyrinato(2 −)]-, 22:160
Dysprosium chloride (DyCl$_3$), 22:39
Dysprosium potassium chloride (KDy$_2$Cl$_7$), 22:2

Erbium:
  porphyrin complexes, 22:156
———, (2,4-pentanedionato)[5,10,15,20-tetrakis(3-fluorophenyl)porphyrinato(2 −)]-, 22:160
———, (2,4-pentanedionato)[5,10,15,20-tetraphenylporphyrinato(2 −)]-, 22:160
Erbium chloride (ErCl$_3$), 22:39
Ethanamine:
  intercalate with hydrogen pentaoxoniobatetitanate(1 −), 22:89
1,2-Ethanediamine:
  cobalt(III) trifluoromethanesulfonate complexes, 22:105
———, N-(2-aminoethyl)-:
  cobalt(III) trifluoromethanesulfonate complexes, 22:106
  cobalt complex, 21:19, 21, 120–126
———, N,N-bis[2-(dimethylamino)ethyl-N',N'-dimethyl-:
  palladium complex, 21:129–132
———, N,N'-bis[2-(dimethylamino)ethyl]-N,N'-dimethyl-:
  palladium complex, 21:133
———, N,N'-bis(1-methylethyl)-:
  platinum complex, 21:87
———, (S,S)-N,N'-bis(1-phenylethyl)-:
  platinum complex, 21:87
———, N,N'-dimethyl-N,N'-bis(1-methylethyl)-:
  platinum complex, 21:87
———, N,N'-dimethyl-N,N'-bis(1-phenylethyl)-, (R,R)-:
  platinum complex, 21:87
———, N,N,N',N'-tetraethyl-:
  platinum complexes, 21:86
———, N,N,N',N'-tetramethyl-:
  palladium complex, 22:168
1,2-Ethanediol:
  iron complex, 22:88
Ethanethiol:
  iron complex, 21:39

———, 2-amino-:
  cobalt complex, 21:19
———, 1,1-dimethyl-:
  iron complex, 21:36, 37
Ethanol:
  uranium complex, 21:165
Ethene:
  iron complex, 21:91
  platinum complexes, 21:86–89
  ruthenium complex, 21:76
Ethylene glycol, see—1,2-Ethanediol, 22:88
Europium:
  porphyrin complexes, 22:156
———, (2,4-pentanedionato)[5,10,15,20-tetrakis(3,5-dichlorophenyl)porphyrinato(2 −)]-, 22:160
———, (2,4-pentanedionato)[5,10,15,20-tetrakis(4-methylphenyl)porphyrinato(2 −)]-, 22:160
———, (2,4-pentanedionato)[5,10,15,20-tetraphenylporphyrinato(2 −)]-, 22:160
Europium chloride (EuCl$_3$), 22:39

Ferrate(1 −), tricarbonylnitrosyl-:
  μ-nitrido-bis(triphenylphosphorus)(1 +), 22:163, 165
———, tridecacarbonylhydridotriruthenium-:
  μ-nitrido-bis(triphenylphosphorus)(1 +), 21:60
Ferrate(2 −), tetrakis(benzenethiolato)tetra-μ-seleno-tetra-:
  bis(tetrabutylammonium), 21:36
———, tetrakis(benzenethiolato)tetra-μ-thio-tetra-:
  bis(tetrabutylammonium), 21:35
———, tetrakis(1,1-dimethylethanethiolato)tetra-μ-seleno-tetra-:
  bis(tetrabutylammonium), 21:37
———, tetrakis(1,1-dimethylethanethiolato)tetra-μ-thio-tetra-:
  bis(tetramethylammonium), 21:36
———, tridecacarbonyltetra-:
  μ-nitrido-bis(triphenylphosphorus)(1 +), 21:66, 68
Ferrate(II), tetrakis(benzenethiolato)-:
  bis(tetraphenylphosphonium), 21:24
Ferrate(II, III), tetrakis(benzenethiolato)tetra-μ$_3$-thio-tetra-:
  bis(tetraphenylphosphonium), 21:27

Ferrate(III), tetrakis(benzenethiolato)di-
μ-thio-di-:
  bis(tetraphenylphosphonium), 21:26
Formic acid, (formyloxy)-:
  iridium complex, 21:102
Furan, tetrahydro-:
  hafnium, niobium, scandium, titanium,
  vanadium, and zirconium complexes,
  21:135–139

Gadolinium:
  porphyrin complexes, 22:156
  ———, (2,4-pentanedionato)[5,10,15,20-tetra-
  phenylporphyrinato-]-, 22:160
Gadolinium chloride ($GdCl_3$), 22:39
Gallate(1 −), tetrabromo-:
  tetrabutylammonium, 22:139
  tetraethylammonium, 22:141
  ———, tetrachloro-:
  tetrabutylammonium, 22:139
  ———, tetraiodo-:
  tetrabutylammonium, 22:140
Gallate(2 −), hexabromodi-:
  bis(tetraphenylphosphonium), 22:139
  bis(triphenylphosphonium), 22:135, 138
  ———, hexachlorodi-:
  bis(triphenylphosphonium), 22:135, 138
  ———, hexaiododi-:
  bis(triphenylphosphonium), 22:135, 138
Gold(I), (L-cysteinato)-, 21:31
Guanidinium (hydrogen difluoride)-
  tetracyanoplatinate (2:0.27:1):
  hydrate (1:1.8), 21:146

Hafnium(IV), tetrachlorobis(tetrahydrofuran)-,
  21:137
3,5-Heptanedione, 2,2,6,6-tetramethyl-:
  actinide and lanthanide complexes, 22:156
4,7,13,16,21,24-Hexaoxa-1,10-diazabicy-
clo[8.8.8]hexacosane:
  potassium complex, 22:151
Hexasulfide:
  palladium complex, nonstoichiometric,
  21:14
Holmium:
  porphyrin complexes, 22:156
  ———, (2,4-pentanedionato)[5,10,15,20-tetra-
  phenylporphyrinato(2 −)]-, 22:160
  ———, (2,2,6,6-tetramethyl-3,5-heptane-
dionato)[5,10,15,20-tetraphenylpor-
phyrinato(2 −)]-, 22:160
Holmium chloride ($HoCl_3$), 22:39
Hydrido complexes:
  chromium, molybdenum and tungsten,
  22:181
  cobalt, iron, osmium, and ruthenium,
  21:58–65
  iron complexes, 21:92
[Hydrogen bis(sulfate)]:
  cesium tetracyanoplatinate (0.46:3:1),
  21:151
Hydrogen difluoride:
  potassium tetracyanoplatinate (0.30:2:1),
  trihydrate, 21:147
Hydrogen pentaoxoniobatetitanate(1 −), 22:89

Intercalation compounds, 22:86, 89
Iridium, (1,3-butanediyl)(η^5-pentamethylcyclo-
pentadienyl)(triphenylphosphine)-, 22:174
Iridium, chloro(η^2-cyclooctene)-
  tris(trimethylphosphine)-, 21:102
  ———, chloro[(formyl-κC-oxy)formato-
κO(2 −)]thio(trimethylphospine)-, 21:102
Iridium(1 +), bis[1,2-ethanediyl-
bis(dimethylphosphine)]-:
  chloride, 21:100
  ———, (carbon dioxide)bis[1,2-ethanediyl-
bis(dimethylphosphine)]-:
  chloride, 21:100
Iridium(I), carbonylchlorobis-
  (dimethylphenylphosphine)-:
  trans-, 21:97
Iridium(III), tris(acetonitrile)-
nitrosylbis(triphenylphosphine)-:
  bis[hexafluorophosphate], 21:104
Iron, bis(η^5-cyclopentadienyl)-μ-(disulfur)bis-
μ-(ethanethiolato)di-, 21:40, 41
  ———, bis[1,2-ethanediyl-
bis(diphenylphosphine)](ethene)-, 21:91
  ———, bis[1,2-ethanediyl-
bis(diphenylphosphine)](trimethyl phosphite)-, 21:93
  ———, [2-[2-(diphenylphosphino)-
ethyl]phenylphosphino]phenyl-
C,P,P'][1,2-ethanediyl-
bis(diphenylphosphine)]hydrido-, 21:92
  ———, [1,2-ethanediolato(2 −)]dioxodi-, 22:88
  ———, [1,2-ethanediyl-

Subject Index

bis(diphenylphosphine)]bis(2,4-pentane-
dionato)-, 21:94
——, methoxyoxo-, 22:87
——, tetrakis(η^5-cyclopentadienyl)-μ_3-(disulfur)-di-μ_3-thiotetra-, 21:42
——, tetrakis(η^5-cyclopentadienyl)-μ_3-(disulfur)-tri-μ_3-thiotetra-, 21:45
——, tridecacarbonyldihydridotriosmium-,
21:63
——, tridecacarbonyldihydridotriruthenium-,
21:58
Iron(2+), bis(acetonitrile)bis(η^5-cyclopenta-
dienyl)bis-μ-(ethanethiolato)-di-:
bis(hexafluorophosphate), 21:39
——, tetrakis(η^5-cyclopentadienyl)-μ_3-(disulfur)tri-μ_3-thiotetra-:
bis(hexafluorophosphate), 21:44
Iron(II), bis(acetonitrile)(2,9-dimethyl-3,10-di-
phenyl-1,4,8,11-tetrazacyclotetradeca-
1,3,8,10-tetraene)-:
bis[hexafluorophosphate(1 −)], 22:108
Iron(III), {[tris[μ-[(1,2-cyclohexanedione dioximato)-$O:O'$]diphenyldiborato(2 −)]-
$N,N',N'',N''',N'''',N'''''$}-, 21:112
Iron chloride oxide (FeClO):
intercalate with 4-aminopyridine (4:1), 22:86
intercalate with pyridine (4:1), 22:86
intercalate with 2,4,6-trimethypyridine,
22:86
Iron oxide (Fe_2O_4):
magnetite, crystal growth of, by skull melting, 22:43
Iron titanium hydride ($FeTiH_{1.94}$), 22:90

Labeling:
of boron compounds, with boron-10, 22:218
Lanthanides:
5,10,15,20-tetraphenylporphyrin complexes,
22:156
Lanthanium iodide (LaI_2), 22:36
Lanthanum, (2,4-pentanedionato)[5,10,15,20-
tetraphenylporphyrinato(2 −)]-, 22:160
——, (2,2,6,6-tetramethyl-3,5-heptane-
dionato)[5,10,15,20-tetraphenylpor-
phyrinato(2 −)]-, 22:160
Lanthanum aluminum nickel hydride
($AlLaNi_4H_4$), 22:96
Lanthanum chloride ($LaCl_3$), 22:39
Lanthanum iodide (LaI_3), 22:31

Lead oxide (PbO_2):
solid solns. with ruthenium oxide (Ru_2O_3),
pyrochlor, 22:69
Lead ruthenium oxide ($Pb_{2.67}Ru_{1.33}O_{6.5}$),
pyrochlore, 22:69
Lithium, (pentafluorophenyl)-, 21:72
Lithium cesium thulium chloride
($Cs_2LiTmCl_6$), 20:10
Lithium nitride (Li_3N), 22:48
Lutetium:
porphyrin complexes, 22:156
——, (2,4-pentanedionato)[5,10,15,20-tetra-
phenylporphyrinato(2 −)]-, 22:160
Lutetium cesium chloride (Cs_2LuCl_5), 22:6
Lutetium cesium chloride (Cs_3LuCl_6), 22:6
Lutetium cesium chloride ($Cs_3Lu_2Cl_9$), 22:6
Lutetium chloride ($LuCl_3$), 22:39

Magnetite (Fe_2O_4):
crystal growth of, by skull melting, 22:43
Manganate(II), tetrakis(benzenethiolato)-:
bis(tetraphenylphosphonium), 21:25
Manganese calcium oxide ($Ca_2Mn_3O_8$), 22:73
Methanamine:
intercalate with hydrogen pentaoxoniobateti-
tanate(1 −), 22:89
Methanamine, N-methyl-:
molybdenum complex, 21:54
Methanesulfonic acid, trifluoro-:
cobalt(III) amine complexes, 22:104, 105
Molecular sieves, 22:61
Molybdate(1 −), pentacarbonylhydrido-:
μ-nitrido-bis(triphenylphosphorus)(1 +),
22:183
Molybdate(V), pentafluorooxo-:
dipotassium, 21:170
Molybdenum:
as substrate for cadmium chalcogenides,
22:80
——, dichlorotetrakis(dimethylamido)di-:
($Mo\equiv Mo$), 21:56
——, hexakis(dimethylamido)di-:
($Mo\equiv Mo$), 21:54

Neodynium, (2,4-pentanedionato)[5,10,15,20-
tetraphenylporphyrinato(2 −)]-, 22:160
——, (2,2,6,6-tetramethyl-3,5-heptane-
dionato)[5,10,15,20-tetraphenylpor-
phyrinato(2 −)]-, 22:160

Neodynium chloride (NdCl₃), 22:39
Nickel(II), chloro(*N*,*N*-diethyldiselenocarbamato)(triethylphosphine)-, 21:9
——, dibromobis(3,3',3''-phosphinidynetripropionitrile)-, 22:113, 115:
  polymer, 22:115
——, dichlorobis(3,3',3''-phosphinidynetripropionitrile)-, 22:113
Nickel aluminum lanthanum hydride (AlLaNi₄H₄), 22:96
Niobium(III), di-μ-chloro-tetrachloro-μ-(dimethyl sulfide)bis(dimethyl sulfide)di-, 21:16
——, hexachlorobis[1,2-ethanediylbis(diphenylphosphine)]di-, 21:18
——, hexachlorobis[[2-[(dimethylarsino)methyl]-2-methyl-1,3-propanediyl]bis(dimethylarsine)]-, 21:18
——, hexachlorobis[*o*-phenylenebis(dimethylarsine)]di-, 21:18
Niobium (IV), tetrachlorobis(tetrahydrofuran)-, 21:138
Nitrogen sulfide (NS), *see*—Sulfur nitride (SN), polymer, 22:143
Nitrosyl complexes:
  iridium, 21:104
Nitrosyl complexes:
  iron and ruthenium, 22:163

cyclo-Octasulfur monoxide, 21:172
Offretite, tetramethylammonium substituted [K₂Na[(CH₃)₄N]Al₄(Si₁₄O₃.₆)].7H₂O, 22:66
Osmium, tridecacarbonyldihydridoirontri-, 21:63
——, tridecacarbonyldihydridorutheniumtri-, 21:64

Palladate(II), bis(hexasulfido)-:
  diammonium, nonstoichiometric, 21:14
Palladium, (2,2'-bipyridine)(1,4-butanediyl)-, 22:170
——, (1,4-butanediyl)bis(triphenylphosphine)-, 22:169
——, (1,4-butanediyl)[1,2-ethanediylbis(diphenylphosphine)]-, 22:167
——, (1,4-butanediyl)(*N*,*N*,*N*',*N*'-tetramethyl-1,2-ethanediamine)-, 22:168
Palladium(I), μ-carbonyl-dichlorobis[methylenebis(diphenylphosphine)]di-, 21:49
——, dichlorobis-μ-[methylenebis(diphenylphosphine)]-di-:
  (*Pd-Pd*), 21:48
Palladium(II), [*N*,*N*'-bis[2-(dimethylamino)ethyl]-*N*,*N*'-dimethyl-1,2-ethanediamine]-:
  bis(hexafluorophosphate), 21:133
——, [*N*,*N*-bis[2-(dimethylamino)ethyl]-*N*',*N*'-dimethyl-1,2-ethanediamine]bromo-:
  bromide, 21:131
——, [*N*,*N*-bis[2-(dimethylamino)ethyl]-*N*',*N*'-dimethyl-1,2-ethanediamine]chloro-:
  chloride, 21:129
——, [*N*,*N*-bis[2-(dimethylamino)ethyl-*N*',*N*'-dimethyl-1,2-ethanediamine]iodo-:
  iodide, 21:130
——, [*N*,*N*-bis[2-(dimethylamino)ethyl]-*N*',*N*'-dimethyl-1,2-ethanediamine](thiocyanato-*N*)-:
  thiocyanate, 21:132
——, chloro(*N*,*N*-diethyldiselenocarbamato)(triphenylphosphine)-, 21:10
2,4-Pentanedione:
  actinide and lanthanide complexes, 22:156
  iron complex, 21:94
Pentasulfide:
  platinum and rhodium complexes, 21:12
1,10-Phenanthroline:
  copper complex, 21:115
Phosphate(1−), hexafluoro-:
  bis(acetonitrile)bis(η⁵-cyclopentadienyl)bis-μ-(ethanethiolato)diiron(2+) (2:1), 21:39
——, (η⁵-cyclopentadienyl)(phenylvinylene)bis(triphenylphosphine)ruthenium(II), 21:80
——, tetrakis(η⁵-cyclopentadienyl)-μ₃-(disulfur)tri-μ₃-thio-tetrairon(2+) (2:1), 21:44
Phosphine:
  iridium complex, 21:104
——, dimethyl-, 21:180
——, dimethylphenyl-, 22:133
  iridium complex, 21:97
——, 1,2-ethanediylbis(dimethyl-:
  iridium complex, 21:100
——, 1,2-ethanediylbis(diphenyl-:
  iron complexes, 21:90–94

niobium complex, 21:18
palladium complex, 22:167
——, methylenebis(diphenyl-:
palladium and rhodium complexes, 21:47–49
——, triethyl-:
nickel complex, 21:9
——, trimethyl-:
iridium complex, 21:102
——, triphenyl-:
cobalt, iridium and rhodium complexes, 22:171, 173, 174
palladium complex, 22:169
ruthenium complexes, 21:29, 78
Phosphines, triaryl, 21:175
Phosphonium, tetraphenyl-:
tetrakis(benzenethiolato)cadmate(II) (2:1), 21:26
tetrakis(benzenethiolato)cobaltate(II) (2:1), 21:24
tetrakis(benzenethiolato)di-μ-thiodifferate(III) (2:1), 21:26
tetrakis(benzenethiolato)ferrate(II) (2:1), 21:24
tetrakis(benzenethiolato)manganate(II) (2:1), 21:25
tetrakis(benzenethiolato)tetra-μ_3-thio-tetraferrate(II, III) (2:1), 21:27
tetrakis(benzenethiolato)zincate(II) (2:1), 21:25
Phosphorotrithious acid:
tributyl ester, 22:131
Phosphorus(1+), μ-nitrido-bis(triphenyl-:
decacarbonyl-μ-nitrosyl-triruthenate(1−), 22:163, 165
hexafluorouranate(V), 21:166
nitrite, 22:164
pentacarbonylhydridochromate(1−), 22:183
pentacarbonylhydridomolybdate(1−), 22:183
pentacarbonylhydridotungstate(1−), 22:182
tricarbonylnitrosylferrate(1−), 22:163, 165
tridecacarbonylcobalttriruthenate(1−), 21:61
tridecacarbonylhydridoirontriruthenate(1−), 21:60
tridecacarbonyltetraferrate(2−) (2:1), 21:66, 68
Platinate, tetracyano-:
cesium azide (1:2:0.25), hydrate, 21:149
cesium chloride (1:2:0.30), 21:142

cesium [hydrogen bis(sulfate)] (1:2:0.46), 21:151
guanidinium (hydrogen difluoride) (1:3:0.27), hydrate (1:1.8), 21:146
potassium (hydrogen difluoride) (1:2:0.30), trihydrate, 21:147
rubidium chloride (1:2:0.30), trihydrate, 21:145
Platinate(II), bis(pentasulfido)-:
bis(tetrapropylammonium)-, 20:13
——, tetracyano-:
dithallium, 21:153
thallium carbonate (1:4:1), 21:153, 154
Platinate(II), trichloro(dimethyl sulfide)-:
tetrabutylammonium, 22:128
Platinate(IV), tris(pentasulfido)-:
diammonium, 21:12, 13
Platinum(II), [N,N'-bis(1-methylethyl)-1,2-ethanediamine]dichloro(ethene)-, 21:87
——, [(S,S)-N,N'-bis(1-phenylethyl)-1,2-ethanediamine]dichloro(ethene)-, 21:87
——, chloro(N,N-diethyldiselenocarbamato)(triphenylphosphine)-, 21:10
——, chlorotris(dimethyl sulfide)-:
tetrafluoroborate(1−), 22:126
——, diammineaquachloro-:
trans-, nitrate, 22:125
——, diamminechloroiodo-:
trans-, chloride, 22:124
——, di-μ-chloro-:
dichlorobis(dimethyl sulfide)di-, 22:128
——, dichloro[N,N'-dimethyl-N,N'-bis(1-methylethyl)-1,2-ethanediamine](ethene)-, 21:87
——, dichloro{(R,R)-N,N'-dimethyl-N,N'-bis(1-phenylethyl)-1,2-ethanediamine](ethene)-, 21:87
——, dichloro(ethene)(N,N,N',N'-tetraethyl-1,2-ethanediamine)-, 21:86, 87
——, (N,N-diethyldiselenocarbamato)methyl(triphenylphosphine)-, 20:10
——, tetraaqua-, 21:192
——, triamminechloro-:
chloride, 22:124
Plumbate(IV), hexachloro-:
dipyridinium, 22:149
Poly(dimethylsiloxane-co-methylphenylsiloxane):
in divanadium stabilization, 22:116

Polythiazyl, *see*—Sulfur nitride (SN) polymer, 22:143
Porphyrin:
  actimide and lanthanide complexes, 22:156
  ——, 5,10,15,20-tetrakis(4-methylphenyl)-:
    actinide and lanthanide complexes, 22:156
  ——, 5,10,15,20-tetraphenyl-:
    actinide and lanthanide complexes, 22:156
Potassium(1+), (4,7,13,16,21,24-hexaoxa-1,10-diazabicyclo[8.8.8]hexacosane)-:
  tetrabismuthide(2−) (2:1), 22:151
Potassium chromium oxide (KCrO$_2$), 22:59
Potassium chromium oxide (K$_{0.5}$CrO$_2$):
  bronze, 22:59
Potassium chromium oxide (K$_{0.6}$CrO$_2$):
  bronze, 22:59
Potassium chromium oxide (K$_{0.7}$CrO$_2$):
  bronze, 22:59
Potassium chromium oxide (K$_{0.77}$CrO$_2$):
  bronze, 22:59
Potassium cobalt oxide (KCoO$_2$), 22:58
Potassium cobalt oxide (K$_{0.5}$CoO$_2$):
  bronze, 22:57
Potassium cobalt oxide (K$_{0.67}$CoO$_2$):
  bronze, 22:57
Potassium dysprosium chloride (KDy$_2$Cl$_7$), 22:2
Potassium hexafluorouranate(V), 21:166
Potassium (hydrogen difluoride) tetracyanoplatinate (2:0.30:1):
  trihydrate, 21:147
Potassium pentafluorooxomolybdate(V) (2:1), 21:170
Potassium pentaoxoniobatetitanate(1−), 22:89
Potassium sodium tetramethylammonium aluminum silicate hydrate [K$_2$Na[(CH$_3$)$_4$N]Al$_4$(Si$_{14}$O$_{36}$)]·7H$_2$O, 22:66
Praesodymium:
  porphyrin complexes, 22:156
  ——, (2,4-pentanedionato)[5,10,15,20-tetrakis(4-methylphenyl)-porphyrinato(2−)]-, 22:160
  ——, (2,4-pentandionato)[5,10,15,20-tetraphenylporphyrinato(2−)]-, 22:160
  ——, [5,10,15,20-tetrakis(4-methylphenyl)porphyrinato(2−)]-, 22:160
Praseodymium cesium chloride (CsPr$_2$Cl$_7$), 22:2
Praseodynium chloride (PrCl$_3$), 22:39

1-Propanamine:
  intercalate with hydrogen pentaoxoniobatetitanate(1−), 22:89
Propionitrile, 3,3′,3″-phosphinidynetri-:
  nickel complexes, 22:113, 115
1*H*-Pyrazole:
  boron-copper complex, 21:108, 110
  ——, 3,5-dimethyl-:
    boron-copper complex, 21:109
Pyridine:
  intercalate with FeClO (1:4), 22:86
  rhenium complex, 21:116, 117
  ——, 4-amino-:
    intercalate with FeClO (1:4), 22:86
  ——, 2,4,6-trimethyl-:
    intercalate with FeClO, 22:86

Rare earth alkali metal bromides and chlorides, 22:1, 10
Rare earth trichlorides, 22:39
Rare earth triiode, 22:31
Rhenium(V), dioxotetrakis(pyridine)-:
  chloride, *trans*-, 21:116
  perchlorate, *trans*-, 21:117
Rhodate(III), tris(pentasulfido)-:
  triammonium, 21:15
Rhodium, (1,4-butanediyl)(η^4-pentamethylcyclopentadienyl)(triphenylphosphine)-, 22:173
Rhodium(1+), bis[*o*-pheylenebis(dimethylarsine)]-:
  chloride, 21:101
  ——, (carbon dioxide)bis[*o*-phenylenebis(dimethylarsine)]-:
    chloride, 21:101
Rhodium(I), tetrakis(1-isocyanobutane)-:
  tetraphenylborate(1−), 21:50
  ——, tetrakis(1-isocyanobutane)-bis[methylenebis(diphenylphosphine)]di-:
    bis[tetraphenylborate(1−)], 21:49
Rubidium chloride tetracyano platinate (2:0.30:1):
  trihydrate, 21:145
Ruthenate(1−), decacarbonyl-μ-nitrosyl-tri-:
  μ-nitrido-bis(triphenylphosphorus)(1+), 22:163, 165
  ——, tridecacarbonylcobalttri-:
    μ-nitrido-bis(triphenylphosphorus)(1+), 21:61

———, tridecacarbonylhydridoirontri-:
  μ-nitrido-bis(triphenylphoshorus)(1+),
  21:60
Ruthenium, (η^6-benzene)(η^4-1,3-cyclohexadiene)-, 22:177
———, bis(η^5-cycloheptanedienyl)-, 22:179
———, bis(η^5-cyclopentadienyl)-, 22:180
———, carbonyltri-μ-chloro-chlorotetrakis(triphenylphosphine)di-:
  compd. wth acetone (1:2), 21:30
———, (η^4-1,5-cyclooctadiene)(η^6-1,3,5-cyclooctatriene)-, 22:178
———, tri-μ-chloro-chloro(thiocarbonyl)tetrakis(triphenylphosphine)di-:
  compd. with acetone (1:1), 21:29
———, tridecacarbonyldihydridoirontri-, 21:58
———, tridecacarbonyldihydridotriosmium-, 21:64
Ruthenium(O), bis(η^2-ethene)(η^6-hexamethylbenzene)-, 21:76
———, (η^4-1,3-cyclohexadiene)(η^6-hexamethylbenzene)-, 21:77
Ruthenium(II), chloro(η^5-cyclopentadienyl)bis(triphenylphoshine)-, 21:78
———, (η^5-cyclopentadienyl)-(phenylethynyl)bis(triphenylphosphine)-, 21:82
———, (η^5-cyclopentadienyl)-(phenylvinylidene)bis-(triphenylphosphine)-:
  hexafluorophosphate(1−), 21:80
———, di-μ-chloro-bis[chloro(η^6-hexamethylbenzene)-, 21:75
———, tris(2,2'-bipyridine)-:
  dichloride, hexahydrate, 21:127
Ruthenium oxide (Ru_2O_3):
  solid solns. with lead oxide (PbO_2), pyrochlor, 22:69
Ruthenocene, see—Ruthenium, bis-:
  (η^5-cyclopentadienyl)-, 22:180

Samarium:
  porphyrin complexes, 22:156
———, (2,4-pentanedionato)[5,10,15,20-tetraphenylporphyrinato(2−)]-, 22:160
———, (2,2,6,6-tetramethyl-3,5-heptanedionato)[5,10,15,20-tetraphenylporphyrinato(2−)]-, 22:160

Saramium chloride ($SmCl_3$), 22:39
Scandium(III), trichlorotris(tetrahydrofuran)-, 21:139
Scandium cesium chloride ($CsScCl_3$), 22:23
Scandium cesium chloride ($Cs_3Sc_2Cl_9$), 22:25
Scandium chloride ($ScCl_3$), 22:39
Selenide:
  iron complex, 21:36, 37
Selenium:
  iron polynuclear complexes, 21:33–37
Selenocarbonyls:
  chromium, 21:1, 2
Serine:
  copper complex, 21:115
Siloxane, dimethyl-:
  copolymer with methylphenylsiloxane, in divanadium stabilization, 22:116
———, methylphenyl-:
  copolymer with dimethylsiloxane, in divanadium stabilization, 22:116
Silver tungstate ($Ag_8(W_4O_{16})$), 22:76
Sodium aluminum silicate hydrate ($NaAlSiO_4 \cdot 2.25H_2O$), 22:61
Sodium aluminum silicate hydrate ($Na_2Al_2Si_5O_{14} \cdot XH_2O$), 22:64
Sodium cobalt oxide ($NaCoO_2$), 22:56
Sodium cobalt oxide ($Na_{0.6}CoO_2$), 22:56
Sodium cobalt oxide ($Na_{0.64}CoO_2$), 22:56
Sodium cobalt oxide ($Na_{0.74}CoO_2$), 22:56
Sodium cobalt oxide ($Na_{0.77}CoO_2$), 22:56
Sodium cyanotri[(^2H)hydro]borate(1−), 21:167
Sodium hexafluorouranate(V), 21:166
Sodium potassium tetramethylammonium aluminum silicate hydrate
  [$K_2Na[(CH_3)_4N]Al_4(Si_{14}O_{36})] \cdot 7H_2O$, 22:66
Sodium tetrapropylammonium aluminum silicate ($Na_{2.4}[(C_3H_7)_4N]_{3.6}Al_{2.6}(Si_{100}O_{207})$), 22:67
Styrene, see—Benzene, vinyl-, 21:80
Sulfur:
  iron cyclopentadienyl complexes, 21:37–46
  iron polynuclear complexes, 21:33–37
Sulfur nitride (SN):
  polymer, 22:143
Sulfur oxide (S_8O), 21:172

Tantalum:
  as high-temp. container for reduced halides, 20:15

Technetate(V), tetrachlorooxo-:
  tetrabutylammonium (1:1), 21:160
Terbium:
  porphyrin complexes, 22:156
  ——, (2,4-pentanedionato)[5,10,15,20-tetraphenylporphyrinato(2−)]-, 22:160
  ——, (2,2,6,6-tetramethyl-3,5-heptanedionato)[5,10,15,10-tetraphenylporphyrinato(2−)]-, 22:160
Terbium chloride (TbCl$_3$), 22:39
1,4,8,11-Tetraazacyclotetradeca-1,3,8,10-tetraene:
  2,9-dimethyl-3,10-diphenyl-, copper, iron, and zinc complexes, 22: 107, 108, 110, 111
2,3,7,8-Teracarbadodecaborane(12):
  2,3,7,8-tetraethyl-, 22:217
Thallium carbonate tetracyanoplatinate(II) (4:1:1), 21:153, 154
Thallium chloride (TlCl), 21:72
Thallium tetracyanoplatinate(II) (2:1), 21:153
Thallium(III), chlorobis(pentafluorophenyl)-, 21:71, 72
  ——, chlorobis(2,3,4,6-tetrafluorophenyl)-, 21:73
  ——, chlorobis(2,3,5,6-tetrafluorophenyl)-, 21:73
  ——, chlorobis(2,4,6-trifluorophenyl)-, 21:73
1-Thia-*closo*-decaborane(9), 22:229
6-Thia-*nido*-decaborane(11), 22:228
Thiocarbonyl complexes:
  ruthenium, 21:29
Thiocyanic acid:
  palladium complex, 21:132
Thorium:
  porphyrin complexes, 22:156
  ——, bis(2,4-pentanedionato)[5,10,15,20-tetraphenylporphyrinato(2−)]-, 22:160
  ——, (2,2,6,6-tetramethyl-3,5-heptanedionato)[5,10,15,20-tetraphenylporphyrinato(2−)]-, 22:160
Thulium cesium lithium chloride (Cs$_2$LiTmCl$_6$), 20:10
Thulium chloride (TmCl$_3$), 22:39
Titanate(1−), pentaoxoniobate-:
  hydrogen, 22:89
  hydrogen, intercalate with 1-butanamine, 22:89

hydrogen, intercalate with ethanamine, 22:89
hydrogen, intercalate with methanamine, 22:89
hydrogen, intercalate with NH$_3$, 22:89
hydrogen, intercalate with 1-propanamine, 22:89
potassium, 22:89
Titanium:
  as substrate for cadmium chalcogenides, 22:80
Titanium(III), chlorobis(η^5-cyclopentadienyl)-, 21:84
  ——, trichlorotris(tetrahydrofuran)-, 21:137
Titanium(IV), tetrachlorobis(tetrahydrofuran)-, 21:135
Titanium iron hydride (FeTiH$_{1.94}$), 22:90
Transition metal alkali metal oxides, 22:56
triars, *see*—Arsine, [2-[(dimethylarsino)methyl]-2-methyl-1,3-propanediyl]bis(dimethyl-, 21:18
Trimethyl phosphite:
  iron complex, 21:93
[^{10}B$_2$]-1,2,4,3,5-Trithiadiborolane:
  3,5-dimethyl-, 22:225
Tungstate(1−), pentacarbonylhydrido-:
  μ-nitrido-bis(triphenylphosphorus)(1+), 22:182

Uranate(V), hexafluoro-:
  μ-nitrido-bis(triphenylphosphorus)(1+), 21:166
  potassium, 21:166
  sodium, 21:166
Uranium(IV) chloride (UCl$_4$), 21:187
Uranium(V), pentaethoxy-, 21:165
Uranium(V) fluoride (UF$_5$), 21:163

Vanadium chloride (VCl$_2$), 21:185
Vanadium(III), chlorobis(η^5-cyclopentadienyl)-, 21:85
  ——, trichlorotris(tetrahydrofuran)-, 21:138

Water:
  cobalt complex, 21:123–126
  platinum complex, 21:192; 22:125
Welding:
  of tantalum, 20:7

Ytterbium:
  porphyrin complexes, 22:156
  ——, (2,4-pentanedionato)[5,10,15,20-tetraphenylporphyrinato(2−)]-, 22:156
  ——, [5,10,15,20-tetrakis(3-fluorophenyl)porphyrinato(2−)](2,2,6,6-tetramethyl-3,5-heptanedionato)-, 22:160
  ——, [5,10,15,20-tetrakis(4-methylphenyl)porphyrinato(2−)](2,2,6,6-tetramethyl-3,5-heptanedionato)-, 22:156
Ytterbium chloride ($YbCl_3$), 22:39
Yttrium:
  porphyrin complexes, 22:156
  ——, (2,4-pentanedionato)[5,10,15,20-tetraphenylporphyrinato(2−)]-, 22:160
Yttrium chloride (YCl_3), 22:39

Zeolite, 22:61
Zeolite A ($NaAlSiO_4 \cdot 2.25H_2O$), 22:63
Zeolite Y ($Na_2Al_2Si_5O_{14} \cdot XH_2O$), 22:64
Zinc(II), chloro(2,9-dimethyl-3,10-diphenyl-1,4,8,11-tetraazacyclotetradeca-1,3,8,10-tetraene)-: hexafluorophosphate(1−), 22:111
Zincate(II), terakis(benzenethiolato)-: bis(tetraphenylphosphonium), 21:25
Zirconium(IV), tetrachlorobis(tetrahydrofuran)-, 21:136
Zirconium bromide (ZrBr), 22:26
Zirconium chloride (ZrCl), 22:26
ZSM-5 ($Na_{2.4}[(C_3H_7)_4N]_{3.6}Al_{2.6}(Si_{100}O_{207})$), 22:67

FORMULA INDEX

The Formula Index, as well as the Subject Index, is a cumulative index for Volumes 21–25. The Index is organized to allow the most efficient location of specific compounds and groups of compounds related by central metal ion or ligand grouping.

The formulas entered in the Formula Index are for the total composition of the entered compound, e.g., F_6NaU for sodium hexafluorouranate (V). The formulas consist solely of atomic symbols (abbreviations for atomic groupings are not used) and arranged in alphabetical order with carbon and hydrogen always given last, e.g., $Br_3CoN_4C_4H_{16}$. To enhance the utility of the Formula Index, all formulas are permuted on the symbols for all metal atoms, e.g., $FeO_{13}Ru_3C_{13}H_{13}$ is also listed at $Ru_3FeO_{13}C_{13}H_{13}$. Ligand groupings are also listed separately in the same order, e.g., $N_2C_2H_8$, 1,2-Ethanediamine, cobalt complexes. Thus individual compounds are found at their total formula in the alphabetical listing; compounds of any metal may be scanned at the alphabetical position of the metal symbol; and compounds of a specific ligand are listed at the formula of the ligand, e.g., NC for Cyano complexes.

Water of hydration, when so identified, is not added into the formulas of the reported compounds, e.g., $Cl_{0.30}N_4PtRb_2C_4 \cdot 3H_2O$.

$Ag_8O_{16}W_4$, Silver tungstate, 22:76
AlH_4LaNi_4, Aluminum lanthanum nickel hydride, 22:96
$AlNaO_4Si \cdot 2.25H_2O$, Sodium aluminum silicate, 22:61
——, Zeolite A, 22:63
$Al_2Na_2O_{14}Si \cdot XH_2O$, Sodium aluminum silicate hydrate, 22:64
——, Zeolite Y, 22:64
$Al_{2.6}N_{3.6}Na_{2.4}O_{207}Si_{100}C_{43}H_{100}$, Sodium tetrapropylammonium aluminum silicate, 22:67
——, ZSM-5, 22:67
$Al_4K_2NNaO_{36}Si_{14}C_4H_{12} \cdot 7H_2O$, Offretite, tetramethylammonium substituted, 22:65
——, Potassium sodium tetramethyl-ammonium aluminum silicate hydrate, 22:65
$As_2C_{10}H_{16}$, Arsine, o-phenylenebis(dimethyl-, rhodium complex, 21:101
$As_4ClO_2RhC_{21}H_{32}$, Rhodium(1+), (carbon dioxide)bis[o-phenylene-bis(dimethylarsine)]-, chloride, 21:101
$As_4ClRhC_{20}H_{32}$, Rhodium(1+), bis[o-phenylenebis(dimethylarsine)]-, chloride, 21:101
$As_4Cl_6Nb_2C_{20}H_{32}$, Niobium(III), hexachlorobis[o-phenylenebis(dimethylarsine)]di-, 21:18

$As_6Cl_6Nb_2C_{22}H_{54}$, Niobium(III), hexachlorobis[[2-[(dimethylarsino)-methyl]-2-methyl-1,3-propanediyl]bis-(dimethylarsine)]-, 21:18
$AuNO_2SC_3H_6$, Gold) (I), (L-cysteinato)-, 21:31
BBr_2CH_3, [^{10}B]Borane, dibromomethyl-, 22:223
BBr_3, [^{10}B]Boron bromide, 22:219
$BClF_4PtS_3C_6H_{18}$, Platinum(II), chlorotris(dimethyl sulfide)-, tetrafluoroborate(1−), 22:126
$BCl_2C_6H_5$, Borane, dichlorophenyl-, 22:207
$BCuN_6OC_{10}H_{10}$, Copper(I), carbonyl[hydrotris(pyrazolato)borato]-, 21:108
$BCuN_6OC_{16}H_{22}$, Copper(I), carbonyl[tris(3,5-dimethylpyrazolato)-hydroborato]-, 21:109
$BCuN_8OC_{13}H_{12}$, Copper(I), carbonyl[tetrakis(pyrazolato)borato]-, 21:110
BKC_8H_{16}, Borate(1−), (cyclooctane-1,5-diyl)dihydro-, potassium, 22:200
$BLiC_8H_{16}$, Borate(1−), (cyclooctane-1,5-diyl)dihydro-, lithium, 22:199
BNC_6H_{16}, Borane, (dimethylamino)diethyl-, 22:209
$BNNaC^2H_3$, Borate(1−), cyanotri[(^2H)-hydro]-, sodium, 21:167

BN$_4$RhC$_{44}$H$_{56}$, Rhodium(I), tetrakis(1-isocyanobutane)-, tetraphenylborate(1 −), 21:50
BN$_6$C$_9$H$_{10}$, Borate(1 −), hydrotris(pyrazolato)-, copper complex, 21:108
BN$_6$C$_{15}$H$_{22}$, Borate(1 −), tris(3,5-dimethylpyrazolato)hydro-, copper complex, 21:109
BN$_8$C$_{12}$H$_{12}$, Borate(1 −), tetrakis(pyrazolato)-, copper complex, 21:110
BNaC$_8$H$_{16}$, Borate(1 −), (cyclooctane-1,5-diyl)dihydro-, sodium, 22:200
BOC$_4$H$_{11}$, Borane, diethylhydroxy-, 22:193
BOC$_5$H$_{13}$, Borane, diethylmethoxy-, 22:190
BOC$_8$H$_{20}$, Diboroxane, tetraethyl-, 22:188
BO$_2$C$_9$H$_{19}$, Borane[(2,2-dimethylpropanoyl)oxy]diethyl-, 22:185
B$_2$FeN$_6$O$_6$C$_{30}$H$_{34}$, Iron(II), {[tris[μ-[(1,2-cyclohexanedione dioximato)-O:O']diphenyldiborat-(2 −)]-$N,N',N'',N''',N'''',N'''''$}-, 21:112
B$_2$N$_4$P$_4$Rh$_2$C$_{118}$H$_{120}$, Rhodium(I), tetrakis-(1-isocyanobutane)bis[methylenebis(diphenylphosphine)]di-, bis[tetraphenylborate(1 −)], 21:49
B$_2$N$_6$O$_6$C$_{30}$H$_{34}$, Borate(2 −), tris[μ-[(1,2-cyclohexanedione dioximato)-O:O']diphenyldi-, iron complex, 21:112
B$_2$O$_5$C$_{14}$H$_{28}$, Boron, bis-μ-(2,2-dimethylpropanoato-O,O'),-diethyl-μ-oxo-di-, 22:196
B$_2$S$_3$C$_2$H$_6$, [^{10}B$_2$]-1,2,4,3,5-Trithiadiborolane, 3,5-dimethyl-, 22:225
B$_4$C$_6$H$_{16}$, 2,3-Dicarba-$nido$-hexaborane(8), 2,3-diethyl-, 22:211
B$_4$FeC$_{12}$H$_{30}$, [1,1'-$commo$-Bis(2,3-dicarba-1-ferra-$closo$-heptaborane)](12), 2,2',3,3'-tetraethyl-1,1-dihydro-, 22:215
B$_7$C$_2$H$_{16}$, 2,6-Dicarba-$nido$-nonaborane, 22:237
B$_8$C$_{12}$H$_{28}$, 2,3,7,8-Tetracarbadodecaborane(12), 2,3,7,8-tetraethyl-, 22:217
B$_9$CoC$_7$H$_{16}$, 1,2-Dicarba-3-cobalta-$closo$-dodecaborane(11), 3-(η^5-cyclopentadienyl)-, 22:235
B$_9$CsH$_{12}$S, Borate(1 −), dodecahydro-6-thia-$arachno$-deca-, cesium, 22:227
B$_9$H$_9$S, 1-Thia-$closo$-decaborane(9), 22:22
B$_9$H$_{11}$S, 6-Thia-$nido$-decaborane(11), 22:228
B$_9$KC$_2$H$_{12}$, Borate(1 −), dodecahydro-7,8-dicarba-$nido$-undeca-, potassium, 22:231
B$_9$SC$_4$H$_{17}$, 7,8-Dicarba-$nido$-undecaborane(11), 9-(dimethyl sulfide)-, 22:239
B$_{10}$H$_{14}$, Decaborane(14), 22:202
B$_{10}$SC$_2$H$_{12}$, 1,2-Dicarba-$closo$-dodecaborane(12)-9-thiol, 22:241

Bi$_4$K$_2$N$_2$O$_{12}$C$_{36}$H$_{72}$, Potassium, (4,7,-13,16,21,24-hexaoxa-1,10-diazabicyclo [8.8.8]hexacosane)-, tetrabismuthide(2 −) (2:1), 22:151
BrCoN$_4$O$_3$C$_5$H$_6$, Cobalt(III), (carbonato)bis(1,2-ethanediamine)-, bromide, 21:120
BrCoN$_4$O$_7$S$_2$C$_4$H$_{18}$ · H$_2$O, Cobalt(III), aquabromobis(1,2-ethanediamine)-, dithionate, $trans$-, monohydrate, 21:124
BrZr, Zirconium bromide, 22:26
Br$_2$N$_4$PdC$_{12}$H$_{30}$, Palladium(II), [N,N-bis[2-(dimethylamino)ethyl]-N',N'-dimethyl-1,2-ethanediamine]bromo-, bromide, 21:131
Br$_2$N$_6$NiP$_2$C$_{18}$H$_{24}$, Nickel(II), dibromobis(3,3',3''-phosphindynetripropionitrile)-, 22:113, 115
(Br$_2$N$_6$NiP$_2$C$_{18}$H$_{24}$)$_x$, Nickel(II), dibromobis(3,3',3''-phosphindynetripropionitrile)-, polymer, 22:115
Br$_3$CoN$_4$C$_4$H$_{16}$, Cobalt(III), dibromobis-(1,2-ethanediamine)-, bromide, $trans$-, 21:120
Br$_3$CoN$_4$C$_4$H$_{16}$ · H$_2$O, Cobalt(III), dibromobis(1,2-ethanediamine)-, bromide, cis-, monohydrate, 21:121
Br$_3$CoN$_4$OC$_4$H$_{18}$ · H$_2$O, Cobalt(III), aquabromobis(1,2-ethanediamine)-, dibromide, cis-, monohydrate, 21:123
Br$_4$GaNC$_8$H$_{20}$, Gallate(1 −), tetrabromo-, tetraethylammonium, 22:141
Br$_4$GaNC$_{16}$H$_{36}$, Gallate(1 −), tetrabromo-, tetrabutylammonium, 22:139
Br$_6$Ga$_2$P$_2$C$_{36}$H$_{32}$, Gallate(2 −), hexabromodi-, bis(triphenylphosphonium), 22:135, 138
Br$_6$Ga$_2$P$_2$C$_{48}$H$_{40}$, Gallate(2 −), hexabromodi-, bis(tetraphenylphosphonium), 22:139

CO, Carbon monoxide:
  chromium complex, 21:1, 2
  cobalt, iron, osmium, and ruthenium complexes, 21:58–65
  copper complex, 21:107–110
  iridium complex, 21:97
  iron complex, 21:66, 68
  iron and ruthenium complexes, 22:163
  palladium complex, 21:49
  ruthenium complex, 21:30
C$_2$H$_4$, Ethene:
  iron complex, 21:91
  platinum complexes, 21:86–89
  ruthenium complex, 21:76

C_4H_{10}, Butane:
cobalt, iridium and rhodium complexes, 22:171, 173, 174
palladium complex, 22:167, 168, 169, 170
C_5H_6, 1,3-Cyclopentadiene:
cobalt complex, 22:171, 235
iron complexes, 21:39–46
ruthenium complex, 21:78; 22:180
titanium and vanadium complexes, 21:84, 85
C_6H_6, Benzene:
chromium complex, 21:1, 2
ruthenium complex, 22:177
C_6H_8, 1,3-Cyclohexadiene:
ruthenium complex, 21:77; 22:177
C_7H_{10}, 1,3-Cycloheptadiene:
ruthenium complex, 22:179
C_8H_6, Benzene, ethynyl-:
ruthenium complex, 21:82
C_8H_8, Benzene, vinyl-:
ruthenium complex, 21:80
——, Styrene, see—Benzene, vinyl-, 21:80
C_8H_{10}, 1,3,5-Cyclooctatriene:
ruthenium complex, 22:178
C_8H_{12}, 1,5-Cyclooctadiene:
ruthenium complex, 22:178
C_8H_{14}, Cyclooctene:
iridium complex, 21:102
C_8H_{16}, Cyclooctane:
boron complex, 22:199
$C_{10}H_4$, Benzene, 1-isopropyl-4-methyl-:
ruthenium complex, 21:75
$C_{10}H_{16}$, 1,3-Cyclopentadiene, 1,2,3,4,5-pentamethyl-, 21:181
iridium and rhodium complex, 22:173, 174
$C_{12}H_{18}$, Benzene, hexamethyl-:
ruthenium complexes, 21:74–77
$Ca_2Mn_3O_8$, Calcium, manganese oxide, 22:73
$CdP_2S_4C_{72}H_{60}$, Cadmate(II), tetrakis(benzenethiolato)-, bis(tetraphenylphosphonium), 21:26
CdSe, Cadmium selenide, 22:82
$CdSe_xTe_{1-x}$, Cadmium selenide telluride, 22:84
$CdSe_{0.65}Te_{0.35}$, Cadmium selenide telluride, 22:81
$CeCl_3$, Cerium chloride, 22:39
$CeN_4O_4C_{54}H_{42}$, Cerium, bis(2,4-pentanedionato)[5,10,15,20-tetraphenylporphyrinato(2−)]-, 22:160
$ClCoN_4O_2SC_6H_{18}$, Cobalt(III), bis(1,2-ethanediamine) (2-mercaptoacetato(2−)-O,S)-, perchlorate, 21:21

$ClCoN_4O_7S_2C_4H_{18} \cdot H_2O$, Cobalt(III), aquachlorobis(1,2-ethanediamine)-, dithionate, trans-, monohydrate, 21:125
$ClF_6N_4PZnC_{24}H_{28}$, Zinc(II), chloro(2,9-dimethyl-3,10-diphenyl-1,4,8,11-tetraazacyclotetradeca-1,3,8,10-tetraene)-, hexafluorophosphate(1−), 22:111
$ClF_8TlC_{12}H_2$, Thallium(III), chlorobis(2,3,4,6-tetrafluorophenyl-), chlorobis(2,3,5,6-tetrafluorophenyl)-, 21:73
$ClF_{10}TlC_{12}$, Thallium(III), chlorobis(pentafluorophenyl)-, 21:71, 72
ClFeO, Iron chloride oxide:
intercalate with 4-aminopyridine (4:1), 22:86
intercalate with pyridine (4:1), 22:86
intercalate with 2,4,6-trimethylpyridine (6:1), 22:86
$ClIrOPC_{17}H_{22}$, Iridium(I), carbonylchlorobis(dimethylphenylphosphine)-, trans-, 21:97
$ClIrO_2P_4C_{13}H_{32}$, Iridium(1+), (carbon dioxide)bis[1,2-ethanediylbis-(dimethylphosphine)]-, chloride, 21:100
$ClIrO_4P_3C_{11}H_{27}$, Iridium, chloro[(formyl-κC-oxy)formato-κO-(2−)]tris(trimethylphosphine)-, 21:102
$ClIrP_3C_{17}H_{41}$, Iridium, chloro(η²-cyclooctene)tris(trimethylphosphine)-, 21:102
$ClIrP_4C_{12}H_{32}$, Iridium(1+), bis[1,2-ethandiylbis(dimethylphosphine)]-, chloride, 21:100
$ClNNiPSe_2C_{11}H_{25}$, Nickel(II), chloro(N,N-diethyldiselenocarbamato)-(triethylphosphine)-, 21:9
$ClNPPdSe_2C_{23}H_{25}$, Palladium, chloro-(N,N-diethyldiselenocarbamato)(triphenylphosphine)-, 21:10
$ClNPPtSe_2C_{23}H_{25}$, Platinum(II), chloro(N,N-diethyldiselenocarbamato)(triphenylphosphine)-, 21:10
$ClN_4O_2ReC_{20}H_{20}$, Rhenium(V), dioxotetrakis(pyridine)-, chloride, trans-, 21:116
$ClN_4O_6ReC_{20}H_{20}$, Rhenium(V), dioxotetrakis(pyridine)-, perchlorate, trans-, 21:117
$ClP_2RuC_{41}H_{35}$, Ruthenium(II), chloro(η⁵-cyclopentadienyl)bis(triphenylphosphine)-, 21:78
$ClTiC_{10}H_{10}$, Titanium(III), chlorobis(η⁵-cyclopentadienyl)-, 21:84

ClVC$_{10}$H$_{10}$, Vanadium(III), chlorobis(η^5-cyclopentadienyl)-, 21:85

ClZr, Zirconium chloride, 22:26

Cl$_{0.30}$Cs$_2$N$_4$PtC$_4$, Platinate, tetracyano-, cesium chloride (1:2:0.30), 21:142

Cl$_{0.30}$N$_4$PtRb$_2$C$_4$ · 3H$_{20}$, Platinate, tetracyano-, rubidium chloride (1:2:0.30), trihydrate, 21:145

Cl$_2$CoN$_5$O$_8$SC$_6$H$_{22}$, Cobalt(III), (2-aminoethanethiolato-N,S-bis(1,2-ethanediamine)-, diperchlorate, 21:19

Cl$_2$H$_6$IN$_2$Pt, Platinum(II), diamminechloroido-, *trans*-, chloride, 22:124

Cl$_2$H$_9$N$_3$Pt, Platinum(II), triamminechloro-, chloride, 22:124

Cl$_2$Mo$_2$N$_4$C$_8$H$_{24}$, Molybdenum, dichlorotetrakis(dimethylamido)di-, ($Mo{\equiv}Mo$), 21:56

Cl$_2$N$_2$PtC$_{10}$H$_{24}$, Platinum(II), [N,N'-bis(1-methylethyl)-1,2-ethanediamine]dichloro(ethene)-, 21:87

Cl$_2$N$_2$PtC$_{12}$H$_{28}$, Platinum(II):
dichloro[N,N'-dimethyl-N,N'-bis(1-methylethyl)-1,2-ethanediamine](ethene)-, 21:87
dichloro(ethene)(N,N,N',N'-tetraethyl-1,2-ethanediamine)-, 21:86, 87

Cl$_2$N$_2$PtC$_{20}$H$_{28}$, Platinum(II), [(S,S)-N,N'-bis(1-phenylethyl)-1,2-ethanediamine]dichloro(ethene)-, 21:87

Cl$_2$N$_2$PtC$_{22}$H$_{32}$, Platinum(II), dichloro[(R,R)-N,N'-dimethyl-N,N'-bis(1-phenylethyl)-1,2-ethanediamine](ethene)-, 21:87

Cl$_2$N$_4$PdC$_{12}$H$_{30}$, Palladium(II), [N,N-bis[2-(dimethylamino)ethyl]-N',N'-dimethyl-1,2-ethanediamine]chloro-, chloride, 21:129

Cl$_2$N$_6$NiP$_2$C$_{18}$H$_{24}$, Nickel(II), dichlorobis(3,3′,3″-phosphinidynetripropionitrile)-, 22:113

Cl$_2$N$_6$RuC$_{30}$H$_{24}$ · 6H$_2$O, Ruthenium(II), tris(2,2′-bipyridine)-, dichloride, hexahydrate, 21:127

Cl$_2$OP$_4$Pd$_2$C$_{51}$H$_{44}$, Palladium(I), μ-carbonyl-dichlorobis[methylenebis(diphenylphosphine)]di-, 21:49

Cl$_2$P$_4$Pd$_2$C$_{50}$H$_{44}$, Palladium(I), dichlorobi-μ-[methylenebis(diphenylphosphine)]-di-, (Pd-Pd), 21:48

Cl$_2$V, Vanadium chloride, 21:185

Cl$_3$CsSc, Cesium sacandium chloride, 22:23

Cl$_3$Dy, Dyprosium chloride, 22:39

Cl$_3$Er, Erbium chloride, 22:39

Cl$_3$Eu, Europium chloride, 22:39

Cl$_3$Gd, Gadolinium chloride, 22:39

Cl$_3$Ho, Holmium chloride, 22:39

Cl$_3$La, Lanthanum chloride, 22:39

Cl$_3$Lu, Lutetium chloride, 22:39

Cl$_3$NPtSC$_{18}$H$_{42}$, Platinate(II), trichloro(dimethyl sulfide)-, tetrabutylammonium, 22:128

Cl$_3$Nd, Neodymium chloride, 22:39

Cl$_3$O$_3$ScC$_{12}$H$_{24}$, Scandium(III), trichlorotris(tetrahydrofuran)-, 21:139

Cl$_3$O$_3$TiC$_{12}$H$_{24}$, Titanium(III), trichlorotris(tetrahydrofuran)-, 21:137

Cl$_3$O$_3$VC$_{12}$H$_{24}$, Vanadium(III), trichlorotris(tetrahydrofuran)-, 21:138

Cl$_3$Pr, Praseodynium chloride, 22:39

Cl$_3$Sc, Scandium chloride, 22:39

Cl$_3$Sm, Saramium chloride, 22:39

Cl$_3$Tb, Terbium chloride, 22:39

Cl$_3$Tl, Thallium chloride, 21:72

Cl$_3$Tm, Thulium chloride, 22:39

Cl$_3$Y, Yttrium chloride, 22:39

Cl$_3$Yb, Ytterbium chloride, 22:39

Cl$_4$HfO$_2$C$_8$H$_{16}$, Hafnium(IV), tetrachlorobis(tetrahydrofuran)-, 21:137

Cl$_4$GaNC$_{16}$H$_{36}$, Gallate(1−), tetrachloro-, tetrabutylammonium, 22:139

Cl$_4$NOTcC$_{16}$H$_{36}$, Technetate(V), tetrachlorooxo-, tetrabutylammonium (1:1), 21:160

Cl$_4$NbO$_2$C$_4$H$_{16}$, Niobium(IV), tetrachlorobis(tetrahydrofuran)-, 21:138

Cl$_4$OP$_4$Ru$_2$C$_{73}$H$_{30}$, Ruthenium, carbonyltri-μ-chloro-chlorotetrakis(triphenylphosphine)di-, compd. with acetone (1:2), 21:30

Cl$_4$O$_2$TiC$_8$H$_{16}$, Titanium(IV), tetrachlorobis(tetrahydrofuran)-, 21:135

Cl$_4$O$_2$ZrC$_8$H$_{16}$, Zirconium(IV), tetrachlorobis(tetrahydrofuran)-, 21:136

Cl$_4$P$_4$Ru$_2$SC$_{73}$H$_{60}$, Ruthenium, tri-μ-chlorochloro(thiocarbonyl)tetrakis-(triphenylphosphine)di-, compd. with acetone, 21:29

Cl$_4$Pt$_2$S$_2$C$_4$H$_{12}$, Platinum(II), di-μ-chlorodichlorobis(dimethyl sulfide)di-, 22:128

Cl$_4$Ru$_2$C$_{20}$H$_{28}$, Ruthenium(II), di-μ-chlorobis[chloro(η^6-1-isopropyl-4-methylbenzene)-, 21:75

Cl$_4$Ru$_2$C$_{24}$H$_{36}$, Ruthenium(II), di-μ-chlorobis[chloro(η^6-hexamethylbenzene)-, 21:75

Cl$_4$U, Uranium(IV) chloride, 21:187
Cl$_5$Cs$_2$Lu, Cesium Lutetium chloride, 22:6
Cl$_6$Cs$_2$LiTm, Cesium lithium thulium chloride, 21:10
Cl$_6$Cs$_3$Lu, Cesium lutetium chloride, 22:6
Cl$_6$Ga$_2$P$_2$C$_{36}$H$_{32}$, Gallate(2−), hexachlorodi-, bis(triphenylphosphonium), 22:135, 138
Cl$_6$N$_2$PbC$_{10}$H$_{12}$, Plumbate(IV), hexachloro-, dipyridinium, 22:149
Cl$_6$Nb$_2$P$_2$C$_{30}$H$_{24}$, Niobium(III), hexachlorobis[1,2-ethanediylbis(diphenylphosphine)]di-, 21:18
Cl$_6$Nb$_2$S$_3$C$_6$H$_{18}$, Niobium(III), di-μ-chlorotetrachloro-μ-(dimethyl sulfide)-bis(dimethyl sulfide)di-, 21:16
Cl$_7$CsPr$_2$, Cesium praseodymium chloride, 22:2
Cl$_7$Dy$_2$K, Potassium dyprosium chloride, 22:2
Cl$_8$EuN$_4$O$_2$C$_{49}$H$_{27}$, Europium, (2,4-pentanedionato)[5,10,15,20-tetrakis(3,5-dichlorophenyl)porphyrinato(2−)]-, 22:160
Cl$_9$Cs$_3$Lu$_2$, Cesium lutetium chloride, 22:6
Cl$_9$Cs$_3$Sc$_2$, Cesium scandium chloride, 22:25
CoB$_9$C$_7$H$_{16}$, 1,2-Dicarba-3-cobalta-*closo*-dodecaborane(11), 3-(η^5-cyclopentadienyl)-, 22:235
CoBrN$_4$O$_3$C$_5$H$_{16}$, Cobalt(III), (carbonato)bis(1,2-ethanediamine)-, bromide, 21:120
CoBrN$_4$O$_7$S$_2$C$_4$H$_{18}$ · H$_2$O, Cobalt(III), aquabromobis(1,2-ethanediamine)-, dithionate, *trans*-, monohydrate, 21:124
CoBr$_3$N$_4$C$_4$H$_{16}$, Cobalt(III), dibromobis(1,2-ethanediamine)-, bromide, *trans*-, 21:120
CoBr$_3$N$_4$C$_4$H$_{16}$ · H$_2$O, Cobalt(III), dibromobis(1,2-ethanediamine)-, bromide, *cis*-, monohydrate, 21:121
CoBr$_3$N$_4$OC$_4$H$_{18}$ · H$_2$O, Cobalt(III), aquabromobis(1,2-ethanediamine)-, dibromide, *cis*-, monohydrate, 21:123
CoClN$_4$O$_2$SC$_6$H$_{18}$, Cobalt(III), bis(1,2-ethandiamine)(2-mercaptoacetato-(2−)-*O*,*S*)-, perchlorate, 21:21
CoClN$_4$O$_7$S$_2$C$_4$H$_{18}$ · H$_2$O, Cobalt(III), aquachlorobis(1,2-ethanediamine)-, dithionate, *trans*-, monohydrate, 21:125
CoCl$_2$N$_5$O$_8$C$_6$H$_{22}$, Cobalt(III), (2-aminoethanethiolato-*N*,*S*)bis(1,2-ethanediamine)-, diperchlorate, 21:19
CoF$_9$N$_3$O$_9$S$_3$C$_7$H$_{16}$, Cobalt(III), [*N*-(2-aminoethyl)-1,2-ethanediamine]-tris-(trifluoromethanesulfonato)-, *fac*-, 22:106
CoF$_9$N$_4$O$_9$S$_3$C$_7$H$_{19}$, Cobalt(III), bis(1,2-ethanediamine)bis(trifluoromethanesulfonato)-, *cis*-, trifluoromethanesulfonate, 22:105
CoF$_9$N$_5$O$_9$S$_3$C$_3$H$_{15}$, Cobalt(III), pentaammine(trifluoromethanesulfonato)-, trifluoromethanesulfonate, 22:104
CoKO$_2$, Potassium cobalt oxide, 22:58
CoK$_{0.5}$O$_2$, Potassium cobalt oxide, 22:57
CoK$_{0.67}$O$_2$, Potassium cobalt oxide, 22:57
CoNO$_{13}$P$_2$Ru$_3$C$_{49}$H$_{30}$, Ruthenate(1−), tridecacarbonylcobalttri-, μ-nitridobis(triphenylphosphorus) (1+), 21:61
CoNaO$_2$, Sodium cobalt oxide, 22:56
CoNa$_{0.6}$O$_2$, Sodium cobalt oxide, 22:56
CoNa$_{0.64}$O$_2$, Sodium cobalt oxide, 22:56
CoNa$_{0.74}$O$_2$, Sodium cobalt oxide, 22:56
CoNa$_{0.77}$O$_2$, Sodium cobalt oxide, 22:56
CoPC$_{28}$H$_{27}$, Cobalt, (1,4-butanediyl)-(η^5-cyclopentadienyl)(triphenylphosphine), 22:171
CoP$_2$S$_4$C$_{72}$H$_{60}$, Cobaltate(II), tetrakis-(benzenethiolato)-, bis(tetraphenylphosphonium), 21:24
CrKO$_2$, Potassium chromium oxide, 22:59
CrK$_{0.5}$O$_2$, Potassium chromium oxide, 22:59
CrK$_{0.6}$O$_2$, Potassium chromium oxide, 22:59
CrK$_{0.7}$O$_2$, Potassium chromium oxide, 22:59
CrK$_{0.77}$O$_2$, Potassium chromium oxide, 22:59
CrNO$_5$P$_2$C$_{41}$H$_{31}$, Chromate(1−), pentacarbonylhydrido-, μ-nitridobis(triphenylphosphorus)(1+), 22:183
CrO$_2$SeC$_9$H$_6$, Chromium(O), (η^6-benzene)dicarbonyl(selenocarbonyl)-, 21:1, 2
CrO$_5$SeC$_6$, Chromium(O), pentacarbonyl(selenocarbonyl)-, 21:1, 4
CsCl$_3$Sc, Cesium scandium chloride, 22:23
CsCl$_7$Pr$_2$, Cesium praseodymium chloride, 22:2
CS$_2$Cl$_{0.30}$N$_4$PtC$_4$, Platinate, tetracyano-, cesium chloride (1:2:0.30), 21:142
Cs$_2$Cl$_5$Lu, Cesium lutetium chloride, 22:6
Cs$_2$Cl$_6$LiTm, Cesium lithium thulium chloride, 20:10
Cs$_2$N$_{4.75}$OPtC$_4$ · XH$_2$O, Platinate, tetracyano-, cesium azide (1:2:0.25), hydrate, 21:149
Cs$_3$Cl$_6$Lu, Cesium lutetium chloride, 22:6
Cs$_3$Cl$_9$Lu$_2$, Cesium lutetium chloride, 22:6
Cs$_3$Cl$_9$Sc$_2$, Cesium scandium chloride, 22:25

Cs$_3$N$_4$O$_{3.68}$PtS$_{0.92}$C$_4$H$_{0.46}$, Platinate, tetracyano-, cesium [hydrogenbis(sulfate)] (1:3:0.46), 21:151

CuBN$_6$OC$_{10}$H$_{10}$, Copper(I), carbonyl[hydrotris(pyrazolato)borato]-, 21:108

CuBN$_6$OC$_{16}$H$_{22}$, Copper(I), carbonyl[tris(3,5-dimethylpyrazolato)-hydroborato]-, 21:109

CuBN$_8$OC$_{13}$H$_{12}$, Copper(I), carbonyltetrakis(pyrazolato)borato]-, 21:110

CuF$_{12}$N$_4$P$_2$C$_{24}$H$_{28}$, Copper(II), (2,9-dimethyl-3,10-diphenyl-1,4,8,11-tetraazacyclotetradeca-1,3,8,10-tetraene)-, bis[hexafluorophosphate(1−)], 22:10

CuI, Copper iodide, 22:101

CuN$_3$O$_7$SC$_{15}$H$_{14}$, Copper(II), (1,10-phenanthroline)[serinato(1−)]-, sulfate, 21:115

DyCl$_3$, Dyprosium chloride, 22:39

DyN$_4$O$_2$C$_{49}$H$_{35}$, Dyprosium, (2,4-pentanedionato)[5,10,15,10-tetraphenylporphyrinato(2−)]-, 22:166

DyN$_4$O$_2$C$_{55}$H$_{47}$, Dysprosium, (2,2,6,6-tetramethyl-3,5-heptanedionato)-[5,10,15,20-tetraphenylporphyrinato(2−)]-, 22:160

Dy$_2$Cl$_7$K, Potassium dysprosium chloride, 22:2

ErCl$_3$, Erbium chloride, 22:39

ErF$_4$N$_4$O$_2$C$_{49}$H$_{31}$, Erbium, (2,4-pentanedionato)-[5,10,15,20-tetrakis(3-fluorophenyl)porphyrinato(2−)]-, 22:160

ErN$_4$O$_2$C$_{49}$H$_{35}$, Erbium, (2,4-pentanedionato)-[5,10,15,20-tetraphenylporphyrinato(2−)]-, 22:160

EuCl$_3$, Europium chloride, 22:39

EuCl$_8$N$_4$O$_2$C$_{49}$H$_{27}$, Europium, (2,4-pentanedionato)[5,10,15,20-tetrakis(3,5-dichlorophenyl)-porphyrinato(2−)]-, 22:160

EuN$_4$O$_2$C$_{49}$H$_{35}$, Europium, (2,4-pentanedionato)-[5,10,15,20-tetraphenylporphyrinato(2−)]-, 22:160

EuN$_4$O$_2$C$_{53}$H$_{43}$, Europium, (2,4-pentanedionato)-[5,10,15,20-tetrakis(4-methylphenyl)porphyrinato(2−)]-, 22:160

F$_{0.54}$N$_{10}$PtC$_6$H$_{12.27}$ · 1.8H$_2$O, Platinate, tetracyano-, guanidinium (hydrogen difluoride) (1:2:0.27), hydrate (1:1:8), 21:146

F$_{0.60}$K$_2$N$_4$PtC$_4$H$_{0.30}$ · 3H$_2$O, Platinate, tetracyano-, potassium (hydrogen difluoride) (1:2:0.30), 21:147

F$_3$C$_6$H$_2$, Benzene, 1,3,5-trifluorothallium complex, 21:73

F$_3$O$_3$SCH, Methranesulfonic acid, trifluoro-, cobalt(III) amine complexes, 22:104, 105

F$_4$C$_6$H$_3$, Benzene, 1,2,3,5-tetrafluorothallium complex, 21.73

———, 1,2,4,5-tetrafluorothalium complex, 21:73

F$_4$N$_4$O$_2$YbC$_{55}$H$_{43}$, Ytterbium, [5,10,15,20-tetrakis(3-fluorophenyl)-porphyrinato(2−)][2,2,6,6-tetramethyl-3,5-heptanedionato)-, 22:160

F$_5$C$_6$H, Benzene, pentafluoro-, lithium and thallium complex, 21:71, 72

F$_5$LiC$_6$, Lithium, (pentafluorophenyl), 21:72

F$_5$MoOK$_2$, Molybdate(V), pentafluorooxo-, dipotassium, 21:170

F$_5$U, Uranium(V) fluoride, β-, 21:163

F$_6$KU, Uranate(V), hexafluoro-, potassium, 21:166

F$_6$NP$_2$UC$_{36}$H$_{30}$, Uranate(V), hexafluoro-, μ-nitrido-bis(triphenylphosphorus)(1+), 21:166

F$_6$NaU, Uranate(V), hexafluoro-, sodium, 21:166

F$_6$P$_3$RuC$_{49}$H$_{41}$, Ruthenium(II), (η^5-cyclopentadienyl)(phenylvinylene)bis-(triphenylphosphine)-, hexafluorophosphate(1−), 21:80

F$_{12}$FeN$_6$PC$_{28}$H$_{34}$, Iron(II), bis(acetonitrile) (2,9-dimethyl-3,10-diphenyl-1,4,8,11-tetraazacyclotetradeca-1,3,8,10-tetraene)-, bis[hexafluorophosphate(1−)], 22:108

F$_{12}$Fe$_2$N$_2$P$_2$S$_2$C$_{18}$H$_{36}$, Iron(2+), bis(acetonitrile)bis(η^5-cyclopentadienyl)bis-μ-(ethanethiolato)di-, bis(hexafluorophosphate), 21:39

F$_{12}$Fe$_4$P$_2$S$_5$C$_{20}$H$_{20}$, Iron(2+), tetrakis(η^5-cyclopentadienyl)-μ$_3$-(disulfur)tri-μ$_3$-thio-tetra-, bis(hexafluorophosphate), 21:44

F$_{12}$IrN$_4$OP$_4$C$_{40}$H$_{46}$, Iridium(III), tris-(acetonitrile)nitrosylbis-(triphenylphosphine)-, bis-[hexafluorophosphate], 21:104

F$_{12}$N$_4$P$_2$PdC$_{12}$H$_{30}$, Palladium(II), [N,N'-bis[2-(dimethylamino)-ethyl]-N,N'-dimethyl-1,2-ethanediamine]-, bis(hexafluorophosphate), 21:133

FeB$_2$N$_6$O$_6$C$_{30}$H$_{34}$, {[tris[μ-[(1,2-cyclohexanedione dioximato)-O:O']diphenyldiborato-

(2−)]-$N,N',N'',N''',N'''',N'''''$}-, 21:112
FeB$_4$C$_{12}$H$_{30}$, [1,1'-*commo*-Bis(2,3-dicarba-1-ferra-*closo*-heptaborane)](12), 2,2',3,3'-tetraethyl-1,1-dihydro-, 22:215
FeClO, Iron chloride oxide:
  intercalate with 4-aminopyridine (4:1), 22:86
  intercalate with pyridine (4:1), 22:86
  intercalate with 2,4,6-trimethylpyridine (6:1), 22:86
FeF$_{12}$N$_6$PC$_{28}$H$_{34}$, Iron(II), bis(acetonitrile)(2,9-dimethyl-3,10-diphenyl-1,4,8,11-tetraazacyclotetradia-1,3,8,10-tetraene)-, bis[hexafluorophosphate(1−)], 22:107, 108
FeH$_{1.94}$Ti, Iron titanium hydride, 22:90
FeNO$_{13}$P$_2$Ru$_3$C$_{49}$H$_{31}$, Ruthenate(1−), tridecacarbonylhydridoirontri-, μ-nitridobis(triphenylphosphorus)(1+), 21:60
FeN$_2$O$_4$P$_2$C$_{39}$H$_{30}$, Ferrate(1−), tricarbonylnitrosyl-, μ-nitrido-bis-(triphenylphosphorus)(1+), 22:163, 165
FeO$_2$CH$_3$, Iron, methoxyoxo-, 22:87
FeO$_3$P$_5$C$_{55}$H$_{57}$, Iron, bis[1,2-ethanediylbis(diphenylphosphine)](trimethyl phosphito)-, 21:93
FeO$_4$P$_2$C$_{36}$H$_{38}$, Iron, [1,2-ethanediylbis(diphenylphosphine)]bis(2,4-pentanedionato)-, 21:94
FeO$_{13}$Os$_3$C$_{13}$H$_2$, Osmium, tridecacarbonyldihydridoirontri-, 21:63
FeO$_{13}$Ru$_3$C$_{13}$H$_2$, Iron, tridecacarbonyldihydridotriruthenium-, 21:58
FeP$_2$S$_4$C$_{72}$H$_{60}$, Bis(tetraphenylphosphonium) tetrakis(benzenethiolato)ferrate(II), 21:24
FeP$_4$C$_{52}$H$_{48}$, Iron, [[2-[2-(diphenylphosphino)ethyl]phenylphosphino]phenyl-C,P,P][1,2-ethanediylbis(diphenylphosphine)]-hydrido-, 21:92
FeP$_4$C$_{54}$H$_{52}$, Iron, bis[1,2-ethanediylbis(diphenylphosphine)]-(ethene)-, 21:91
FeS$_4$C$_{14}$H$_{20}$, Iron, bis(η^5-cyclopentadienyl)-μ-(disulfur)bis-μ-(ethanethiolato)-di-, 21:40, 41
Fe$_2$F$_{12}$N$_2$P$_2$S$_2$C$_{18}$H$_{36}$, Iron(2+), bis(acetonitrile)bis(η^5-cyclopentadienyl)bis-μ-(ethanethiolato)-di-, bis(hexafluorophosphate), 21:39
Fe$_2$O$_4$, Iron oxide, 22:43
———, Magnetite, 22:43

Fe$_2$O$_4$C$_2$H$_4$, Iron, [1,2-ethanediolato-(2−)]dioxodi-, 22:88
Fe$_2$P$_2$S$_6$C$_{72}$H$_{60}$, Ferrate(III), tetrakis(benzenethiolato)di-μ-thiodi-, bis(tetraphenylphosphonium), 21:26
Fe$_4$F$_{12}$P$_2$S$_5$C$_{20}$H$_{20}$, Iron(2+), tetrakis(η^5-cyclopentadienyl)-μ$_3$-(disulfur)tri-μ$_3$-thio-tetra-, bis(hexafluorophosphate), 21:44
Fe$_4$N$_2$O$_{13}$P$_4$C$_{85}$H$_{60}$, Ferrate(1−), tridecacarbonyltetra-, μ-nitridobis(triphenylphosphorus)(1+) (1:2), 21:66, 68
Fe$_4$N$_2$S$_4$Se$_4$C$_{48}$H$_{108}$, Ferrate(2−), tetrakis(1,1-dimethylethanethiolato)tetra-μ-selenotetra-, bis(tetrabutylammonium), 21:37
Fe$_4$N$_2$S$_4$Se$_4$C$_{56}$H$_{92}$, Ferrate(2−), tetrakis-(benzenethiolato)tetra-μ-seleno-tetra, bis(tetrabutylammonium), 21:36
Fe$_4$N$_2$S$_8$C$_{24}$H$_{60}$, Ferrate(2−), tetrakis(1,1-dimethylethanethiolato)tetra-μ-thio-tetra-, bis(tetramethylammonium), 21:30
Fe$_4$N$_2$S$_8$C$_{56}$H$_{92}$, Ferrate(2−), tetrakis-(benzenethiolato)tetra-μ-thio-tetra-, bis(tetrabutylammonium), 21:35
Fe$_4$P$_2$S$_8$C$_{72}$H$_{60}$, Ferrate(II, III), tetrakis-(benzenethiolato)-μ$_3$-thio-tetra-, bis(tetraphenyl-phosphonium), 21:27
Fe$_4$S$_5$C$_{20}$H$_{20}$, Iron, tetrakis(η^5-cyclo-pentadienyl)-μ$_3$-(disulfur)tri-μ$_3$-thio-tetra-, 21:45
Fe$_4$S$_6$C$_{20}$H$_{20}$, Iron, tetrakis(η^5-cyclopentadienyl)bis-μ$_3$-(disulfur)-di-μ$_3$-thio-tetra-, 21:42

GaBr$_4$NC$_8$H$_{20}$, Gallate(1−), tetrabromo-tetraethylammonium, 22:141
GaBr$_4$NC$_{16}$H$_{36}$, Gallate(1−), tetrabromo-, tetrabutylammonium, 22:139
GaCl$_4$NC$_{16}$H$_{36}$, Gallate(1−), tetrachloro-, tetrabutylammonium, 22:139
GaI$_4$NC$_{16}$H$_{36}$, Gallate(1−), tetraiodo-, tetrabutylammonium, 22:140
Ga$_2$Br$_2$P$_2$C$_{36}$H$_{32}$, Gallate(2−), hexabromodi-, bis(triphenylphosphonium), 22:135, 138
Ga$_2$Br$_6$P$_2$C$_{48}$H$_4$, Gallate(2−), hexabromodi-, bis(tetraphenylphosphonium), 22:139
Ga$_2$Cl$_6$P$_2$C$_{36}$H$_{32}$, Gallate(2−), hexachlorodi-, bis(triphenylphosphonium), 22:135, 138
Ga$_2$I$_6$P$_2$C$_{36}$H$_{32}$, Gallate(2−), hexaiododi-, bis(triphenylphosphonium), 22:135, 138
GdCl$_3$, Gadolinium chloride, 22:39

GdN$_4$O$_2$C$_{49}$H$_{35}$, Gadolinium, (2,4-pentanedianato) [5,10,15,20-tetraphenylporphyrinato(2−)]-, 22:160

HCrNO$_5$P$_2$C$_{41}$H$_{30}$, Chromate(1−), pentacarbonylhydrido-, μ-nitridobis(triphenylphosphorus) (1+), 22:183
HFeNO$_{13}$P$_2$Ru$_3$C$_{49}$H$_{30}$, Ruthenate(1−), tridecacarbonylhydridoirontri-, μ-nitrido-(triphenylphosphorus)(1+), 21:60
HFeP$_4$C$_{52}$H$_{48}$, Iron, [[2-[2-(diphenylphosphino)ethyl]phenylphosphino]phenyl-C,P,P']-[1,2-ethanediylbis(diphenylphosphine)]hydrido-, 21:92
HMoNO$_5$P$_2$C$_{41}$H$_{30}$, Molybdate(1−), pentacarbonylhydrido-, μ-nitridobis(triphenylphosphorus)(1+), 22:183
HNO$_5$P$_2$WC$_{41}$H$_{30}$, Tungstate(1−), pentacarbonylhydrido-, μ-nitridobis(triphenylphosphorus)(1+), 22:182
HNbO$_5$Ti, Hydrogen pentaoxoniobatetitanate(1−), 22:89
   intercalate with 1-butanamine, 22:89
   intercalate with ethanamine, 22:89
   intercalate with methanamine, 22:89
   intercalate with NH$_3$, 22:89
   intercalate with 1-propanamine, 22:89
HO$_8$S$_2$, [hydrogen bis(sulfate)] platinum chain complexes, 21:151
H$_{1.94}$FeTi, Iron titanium hydride, 22:90
H$_2$FeO$_{13}$Os$_3$C$_{13}$, Osmium, tridecacarbonyldihydridoirontri-, 21:63
H$_2$FeO$_{13}$Ru$_3$C$_{13}$, Iron, tridecacarbonyldihydridotriruthenium-, 21:58
H$_2$O$_{13}$Os$_3$RuC$_{13}$, Osmium, tridecacarbonyldihydridorutheniumtri-, 21:64
H$_3$N, Ammonia intercalate with HNbTiO$_5$, 22:89
H$_4$AlLaNi$_4$, Aluminum lanthanum nickel hydride, 22:96
H$_8$O$_4$Pt, Platinum(II), tetraaqua-, 21:192
H$_{16}$N$_4$O$_7$P$_2$, Ammonium diphosphate, 21:157
^{2}H$_3$BNNaC, Borate(1−), cyanotri[(^2H)hydro]-, sodium, 21:167
HfCl$_4$O$_2$C$_8$H$_{16}$, Hafnium(IV), tetrachlorobis(tetrahydrofuran)-, 21:137
HoCl$_3$, Holmium chloride, 22:39
HoN$_4$O$_2$C$_{49}$H$_{35}$, Holmium, (2,4-pentanedionato)-[5,10,15,20-tetraphenylporphyrinato(2−)], 22:160

HoN$_4$O$_2$C$_{55}$H$_{47}$, Holmium, (2,2,6,6-tetrametyl-3,5-heptanedionato)[5,10,15,20-tetraphenylporphyrinato(2−)]-, 22:160

I$_2$La, Lanthanium iodide, 22:36
I$_2$N$_4$PdC$_{12}$H$_{30}$, Palladium(II), [N,N-bis[2-(dimethylamino)ethyl]-N',N'-dimethyl-1,2-ethanediamine]iodo-, iodide, 21:130
I$_3$La, Lanthanum iodide, 22:31
IrClOPC$_{17}$H$_{22}$, Iridium(I), carbonylchlorobis(dimethylphenylphosphine)-, trans-, 21:97
IrClO$_2$P$_4$C$_{13}$H$_{32}$, Iridium(1+), (carbondioxide)bis[1,2-ethanediylbis(dimethylphosphine)]-, chloride, 21:100
IrClO$_4$P$_3$C$_{11}$H$_{27}$, Iridium, chloro[(formyl-κC-oxy)formato-κO-(2−)]tris(trimethylphosphine)-, 21:102
IrClP$_3$C$_{17}$H$_{41}$, Iridium, chloro(η^2-cyclooctene)tris(trimethylphosphine)-, 21:102
IrClP$_4$C$_{12}$H$_{32}$, Iridium(1+), bis[1,2-ethanediylbis(dimethylphosphine)]-, chloride, 21:100
IrF$_{12}$N$_4$OP$_4$C$_{40}$H$_{36}$, Iridium(III), tris(acetonitrile)nitrosylbis(triphenylphosphine)-, bis[hexafluorophosphate], 21:104
IrPC$_{32}$H$_{38}$, Iridium, (1,3-butanediyl)-(η^5-pentamethylcyclopentadienyl)(triphenylphosphine), 22:174

KCoO$_2$, Potassium cobalt oxide, 22:58
KCl$_7$Dy$_2$, Potassium dysprosium chloride, 22:2
KCrO$_2$, Potassium chromium oxide, 22:59
KF$_6$U, Uranate(V), hexafluoro-, potassium, 21:166
KNbO$_5$Ti, Potassium, pentaoxoniobatetitanate(1−), 22:89
K$_{0.5}$CoO$_2$, Potassium cobalt oxide, 22:57
K$_{0.5}$CrO$_2$, Potassium chromium oxide, 22:59
K$_{0.67}$CoO$_2$, Potassium cobalt oxide, 22:57
K$_{0.6}$CrO$_2$, Potassium chromium oxide, 22:59
K$_{0.7}$CrO$_2$, Potassium chromium oxide, 22:59
K$_{0.77}$CrO$_2$, Potassium chromium oxide, 22:59
K$_2$Al$_4$NNaO$_{36}$Si$_{14}$C$_4$H$_{12}$ · 7H$_2$O, Potassium sodium tetramethylammonium aluminum silicate hydrate, 22:65
K$_2$Bi$_4$N$_2$O$_{12}$C$_{36}$H$_{72}$, Potassium, (4,7,13,16,21,24-hexaoxa-1,10-diazabicy-

clo[8.8.8]hexacosane)-, tetrabismuthide, 22:151

$K_2F_{0.60}N_4PtC_4H_{0.30} \cdot 3H_2O$, Platinate, tetracyano-, potassium (hydrogen difluoride) (1:2:0.30), trihydrate, 21:147

K_2F_5MoO, Molybdate(V), pentafluorooxo-, dipotassium, 21:170

$LaAlH_4Ni_4$, Aluminum lanthanum nickel hydride, 22:96

$LaCl_3$, Lanthanum chloride, 22:39

LaI_2, Lanthanum iodide, 22:36

LaI_3, Lanthanum iodide, 22:31

$LaN_4O_2C_{49}H_{35}$, Lanthanum, (2,4-pentanedionato)[5,10,15,20-tetraphenylporphyrinato(2−)], 22:160

$LaN_4O_2C_{55}H_{47}$, Lanthanum, (2,2,6,6-tetraphenylporphyrinato(2−)]-, 22:160

$LiCl_6Cs_2Tm$, Cesium lithium thulium chloride, 22:7

LiF_5C_6, Lithium, (pentafluorophenyl)-, 21:72

Li_3N, Lithium nitride, 22:48

$LuCl_3$, Lutetium chloride, 22:39

$LuCl_5Cs_2$, Cesium lutetium chloride, 22:6

$LuCl_6Cs_3$, Cesium lutetium chloride, 22:6

$LuN_4O_2C_{49}H_{35}$, Lutetium, (2,4-pentanedionato)-[5,10,15,20-tetraphenylporphyrinato(2−)]-, 22:160

Lu_2Cl_9Cs, Cesium lutetium chloride, 22:6

$Me_2Ph_2[14]$-1,3,8,10-tetraeneN, see —1,4,8,11-Tetraazacyclotetradeca-1,3,8,10-tetraene, 2,9-dimethyl-3,10-diphenyl-, 22:107

$MnP_2S_4C_{72}H_{60}$, Bis(tetraphenylphosphonium) tetrakis(benzenethiolato)manganate(II), 21:25

$Mn_3Ca_2O_8$, Calcium manganese oxide, 22:73

MoF_5OK_2, Molybdate(V), pentafluorooxo-, dipotassium, 21:170

$MoNO_5P_2C_{41}H_{31}$, Molybdate(1−), pentacarbonylhydrido-, μ-nitridobis-(triphenylphosphorus)(1+), 22:183

$Mo_2Cl_2N_4C_8H_{24}$, Molybdenum, dichlorotetrakis(dimethylamido)di-, ($Mo \equiv Mo$), 21:56

$Mo_2N_6C_{12}H_{36}$, Molybdenum, hexakis-(dimethylamido)di-, ($Mo \equiv Mo$), 21:54

NC, Cyano:
  boron complex, 21:167

platinum chain complexes, 21:142–156

NCH_5, Methanamine
  intercalate with $HNbTiO_5$, 22:89

NC_2H_3, Acetonitrile:
  copper, iron, and zinc complexes, 22:108, 110, 111
  iridium complex, 21:104
  iron complexes, 21:39–46

NC_2H_7, Ethanamine intercalate with $HNbTiO_5$, 22:89

———, Methaneamine, N-methylmolybdenum complex, 21:54

NC_3H_9, 1-Propanamine
  intercalate with $HNbTiO_5$, 22:89

NC_4H_{11}, 1-Butanamine
  intercalate with $HNbTiO_5$, 22:89

NC_5H_5, Pyridine
  intercalate with FeClO (1:4), 22:86
  rhenium complex, 21:116, 117

NC_5H_9, Butane, 1-isocyano-,
  rhodium complex, 21:49

NC_8H_{11}, Pyridine, 2,4,6-trimethylintercalate with FeClO (1:6), 22:86

NO, Nitrogen oxide
  iridium complex, 21:104
  iron and ruthenium, 22:163

$NO_2SC_3H_7$, L-Cysteine
  gold complex, 21:31

$NO_3C_3H_7$, Serine
  copper complex, 21:115

$NO_5P_2WC_{41}H_{31}$, Tungstate(1−), pentacarbonylhydrido-, μ-nitridobis(triphenylphosphorus) (1+), 22:182

$NPPtSe_2C_{24}H_{28}$, Platinum(II), (N,N-diethyldiselenocarbamato)methyl-(triphenylphosphine)-, 21:10

$(NS)_x$, Sulfur nitride, polymer, 22:143

NSCH, Thiocyanic acid,
  palladium complex, 21:132

NSC_2H_7, Ethanethiol, 2-aminocobalt complex, 21:19

$NSe_2C_5H_{11}$, Diselenocarbamic acid, N,N-diethyl-
  nickel, palladium, and platinum complexes, 21:9

$N_2C_2H_8$, 1,2-Ethanediamine cobalt complexes, 21:19, 21, 120–126
  cobalt(III) trifluoromethanesulfonate complexes, 22:105

$N_2C_3H_4$, 1H-Pyrazole
  boron-copper complex, 21:108, 110

N₂C₅H₆, Pyridine, 4-amino-intercalate with FeClO, 22:86

N₂C₅H₈, 1H-Pyrazole, 3,5-dimethylboron-copper complex, 21:109

N₂C₆H₁₆, 1,2-Ethanediamine, N,N,-N',N'-tetramethyl-
palladium complex, 22:168

N₂C₈H₂₀, 1,2-Ethanediamine, N,N'-bis(1-methylethyl)-,
platinum complex, 21:87

N₂C₁₀H₈, 2,2'-Bipyridine:
palladium complex, 22:170
ruthenium complex, 21:127

N₂C₁₀H₂₄, 1,2-Ethanediamine, N,N'-dimethyl-N,N'-bis(1-methylethyl)-,
platinum complex, 21:88
——, N,N,N',N'-tetraethylplatinum complex, 21:86, 87

N₂C₁₂H₈, 1,10-Phenathroline copper complex, 21:115

N₂C₁₈H₂₄, 1,2-Ethanediamine, N,N-bis(1-phenylethyl)-, (S,S)-
platinum complex, 21:87

N₂C₂₀H₂₈, 1,2-Ethanediamine, N,N'-dimethyl-N,N'-bis(1-phenylethyl)-, (R,R)-,
platinum complex, 21:87

N₂O₂C₆H₁₀, 1,2-Cyclohexanedione, dioxime boron-iron complex, 21:112

N₂O₂P₂C₃₆H₃₀, Phosphorus(1+), μ-nitrido-bis(triphenyl-, nitrite, 22:164

N₂O₁₁P₂Ru₃C₄₆H₃₀, Ruthenate(1−), decacarbonyl-μ-nitrosyl-tri-, μ-nitrido-bis(triphenylphosphorus) (1+), 22:163, 165

N₂PdC₁₀H₂₄, Palladium, (1,4-butanediyl)-(N,N,N',N'-tetramethyl-1,2-ethanediamine)-, 22:168

N₂PdC₁₄H₁₆, Palladium, (2,2'-bipyridine)(1,4-butanediyl)-, 22:170

N₂PdS₁₁H₈, Palladate(II), bis(hexasulfido)-, diammonium, nonstoichiometric, 21:14

N₂PtS₁₀H₈, Platinate(II), bis(pentasulfido)-, bis(tetrapropylammonium), 21:13

N₂PtS₁₅H₈, Platinate(IV), tris(pentasulfido)-, diammonium, 21:12, 13

N₂S₃H₈, Ammonium pentasulfide, 21:12

N₃, Azido
platinum chain complexes, 21:149

N₃C₄H₁₃, 1,2-Ethanediamine, N-(2-aminoethyl)-
cobalt(III) trifluoromethanesulfonate complexes, 22:106

N₃PC₉H₁₂, Propionitrile, 3,3',3"-phosphinidynetri-
nickel complex, 22:113, 115

N₃RhS₁₅H₁₂, Rhodate(III), tris(pentasulfido)-, triammonium, 21:15

N₄C₁₂H₃₀, 1,2-Ethanediamine, N,N-bis[2-(dimethylamino)ethyl]-N',-N'-dimethyl-palladium complex, 21:129–132
——, N,N'-bis[2-(dimethylamino)-ethyl]-N,N'-dimethyl-,
palladium complex, 21:133

N₄C₂₀H₁₄, Porphyrin
actinide and lanthanide complexes, 22:156
——, 5,10,15,20-tetraphenyl-
actinide and lanthanide complexes, 22:156

N₄C₂₄H₂₈, 1,4,8,11-Tetraazacyclotetradeca-1,3,8,10-tetraene, 2,9-dimethyl-3,10-diphenyl-
copper, iron, and zinc complexes, 22:108, 110, 111

N₄C₄₈H₃₈, Porphyrine, 5,10,15,20-tetrakis(4-methylphenyl)-
actinide and lanthanide complexes, 22:156

N₄NdO₂C₁₅H₄₇, Neodymium, (2,2,6,6-tetramethyl-3,5-heptanedionato)[5,10,15,20-tetraphenylporphyrinato(2−)]-, 22:160

N₄NdO₂C₄₉H₃₅, Neodymium, (2,4-pentanedionato)[5,10,15,20-tetraphenylporphyrinato(2−)]-, 22:160

N₄O₂PrC₄₉H₃₅, Praseodymium, (2,4-pentanedionato)[5,10,15,20-tetraphenylporphyrinato(2−)]-, 22:160

N₄O₂PrC₅₃H₄₃, Praseodymium, (2,4-pentanedionato)[5,10,15,20-tetrakis(4-methylphenyl)porphyrinato(2−)]-, 22:160

N₄O₂SmC₄₉H₃₅, Samarium, (2,4-pentanedionato)[5,10,15,20-tetraphenylporphyrinato(2−)]-, 22:160

N₄O₂SmC₅₅H₄₇, Samarium, (2,2,6,6-tetramethyl-3,5-heptanedionato)[5,10,15,20-tetraphenylporphyrinato(2−)]-, 22:160

N₄O₂TbC₄₉H₃₅, Terbium, (2,4-pentanedionato)[5,10,15,20-tetraphenylporphyrinato(2−)]-, 22:160

N₄O₂TbC₅₅H₄₇, Terbium, (2,2,6,6-tetramethyl-3,5-heptanedionato)[5,20,25,20-tetraphenylporphyrinato(2−)]-, 22:160

N₄O₂TmC₅₅H₄₇, Thulium, (2,2,6,6-tetramethyl-

3,5-heptanedionato)[5,10,15,20-tetraphenylporphyrinato(2−)]-, 22:160
$N_4O_2YC_{49}H_{35}$, Yttrium, (2,4-pentanedionato)[5,10,15,20-tetraphenylporphyrinato(2−)]-, 22:160
$N_4O_2YbC_{53}H_{43}$, Ytterbium, (2,4-pentanedionato)[5,10,15,20-tetrakis(4-methylphenyl)porphyrinato(2−)]-, 22:156
$N_4O_2YbC_{59}H_{55}$, Ytterbium, [5,10,15,20-tetrakis(4-methylphenyl)porphyrinato(2−)](2,2,6,6-tetramethyl-3,5-heptanedionato)-, 22:156
$N_4O_3PtTl_4C_5$, Platinate(II), tetracyano-, thallium carbonate(1:4:1), 21:153, 154
$N_4O_4ThC_{54}H_{42}$, Thorium, bis(2,4-pentanedionato)[5,10,15,20-tetraphenylporphyrinato(2−)]-, 22:160
$N_4PrC_{48}H_{36}$, Praseodymium, [5,10,15,20-tetrakis(4-methylphenyl)porphyrinato(2−)]-, 22:160
$N_4PtTl_2C_4$, Platinate(II), tetracyano-, dithallium, 21:153
$N_6PdS_2C_{14}H_{30}$, Palladium(II), [N,N-bis[2-(dimethylamino)ethyl]-N',N'-dimethyl-1,2-ethanediamine](thiocyanato-N)-, thiocyanate, 21:132
$NaAlO_4Si \cdot 2.25H_2O$, Sodium aluminum silicate, 22:61
$NaAl_4K_2NO_{36}Si_{14}C_4H_{12} \cdot 7H_2O$, Potassium, sodium tetramethyl ammonium aluminum silicate hydrate, 22:65
$NaBNC^2H_3$, Borate(1−), cyanotri[(^2H)hydro]-, sodium, 21:167
$NaCoO_2$, Sodium cobalt oxide, 22:56
NaF_6U, Uranate(V), hexafluoro-, sodium, 21:166
$Na_{0.6}CoO_2$, Sodium cobalt oxide, 22:56
$Na_{0.64}CoO_2$, Sodium cobalt oxide, 22:56
$Na_{0.74}CoO_2$, Sodium cobalt oxide, 22:56
$Na_{0.77}CoO_2$, Sodium cobalt oxide, 22:56
$Na_2Al_2O_{14}Si \cdot XH_2O$, Sodium aluminum silicate hydrate, 22:64
$Na_{2.4}Al_{2.6}N_{3.6}O_{207}Si_{100}C_{43}H_{100}$, Sodium tetrapropylammonium aluminum silicate, 22:67
$NbCl_4O_2C_4H_{16}$, Niobium(IV), tetrachlorobis(tetrahydrofuran)-, 21:138
$NbHO_5Ti$, Hydrogen, pentaoxoniobatetitanate(1−), 22:89
$NbKO_5Ti$, Potassium, pentaoxoniobatetitanate(1−), 22:89

$Nb_2As_4Cl_6C_{20}H_{32}$, Niobium(III), hexachlorobis[o-phenylenebis(dimethylarsine)]di-, 21:18
$Nb_2As_6Cl_6C_{22}H_{54}$, Niobium(III), hexachlorobis[[2-(dimethylarsino)-methyl]-2-methyl-1,3-propanediyl]bis(dimethylarsine)]-, 21:18
$Nb_2Cl_6P_2C_{30}H_{24}$, Niobium(III), hexachlorobis[1,2-ethanediylbis(diphenylphosphine)]di-, 21:18
$Nb_2Cl_6S_3C_{16}H_{18}$, Niobium(III), di-μ-chlorotetrachloro-μ-(dimethylsulfide)-bis(dimethyl sulfide)-di-, 21:16
$NdCl_3$, Neodymium chloride, 22:39
$NdN_4O_2C_{15}H_{47}$, Neodymium, (2,2,6,6-tetramethyl-3,5-heptanedionato)[5,10,15,20-tetraphenylporphyrinato(2−)]-, 22:160
$NdN_4O_2C_{49}H_{35}$, Neodymium, (2,4-pentanedionato)[5,10,15,20-tetraphenylporphyrinato(2−)]-, 22:16
$NiBr_2N_6P_2C_{18}H_{24}$, Nickel(II), dibromobis(3,3',3''-phosphinidynetripropionitrile)-, 22:113, 115
$(NiBr_2N_6P_2C_{18}H_{24})_x$, Nickel(II), dibromobis(3,3',3''-phosphinidyne tripropionitrile)-, 22:115
$NiClNPSe_2C_{11}H_{25}$, Nickel(II), chloro(N,N-diethyldiselenocarbamato)(triethylphosphine)-, 21:9
$NiCl_2N_6P_2C_{18}H_{24}$, Nickel(II), dichlorobis(3,3',3''-phosphinidynetripropionitrile)-, 22:113
Ni_4AlH_4La, Aluminum lanthanum nickel hydride, 22:96

OC, Carbon monoxide:
  chromium complex, 21:1, 2
  cobalt, iron, osmium, and ruthenium complexes, 21:58–65
  iron complex, 21:66, 68
  palladium complex, 21:49
  ruthenium complex, 21:30
OC_3H_6, Acetone, compd. with carbonyltri-μ-chloro-chlorotetrakis(triphenylphosphine)diruthenium (1:2), 21:30
——, compd. with tri-μ-chloro(thiocarbonyl)tetrakis(triphenylphosphine)diruthenium, 21:29

OC$_4$H$_8$, Furan, tetrahydrohafnium, niobium, scandium, titanium, vanadium, and zirconium complexes, 21:135–139
OH$_2$, Water:
cobalt complex, 21:123–126
platinum complex, 21:192
OPC$_{19}$H$_{15}$, Benzaldehyde, 2-(diphenylphosphino)-, 21:176
OS$_8$, cyclo-Octasulfur monoxide, 21:172
O$_2$C$_2$H$_6$, 1,2-Ethandiol, 22:86
O$_2$C$_5$H$_8$, 2,4-Pentanedione:
actinide and lanthanide complexes, 22:156
iron complex, 21:94
O$_2$C$_{11}$H$_{20}$, 3,5-Heptanedione, 2,2,6,6-tetramethyl-
actinide and lanthanide complexes, 22:156
O$_2$PC$_{19}$H$_{15}$, Benzoic acid, 2-(diphenylphosphino)-, 21:178
O$_2$SC$_2$H$_4$, Acetic acid, 2-mercaptocobalt complex, 21:21
O$_3$CH$_2$, Carbonic acid
platinum chain complex, 21:153, 154
cobalt complex, 21:120
O$_3$PC$_3$H$_9$, Trimethyl phosphite
iron complex, 21:93
O$_4$C$_2$H$_2$, Formic acid, (formyloxy)-iridium complex, 21:102
O$_5$UC$_{10}$H$_{25}$, Uranium(V), pentaethoxy-, 21:165
O$_7$P$_2$, Diphosphate, 21:157
O$_{13}$Os$_3$RuC$_{13}$H$_2$, Osmium, tridecacarbonyldihydridorutheniumtri-, 21:64
Os$_3$FeO$_{13}$C$_{13}$H$_3$, Osmium, tridecacarbonyldihydridoirontri-, 21:63
Os$_3$O$_{13}$RuC$_{13}$H$_2$, Osmium, tridecacarbonyldihydridorutheniumtri-, 21:64

PC$_2$H$_7$, Phosphine, dimethyl-, 21:180
PC$_3$H$_9$, Phosphine, trimethyl-
iridium complex, 21:102
PC$_6$H$_{15}$, Phosphine, triethyl-
nickel complex, 21:9
PC$_8$H$_{11}$, Phosphine, dimethylphenyl-, 22:133
iridium complex, 21:97
PC$_{18}$H$_{15}$, Phosphine, triphenyl-, 21:78
cobalt, iridium, and rhodium complexes, 22:171, 173, 174
iridium complex, 21:104

palladium complex, 22:169
palladium and platinum complexes, 21:10
ruthenium complex, 21:29
PN$_3$C$_9$H$_{12}$, Propionitrile, 3,3′,3″-phosphinidynetri-
nickel complex, 22:113, 115
PRhC$_{32}$H$_{38}$, Rhodium, (1,4-butanediyl)-(η^5-pentamethylcyclopentadienyl)-(triphenylphosphine)-, 22:173
PS$_3$C$_{12}$H$_{27}$, Phosphorotrithious acid, tributyl ester, 22:131
P$_2$C$_2$H$_{16}$, Phosphine, 1,2-ethanediylbis(dimethyl-
iridium complex, 21:100
P$_2$C$_{25}$H$_{22}$, Phosphine, methylenebis(diphenylpalladium and rhodium complexes, 21:47–49
P$_2$C$_{26}$H$_{24}$, Phosphine, 1,2-ethanediylbis(diphenyl-:
iron complexes, 21:91–94
palladium complex, 22:167
P$_2$PdC$_{30}$H$_{32}$, Palladium, (1.4-butanediyl)[1,2-ethanediylbis(diphenylphosphine)]-, 22:167
P$_2$PdC$_{40}$H$_{38}$, Palladium, (1,4-butanediyl)bis(triphenylphosphine)-, 22:169
P$_2$RuC$_{49}$H$_{40}$, Ruthenium(II), (η^5-cyclopentadienyl)(phenylethynyl)bis(triphenylphosphine)-, 21:82
P$_2$S$_4$ZnC$_{72}$H$_{60}$, Zincate(II), tetrakis-(benzenethiolato)-, bis(tetraphenylphosphonium), 21:25
PbCl$_6$N$_2$C$_{10}$H$_{12}$, Plumbate(IV), hexachloro-, dipyridinium, 22:149
PbO$_2$, Lead oxide
solid solns. with ruthenium oxide (Ru$_2$O$_3$), pyrochlore, 22:69
Pb$_{2.67}$Ru$_{1.33}$O$_{6.5}$, Lead ruthenium oxide, pyrochlore, 22:69
PdBr$_2$N$_4$C$_{12}$H$_{30}$, Palladium(II), [N,N-bis[2-(dimethylamino)ethyl]-N′,N′-dimethyl-1,2-ethandiamine]bromo-, bromide, 21:131
PdClNPSe$_2$C$_{23}$H$_{25}$, Palladium, chloro(N,N-diethyldiselenocarbamato)-(triphenylphosphine)-, 21:10
PdCl$_2$N$_4$C$_{12}$H$_{30}$, Palladium(II), [N,N-bis[2-(dimethylamino)ethyl]-N′,N′-dimethyl-1,2-ethanediamine]chloro-, chloride, 21:129
PdCl$_2$P$_4$C$_{50}$H$_{44}$, Palladium(I), dichlorobis-μ-[methylenebis(diphenylphosphine)]-di-, (Pd-Pd), 21:48

$PdF_{12}N_4P_2C_{12}H_{30}$, Palladium(II), [$N,N'$-bis[2-(dimethylamino)-ethyl]-N,N'-dimethyl-1,2-ethanediamine]-, bis(hexafluorophosphate), 21:133

$PdI_2N_4C_{12}H_{30}$, Palladium(II), [N,N-bis[2-(dimethylamino)ethyl-N'-N'-dimethyl-1,2-ethanediamine]-iodo-, iodide, 21:130

$PdN_2C_{10}H_{24}$, Palladium, (1,4-butanediyl)(N,N,N',N'-tetramethyl-1,2-ethanediamine)-, 22:168

$PdN_2C_{14}H_{16}$, Palladium, (2,2'-bipyridine)(1,4-butanediyl)-, 22:170

$PdN_2S_{11}H_8$, Palladate(II), bis(hexasulfido)-, diammonium, nonstoichiometric, 21:14

$PdN_6S_2C_{14}H_{30}$, Palladium(II), [N,N-bis[2-(dimethylamino)ethyl]-N',N'-dimethyl-1,2-ethanediamine](thiocyanato-N)-, thiocyanate, 21:132

$PdP_2C_{30}H_{32}$, Palladium, (1,4-butanediyl)[1,2-ethanediylbis(diphenylphosphine)]-, 22:167

$PdP_2C_{40}H_{38}$, Palladium, (1,4-butanediyl)bis(triphenylphosphine)-, 22:169

$Pd_2Cl_2P_4C_{51}H_{44}$, Palladium(I), μ-carbonyl dichlorobis[methylenebis(diphenylphosphine)]di-, 21:49

$PrCl_3$, Praseodymium chloride, 22:39

$PrN_4C_{48}H_{36}$, Praseodymium, [5,10,15,20-tetrakis(4-methylphenyl)porphyrinato(2−)]-, 22:160

$PrN_4O_2C_{49}H_{35}$, Praseodymium, (2,4-pentanedionato)[5,10,15,20-tetraphenylporphyrinato(2−)]-, 22:160

$PrN_4O_2C_{53}H_{43}$, Praseodymium, (2,4-pentanedionato)[5,10,15,20-tetrakis(4-methylphenyl)porphyrinato(2−)]-, 22:160

Pr_2Cl_7Cs, Cesium praseodymium chloride, 22:2

$PtBClF_4S_3C_6H_{18}$, Platinum(II), chlorotris(dimethyl sulfide)-, tetrafluoroborate(1−), 22:126

$PtClH_8N_3O_4$, Platinum(II), diammineaquachloro- trans-, nitrate, 22:125

$PtCl_{0.30}Cs_2N_4C_4$, Platinate, tetracyano-, cesium chloride (1:2:0.30), 21:142

$PtCl_{0.30}N_4Rb_2C_4 \cdot 3H_2O$, Platinate, tetracyano-, rubidium chloride (1:2:0.30), trihydrate, 21:145

$PtClNPSe_2C_{23}H_{25}$, Platinum(II), chloro(N,N-diethyldiselenocarbamato)(triphenylphosphine)-, 21:10

$PtCl_2H_6IN_2Pt$, Platinum(II), diamminechloroiodo-, trans-, chloride, 22:124

$PtCl_2H_9N_3$, Platinum(II), triamminechloro-, chloride, 22:124

$PtCl_2N_2C_{10}H_{24}$, Platinum(II), [N,N'-bis(1-methylethyl)-1,2-ethanediamine)dichloro(ethene)-, 21:87

$PtCl_2N_2C_{12}H_{28}$, Platinum(II), dichloro(ethene)(N,N,N',N'-tetraethyl-1,2-ethanediamine)-, 21:86, 87

$PtCl_2N_2C_{20}H_{28}$, Platinum(II), [(S,S)-N,N'-bis(phenylethyl)-1,2-ethanediamine]dichloro(ethene)-, 21:87

$PtCl_2N_2C_{22}H_{32}$, Platinum(II), dichloro[(R,R)-N,N'-dimethyl-N,N'-bis(1-phenylethyl)-1,2-ethanediamine](ethene)-, 21:87

$PtCl_2N_2PtC_{12}H_{28}$, Platinum(II), dichloro[N,N'-dimethyl-N,N'-bis(1-methylethyl)-1,2-ethanediamine)(ethene)-, 21:87

$PtCl_3NSC_{18}H_{42}$, Platinate(II), trichloro(dimethyl sulfide)-, tetrabutylammonium, 22:128

$PtCs_2N_{4.75}C_4 \cdot XH_2O$, Platinate, tetracyano-, cesium azide (1:2:0.25), hydrate, 21:149

$PtCs_3N_4O_{3.68}S_{0.92}C_4H_{0.46}$, Platinate, tetracyano-, cesium [hydrogen bis(sulfate)] (1:3:0.46), 21:151

$PtF_{0.54}N_{10}C_6H_{12.27} \cdot 1.8H_2O$, Platinate, tetracyano-, guanidinium (hydrogen difluoride) (1:1:0.27), hydrate (1:1.8), 21:146

$PtF_{0.60}K_2N_4PtC_4H_{0.30} \cdot 3H_2O$ Platinate, tetracyano-, potassium (hydrogen difluoride) (1:2:0.30), trihydrate, 21:147

PtH_8O_4, Platinum(II), tetraaqua-, 21:192

$PtNPSeC_{24}H_{48}$, Platinum(II), (N,N-diethyldiselenocarbamato)methyl(triphenylphosphine)-, 20:10

$PtN_2S_{10}H_8$, Platinate(II), bis(pentasulfido)-, bis(tetrapropylammonium), 20:13

$PtN_2S_{15}H_8$, Platinate(IV), tris(pentasulfido)-, diammonium, 21:12, 13

$PtN_4O_3Tl_4C_5$, Platinate(II), tetracyano-, thallium carbonate (1:4:1), 21:153, 154

$PtN_4Tl_2C_4$, Platinate(II), tetracyano-, dithallium, 21:153

$Pt_2Cl_4S_2C_4H_{12}$, Platinum(II), di-chloro-dichlorobis(dimethyl sulfide)di-, 22:128

$Rb_2Cl_{0.30}N_4PtC_4 \cdot 3H_2O$, Platinate, tetracyano-, rubidium chloride (1:2:0.30), trihydrate, 21:145

ReClN$_4$O$_2$C$_{20}$H$_{20}$, Rhenium(V), dioxotetrakis(pyridine)-, chloride trans-, 21:116
ReClN$_4$O$_6$C$_{20}$H$_{20}$, Rhenium(V), dioxotetrakis(pyridine)-, perchlorate, trans-, 21:117
RhAs$_4$ClC$_{20}$H$_{32}$, Rhodium(1 +), bis[o-phenylenebis(dimethylarsine)]-, chloride, 21:101
RhAs$_4$ClO$_2$C$_{21}$H$_{32}$, Rhodium(1 +), (carbon dioxide)bis[o-phenylenebis(dimethylarsine)]-, chloride, 21:101
RhBN$_4$C$_{44}$H$_{56}$, Rhodium(I), tetrakis(1-isocyanobutane)-, tetraphenylborate(1 −), 21:50
RhN$_3$S$_{15}$H$_{12}$, Rhodate(III), tris(pentasulfido)-, triammonium, 21:15
RhPC$_{32}$H$_{38}$, Rhodium, (1,4-butanediyl)-(η^5-pentamethylcyclopentadienyl)-(triphenylphosphine)-, 22:173
Rh$_2$B$_2$N$_4$P$_4$C$_{118}$H$_{120}$, Rhodium(I), tetrakis(1-isocyanobutane)bis[methylene-(diphenylphosphine)]di-, bis[tetraphenylborate(1 −)], 21:49
RuC$_{10}$H$_{10}$, Ruthenium, bis(η^5-cyclopentadienyl)-, 22:180
RuC$_{12}$H$_{14}$, Ruthenium, (η^6-benzene)(η^4-1,3-cyclohexadiene)-, 22:177
RuC$_{14}$H$_{18}$, Ruthenium, bis(η^5-cycloheptadienyl)-, 22:179
RuC$_{16}$H$_{22}$, Ruthenium, (η^4-1,5-cyclooctadiene)(η^6-1,3,5-cyclooctatriene)-, 22:178
RuC$_{16}$H$_{26}$, Ruthenium(O), bis(η^2-ethylene)(η^6-hexamethylbenzene)-, 21:76
RuC$_{18}$H$_{26}$, Ruthenium(O), (η^4-1,3-cyclohexadiene)(η^6-hexamethylbenzene)-, 21:77
RuClP$_2$C$_{41}$H$_{35}$, Ruthenium(II), chloro(η^5-cyclopentadienyl)bis(triphenylphosphine)-, 21:78
RuCl$_2$N$_6$C$_{30}$H$_{24}$ · 6H$_2$O, Ruthenium(II), tris(2,2′-bipyridine)-, dichloride, hexahydrate, 21:127
RuF$_6$P$_3$C$_{49}$H$_{41}$, Ruthenium(II), (η^5-cyclopentadienyl)(phenylvinylene)-bis(triphenylphosphine)-, hexafluorophosphate(1 +), 21:80
RuO$_{13}$Os$_3$C$_{13}$H$_2$, Osmium, tridecacarbonyldihydridorutheniumtri-, 21:64
RuP$_2$C$_{49}$H$_{40}$, Ruthenium(II), (η^5-cyclopentadienyl) (phenylethynyl)bistriphenylphosphine)-, 21:82
Ru$_{1.33}$Pb$_{2.67}$O$_{6.5}$, Lead ruthenium oxide, pyrochlore, 22:69
Ru$_2$Cl$_4$C$_{20}$H$_{28}$, Ruthenium(II), di-μ-chlorobis[chloro(η^6-1-isopropyl-4-methylbenzene)-, 21:75
Ru$_2$Cl$_4$C$_{24}$H$_{36}$, Ruthenium(II), di-μ-chlorobis[chloro(η^6-hexamethylbenzene)-, 21:75
Ru$_2$O$_3$, Ruthenium oxide, solid solns. with lead oxide PbO$_2$, pyrochlore, 22:69
Ru$_3$CoNO$_{13}$P$_2$C$_{49}$H$_{30}$, Ruthenate(1 −), tridecacarbonylcobalttri-, μ-nitridobis(triphenylphosphorus)(1 +), 21:63
Ru$_3$FeNO$_{13}$P$_2$C$_{49}$H$_{31}$, Ruthenate(1 −), tridecacarbonylhydridoirontri-, μ-nitridobis(triphenylphosphorus)(1 +), 21:60
Ru$_3$FeO$_{13}$C$_{13}$H$_2$, Iron, tridecacarbonyldihydridotriruthenium-, 21:58
Ru$_3$N$_2$O$_{11}$P$_2$C$_{46}$H$_{30}$, Ruthenate(1 −), decacarbonyl-μ-nitrosyl-tri-, μ-nitridobis(triphenylphosphorus)(1 +), 22:163, 165

S, Sulfur iron cyclopentadienyl complexes, 21:42
SC, Carbon monosulfide ruthenium complex, 21:29
SC$_2$H$_6$, Dimethyl sulfide:
  boron complex, 22:239
  platinum(II) complexes, 22:126, 128
——, Ethanethiol
  iron complexes, 21:39–45
SC$_2$H$_8$, Dimethyl sulfide
  niobium complex, 21:16
SC$_4$H$_{10}$, Ethanethiol, 1,1-dimethyl-
  iron complex, 21:30, 37
SC$_6$H$_6$, Benzenethiol:
  iron complex, 21:35, 36
  cadmium, cobalt, iron, manganese, and zinc complexes, 21:24–27
(SN)$_x$, Sulfur nitride, polymer, 22:143
SNC$_2$H$_7$, Ethanethiol, 2-amino-
  cobalt complex, 21:19
SO$_2$C$_2$H$_4$, Acetic acid, 2-mercapto-,
  cobalt complex, 21:21
S$_2$, Sulfur
  iron cyclopentadienyl complexes, 21:40–46
S$_3$PC$_{12}$H$_{27}$, Phosphorotrithious acid, tributyl ester, 22:131
S$_5$, Pentasulfide
  platinum and rhodium complexes, 21:12, 13
S$_6$, Hexasulfide
  palladium complex, 21:172
S$_8$O, cyclo-Octasulfur monoxide, 21:172
ScCl$_3$, Scandium chloride, 22:39

Formula Index 277

ScCl$_3$Cs, Cesium scandium chloride, 22:23
ScCl$_3$O$_3$C$_{12}$H$_{24}$, Scandium(III), trichloro-tris(tetrahydrofuran)-, 21:139
Sc$_2$Cl$_9$Cs$_3$, Cesium scandium chloride, 22:25
SeC, Carbon selenide
  chromium complex, 21:1, 2
Se$_2$C, Carbon diselenide, 21:6, 7
Se$_2$NC$_5$H$_{11}$, Diselenocarbamic acid, N,N-diethyl-
  nickel, palladium, and platinum complexes, 21:9
SmCl$_3$, Saramium chloride, 22:39
SmN$_4$O$_2$C$_{49}$N$_{35}$, Samarium, (2,4-pentanedionato)[5,10,15,20-tetraphenylporphyrinato(2−)]-, 22:160
SmN$_4$O$_2$C$_{55}$H$_{47}$, Samarium, (2,2,6,6-tetramethyl-3,5-heptanedionato)[5,10,15,20-tetraphenylporphyrinato(2−)]-, 22:160

TbCl$_3$, Terbium chloride, 22:39
TbN$_4$O$_2$C$_{49}$H$_{35}$, Terbium, (2,4-pentanedionato)[5,10,15,20-tetraphenylporphyrinato(2−)]-, 22:160
TbN$_4$O$_2$C$_{55}$H$_{47}$, Terbium, (2,2,6,6-tetramethyl-3,5-heptanedionato)(5,10,15,20-tetraphenylporphyrinato(2−)]-, 22:160
TcCl$_4$NOC$_{16}$H$_{36}$, Technetate(V), tetrachlorooxo-, tetrabutylammonium (1:1), 21:160
ThN$_4$O$_4$C$_{54}$H$_{42}$, Thorium, bis(2,4-pentanedionato)[5,10,25,20-tetraphenylporphyrinato(2−)]-, 22:160
TiClC$_{10}$H$_{10}$, Titanium(III), chlorobis(η^5-cyclopentadienyl)-, 21:84
TiCl$_3$O$_3$C$_{12}$H$_{24}$, Titanium(III), trichloro-tris(tetrahydrofuran), 21:137
TiCl$_4$O$_2$C$_8$H$_{16}$, Titanium(IV), tetrachlorobis(tetrahydrofuran)-, 21:135
TiFeH$_{1.94}$, Iron titanium hydride, 22:90
TiHNbO$_5$, Hydrogen, pentaoxoniobatetitanate(1−), 22:89
TiKNbO$_5$, Potassium, pentaoxoniobatetitanate(1−), 22:89
TlClF$_4$C$_6$H, Thallium(III), chlorobis(2,3,4,6-tetrafluorophenyl)-, 21:73
——, chlorobis(2,3,5,6-tetrafluorophenyl)-, 21:73
TlClF$_{10}$C$_{12}$, Thallium(III), chlorobis(pentafluorophenyl)-, 21:71, 72
TlCl$_3$, Thallium chloride, 21:72
Tl$_2$N$_4$PtC$_4$, Platinate(II), tetracyano-, dithallium, 21:153

Tl$_4$N$_4$O$_3$PtC$_5$, Platinate(II), tetracyano-, thallium carbonate (1:4:1), 21:153, 154
TmCl$_3$, Thulium chloride, 22:39
TmCl$_6$Cs$_2$Li, Cesium lithium thulium chloride, 22:10
TmN$_4$O$_2$C$_{55}$H$_{47}$, Thulium, (2,2,6,6-tetramethyl-3,5-heptanedionato)[5,10,15,20-tetraphenylporphyrinato(2−)]-, 22:160

UCl$_4$, Uranium(IV) chloride, 21:187
UF$_5$, Uranium(V) fluoride, β-, 21:163
UF$_6$K, Uranate(V), hexafluoro-, potassium, 21:166
UF$_6$NP$_2$C$_{36}$H$_{30}$, Uranate(V), hexafluoro-, μ-nitrido-bis(triphenylphosphorus)(1+), 21:166
UF$_6$Na, Uranate(V), hexafluoro-, sodium, 21:166
UO$_5$C$_{10}$H$_{25}$, Uranium(V), pentaethoxy-, 21:165

VC1C$_{10}$H$_{10}$, Vanadium(III), chlorobis(η^5-cyclopentadienyl)-, 21:85
VCl$_2$, Vanadium chloride, 21:185
VCl$_3$O$_3$C$_{12}$H$_{24}$, Vanadium(III), trichlorotris(tetrahydrofuran)-, 21:138
V$_2$, Divanadium, 22:116

WNO$_5$P$_2$C$_{41}$H$_{31}$, Tungstate(1−), pentacarbonylhydrido-, μ-nitridobis(triphenylphosphorus)(1+), 22:182
W$_4$Ag$_8$O$_{16}$, Silver tungstate, 22:76

YCl$_3$, Yttrium chloride, 22:39
YN$_4$O$_2$C$_{49}$H$_{35}$, Yttrium, (2,4-pentanedionato)[5,10,15,20-tetraphenylporphyrinato(2−)]-, 22:160
YbCl$_3$, Ytterbium chloride, 22:39
YbF$_4$N$_4$O$_2$C$_{55}$H$_{43}$, Ytterbium, [5,10,15,20-tetrakis(3-fluorophenyl)-porphyrinato(2−)](2,2,6,6-tetramethyl-3,5-heptanedionato)-, 22:160
YbN$_4$O$_2$C$_{53}$H$_{43}$, Ytterbium, (2,4-pentanedionato)[5,10,15,20-tetrakis(4-methylphenyl)porphyrinato(2−)]-, 22:156
YbN$_4$O$_2$C$_{59}$H$_{55}$, Ytterbium, [5,10,15,20-tetrakis(4-methylphenyl)-porphyrinato(2−)](2,2,6,6-tetramethyl-3,5-heptanedionato)-, 22:156

ZnClF$_6$N$_4$PC$_{24}$H$_{28}$, Zinc(II), chloro(2,9-dimethyl-3,10-diphenyl-1,4,8,11-tetraazacyclotetradeca-1,3,8,10-tetraene)-, hexafluorophosphate(1−), 22:111

ZnP$_2$S$_4$C$_{72}$H$_{60}$, Zincate(II), tetrakis(benzenethiolato)-, bis(tetraphenylphosphonium), 21:25

ZrBr, Zirconium bromide, 22:26

ZrCl, Zirconium chloride, 22:26

ZrCl$_4$O$_2$C$_8$H$_{16}$, Zirconium(IV), tetrachlorobis(tetrahydrofuran)-, 21:136